T0299897

Fundamentals of Plant Pathology

Fundamentals of Plant Pathology

Dr. S. Parthasarathy

Assistant Professor (Plant Pathology)
Amrita School of Agricultural Sciences
Amrita Vishwa Vidyapeetham
Arasampalayam, Coimbatore - 642109

CRC Press is an imprint of the
Taylor & Francis Group, an **informa** business

Elite Publishing House

First published 2024
by CRC Press
4 Park Square, Milton Park, Abingdon, Oxon, OX14 4RN

and by CRC Press
2385 NW Executive Center Drive, Suite 320, Boca Raton FL 33431

CRC Press is an imprint of Informa UK Limited

British Library Cataloguing-in-Publication Data
A catalogue record for this book is available from the British Library

ISBN: 9781032711843 (hbk)
ISBN: 9781032711874 (pbk)
ISBN: 9781032711881 (ebk)

DOI: 10.4324/9781032711881

Typeset in Adobe Caslon Pro
by Elite Publishing House, Delhi

Contents

	Preface	vii
	About the Author	ix
1	Introduction	1
2	History of Plant Pathology	8
3	Classification of Plant Diseases	19
4	Eukaryotic Fungi and Related Pathogens: Characters and Structures	31
5	Asexual Reproduction in Fungi and Chromista	47
6	Sexual Reproduction in Fungi and Chromista	58
7	Taxonomy of Plant Pathogenic Fungi and Related Eukarya	66
8	Protozoa	74
9	Chromista	79
10	Fungi: *Chitridiomycota*	106
11	Fungi: *Zygomycota*	112
12	Fungi: *Ascomycota*	118
13	Fungi: *Basidiomycota*	155
14	Prokaryotic Bacterial Plant Pathogens	181
15	Prokaryotic Mollicutes	195
16	Virus	203
17	Phanerogamic Parasites	227
18	Algal Plant Parasite	235
19	Nematodes: Morphology and Reproduction	239

20	Ecological Classification of Plant Parasitic Nematodes	252
21	Nematode Disease Symptoms on Crop Plants	256
22	Liberation and Dispersal of Plant Pathogens	260
23	Survival of Plant Pathogens	272
24	Types of Parasitism	277
25	Variability and Resistance in Plant Pathogens	284
26	Pathogenesis	295
27	Enzymes in Disease Development	304
28	Toxins in Disease Development	310
29	Plant Growth Regulators in Disease Development	318
30	Effect of Pathogens on Physiological Functions of the Plants	323
31	Plant Disease Epidemiology	329
32	Defense Mechanism in Plants	345
33	Principles of Plant Disease Management	355
34	Chemical Fungicides and Antibiotics	375
35	Application of Fungicides	391
36	Principles of Integrated Nematode Management	398
	Glossary	410

Preface

The book introduces the nature, causes, and impact of plant diseases, briefly describes the history of plant pathology as a scientific discipline, and introduces the disease cycle as a critical tool for understanding disease development and devising appropriate management strategies. The book describes the diverse organisms and agents that cause disease - plant pathogens. A few chapters describe true fungi, other eukaryotic fungal-like organisms, bacteria, viruses, and nematodes.

The book concerns the approaches we can take to alleviate the effects of plant pathogens. It introduces the principal strategies for plant disease and nematode management (or control). The chapter concerns plant health legislation, which aims to prevent the spread of plant pathogens and pests between countries and regions. The following chapters address cultural management practices, i.e., physical manipulations of the environment to reduce diseases, chemical control, and biological control, where the latter concerns the use of friendly microorganisms to control the foe. Disease resistance is traditionally achieved by plant breeding. However, sources of resistance are not always available, so the status of developing biotechnological approaches for disease management is the subject of the next chapter. The following chapter addresses disease assessment and forecasting, which are essential for managing diseases. Plant disease management is most important for the successful and profitable cultivation of horticultural crops.

This book is mainly intended for the Plant Pathology courses of graduate students in Agriculture, Horticulture, Botany, Forestry, and Mycology. In semester system of education of B.Sc./M.Sc. (Ag.), B.Sc./M.Sc. (Horti.), B.Sc./M.Sc. (Forestry), B.Sc./M.Sc. (Botany), and B. Tech. (Horti.), students are pretty dynamic, and this book will precisely help with the changeover in young minds.

Dr. S. Parthasarathy

About the Author

Dr. S. Parthasarathy is working as Assistant Professor (Plant Pathology), in Department of Plant Pathology, Amrita School of Agricultural Sciences. He has completed his PG and Ph.D. from Tamil Nadu Agricultural University, Coimbatore. He is a recipient of innovative fellowships such as BIRAC Post Master's Innovation Fellow in DBT during 2015-17, Canadian International Development Research Centre - Nano Project JRF during 2013-14, UGC Non-Special Assistance Programme during 2012 and Ministry of Human Resource Development Fellowship for UG during 2009 and he also received medals during his UG program and several meritorious awards in International and National conferences for his research contributions. A budding expert in Plant Pathology, he has published over 60 research articles in refereed journals, authored 12 books and contributed about 40 chapters to different books in National and International repute. He also served as a member of editorial board in national journals and several international scientific societies. His areas of specialization are Biological Control and Molecular Plant Pathology.

About the Author

Chapter - 1

Introduction

Nowadays, crop failure due to adverse climate, pests, weeds, or diseases, is erratic in developed agriculture, and instead, there are surpluses of some foods. Nevertheless, the disease still takes a toll, and much time, effort, and money is spent on protecting crops from harmful agents. Further, plant diseases occur in one or more regions on a large scale and cause severe epidemics and losses of produce from crops. Newer plant diseases also occur in crops and share the losses. Most famines and social upheaval of historical significance have resulted from destruction of food crops by environmental extremes or from damage caused by pests and diseases. The vital role of plant pathology in attaining food security and food safety for the world cannot be overemphasized. Plant pathologists attempt to improve plant health and crop productivity through the study of plant diseases, so that the severity and impact of diseases (suffering) can be alleviated.

Definition: Plant Pathology or Phytopathology or *Fitopathologia* is the branch of agricultural or biological science which deals with the plant diseases, their causes, etiology and their management.

» The science of plant pathology is the study of all aspects of disease in plants, including causal agents, their diagnosis, physiological effects, population dynamics and control.

» Phytopathology (Greek *Phyton* = plant + *pathos* - disease, ailments + *logos* = discourse, knowledge) is study of suffering plants.

Plant Pathology has the following major objectives

1. To study biotic (living) and abiotic (non-living and environmental) causes of diseases or disorders or deficiencies.
2. To study the mechanism of disease development by pathogens.

3. To study the plant (host) pathogen interaction in relation to environment.
4. To develop methods of management of plant diseases.

Plant Pathology has the following scope

Plant Pathology (phytopathology) is defined as the study of the organisms (infectious organisms) and environmental conditions (physiological factors) that cause disease in plants, the mechanisms by which disease occurs, the interactions between these causal agents and the plant (effects on plant growth, yield and quality). Plant pathology also involves the study of pathogen identification, disease etiology, disease cycles, economic impact, plant disease epidemiology, plant disease resistance, how plant diseases affect humans and animals, pathosystem genetics, and management of plant diseases. It also interfaces knowledge from other scientific fields such as mycology, microbiology, virology, biochemistry, bioinformatics, etc. Since plant pathology is directly relevant to man's need to grow enough food and fiber to sustain civilization. In this aspect, major scopes are

1. Plant disease diagnosis
2. Plant disease management
3. Biopesticide production
4. Mushroom cultivation

Plant Disease: Any malfunctioning of host cells and tissues those results from continuous irritation by a pathogenic agent or environmental factor that leads to development of symptoms.

» A plant is said to be 'diseased' when there is a harmful deviation from normal functioning of physiological process

» **Dis and ease** - uneasiness

» Any abnormal condition that alters the normal appearance or physiochemical function of a plant.

Significance of Plant Diseases

The study of plant diseases is important as they cause plant and plant produce loss. The various types of losses occur in the field, in storage or any time between sowing and consumption of produce. The diseases are responsible for direct monitory loss and material loss. Plant diseases still inflect suffering on untold millions of people worldwide causing an estimated annual yield loss of 14% globally with an estimated economic loss of 220 billion U.S. dollars. Fossil evidence indicates that plants were affected by different diseases 250 million year ago. The Plant disease has been associated with many important events in the history of mankind of the earth.

Historical Epidemic Outbreak

1. St. Antony fire or Holyfire or Ergotism in Rye

Ergot derives from the Latin word articulum (articulation or join) via the Old French argot. It was first reported in Rhine Valley, during 857 A.D. In the European Middle Ages, diseases caused by ergot were referred to as St. Anthony's fire or holy fire. Thousands of people have died of ergotism, and mortality rates averaged 40% in some documented epidemics in the 1800s. At the end of the nineteenth century, ergot was still regarded as a 'glorious chemical mess' and people began to use ergot as an important therapeutic agent and chemists have intensively studied more than 40 alkaloids produced by the fungus, including the infamous hallucinogen, lysergic acid diethylamide (LSD). Ingestion of the alkaloids produced by the *Claviceps purpurea* fungus that infects rye and other cereals causes ergotism.

2. The Great Irish Potato Famine

The Irish famine of 1846, in which two million people died and many more emigrated from their home land. It swept the whole of Europe and USA, but it was catastrophic in Ireland. Over one million people died and two million population migrated between 1845 and 1850. The famine was caused by the destruction of the potato crop, the staple food in Ireland at that time, by *Phytophthora infestans*, the causal agent of late blight disease.

Other Major Disease Outbreaks

Most rusts are notorious for the economic harm they bring to vital agricultural and forest plants. After the first mention of the coffee rust in Ceylon (previously Sri Lanka) in 1867, the entire coffee production of the island was eradicated in less than three decades, leaving the British society to drink solely tea. In 1870, Ceylon exported 100 million pounds (45 million kgs) of coffee annually, but by 1889, coffee production had decreased to 5 million pounds (2.3 million kgs). The introduction of the poplar rusts *Melampsora medusae* in 1972 and *M. larici-populina* approximately one year later destroyed the Australian poplar industry in its infancy. Cereal rust diseases cause annual losses of $5 billion and pose a possible threat to the global food supply. Stem or black rust of wheat was traditionally known to produce periodic extreme devastation and was the most feared disease across all continents where wheat was farmed. During the Cold War, it gained popularity due to the interest of biological warfare researchers from both superpowers. In the 21st century Ug99, a virulent form of stem rust has moved from Africa to the Middle East, endangering the wheat fields of South Asia and the high-yielding wheat types created during the Green Revolution.

In France, grape downy mildews occurred epiphytotically between 1878 and 1882. In 1910, Panama wilt of banana harmed the harvest in Panama and Surinam. In 1930, the Sigatoka leaf spot of bananas harmed the sector in Central and South America, while the

Tristeza virus caused a reduction in the citrus harvest in Argentina and Brazil. The downy mildew of maize in Java in 1912, the powdery mildew of rubber in Malaya in 1938, the bunchy top of banana in Ceylon in 1940, and the wart disease of potato in the Netherlands in 1953 are also notable damaging diseases. Cassava mosaic disease frequently results in 15-24% storage root production losses of cassava yearly across sub-Saharan Africa, which is comparable to a loss of US$ 1.2 to US$ 2.3 billion annually, and this resulted in the deaths of over 3,000 people in Uganda due to malnutrition in 1994. As a result of witches' broom (*Moniliophthora perniciosa* -previously *Crinipellis perniciosa*) epidemics during 1988-1990, Brazil became a net importer of 141,000 tonnes in 1998-2000.

The Coffee Wilt Disease (CWD) *Gibberella xylarioides* (*Fusarium xylarioides*), a fungal infection that wiped out more than 12 million robusta coffee trees in central and western Uganda regions at the end of the 20th century, caused an estimated loss of $1 billion in 2009, and continues to spread, is a more recent example of one of the longest-lasting crop diseases in Uganda.

Major Disease Outbreaks in India

The Great Bengal Famine

In the last year of Second World War (1943) Bengal had to face a serious rice famine. One of the reasons to which this famine has been attributed was the loss in yield of the rice crop due to attack of Helminthosporium leaf spot which had been affecting the crop for the last several years. Most of the population was dependent on rice as a single crop. Situation was similar to the Irish potato famine but not so catastrophic. It was due to brown spot of rice caused by the fungus *Cochliobolus miyabeanus*, in which at least 1.5 million people died. Favourable weather conditions for the fungus resulted in an epidemic which reduced the yield of rice crops by 40-90%.

In India, wheat rust had been considered to cause a loss of over Rs. 40 million annually. In the year of epidemics there have been a losses amounting to Rs. 500 million or more. Although introduction of dwarf high yielding varieties has reduced the losses to a great extent even now thefarmers lose 8-10% of the expected yield due to rusts. First stem rust epidemic record goes back to 1786 A.D. in central India. In 1946-47 wheat rust epidemic caused food shortage and an estimated loss of about 200 million rupees annually. Widespread occurrence of leaf rust was observed during 1971-73 in popular cultivar Kalyansona in northern plains and during 1993-94 in HD2285 and HD2329 covering approximately 4 million hectares in North Eastern Plain Zone. Wheat rust research in India was initiated in early 1920's, systematic work on race analysis and testing of varieties against wheat rusts in India was started in 1931 and so far, 32 races of *Puccinia graminis* var. *tritici* have been identified.

The loose smut of wheat is estimated to cause an average loss of 3% every year. The

'Molya' disease, caused by a nematode is another example. The disease of wheat and barley prevalent in most parts of Rajasthan causes a loss of Rs. 30 million in barley and 40 million in wheat every year. Different smuts of sorghum are responsible for an annual loss of Rs. 100 million. 5% to 75% loss in chickpea due to *Ascochyta* blight was reported from Rajasthan during 1982. Wilt of pigeon pea causes 5-10% loss every year in U.P. and Bihar.

Global status of severe losses caused by plant diseases

Year	Diseases	Location
1800s	Ergot of rye and wheat	Europe
1840	Late blight of potato	Northern Europe
1845	Late blight of potato	Ireland Great Irish Famine
1854	Grape powdery mildew	France
1867-68	Coffee rust	Ceylon
1878-85	Grape downy mildew	France
1894	False smut (Inakoji Byo) of rice	Japan
1910	Panama wilt of banana	Panama and Surinam
1916	Wheat stem rust	North America
1920	Chestnut blight	North America
1930	Dutch elm disease	North America and Holland
1930	Sigatoka leaf spot of banana	Central and South America
1930s	*Citrus tristeza virus*	Argentina and Brazil
1938	Powdery mildew of rubber	Malaysia
1940	Bunchy top of banana	Ceylon
1960	Bacterial blight of rice	Southeast Asia
1970	Southern corn leaf blight	United States
1973	Apple scab	India
1979	Root knot nematode	Worldwide
1980's	Tomato spotted wilt virus	United States and Worldwide
1990	Panama disease tropical race 4	Southeast Asia
1994	*Cassava mosaic virus*	Sub-saharan Africa (Uganda)
1998	Fire blight of apple	North western US and worldwide
1999	Wheat black (stem) rust - Ug99	East Africa and Middle East
2012	Coffee rust	Central America and Caribbean
2014	Cassava mosaic disease	West and Central Africa, Srilanka
2016	Wheat Blast	India and Bangladesh

National status of severe losses caused by plant diseases

Year	Diseases	Location
1807	Cereal rust	Punjab
1908	Tundu disease of wheat	Punjab
1917	Spike disease of sandalwood	Karnataka
1918	Black arm of cotton	Madras
1931	Green ear of sorghum	Central India
1930s-40s	Karnal bunt of wheat	Haryana
1942-43	Brown spot of rice	Bengal
1946-47	Wheat rust	Madhya Pradesh
1938-42	Red rot of sugarcane	Uttar Pradesh, Punjab and Bihar
1957	Downy mildew of pearl millet	Maharashtra, Gujarat and Rajasthan
1957	Ergot of pearl millet	Maharashtra
1960	Gummosis of sugarcane	Madras
1963	Bacterial blight of rice	Bihar
1967	Mango malformation	Bihar
1970s	Apple scab	Himachal and Kashmir
1984	Apple scab	Jammu and Kashmir
1985-87	Rice tungro virus	Tamil Nadu and Orissa
1985-88	Rice blast	Tamil Nadu
1985-2001	Coconut root (wilt) disease	Kerala
1990	Rice tungro virus	Andhra Pradesh
1990s	Bunchy top of banana	Tamil Nadu
1990	Bacterial blight of tomato	Maharastra
1999	Rice tungro virus	Punjab
2003	Bacterial blight of pomegranate	Maharashtra, Karnataka and Andhra Pradesh
2004	Yellow rust of wheat	Northern India

Practice Questions

1. ....is the pathogenic cause of the bengal famine in India
2. .... year(s) Ireland faced a famine due to failure of the potato crop
3. .... wheat black (stem) rust strain or race is spreading pandamically in the present century

4. The word phytopathology is derived from...........language
5. Expand LSD............
6. What are the main objectives of the Subject discipline of Plant Pathology?
7. Attempt a brief Note on Irish Famine and Bengal Famine including year of occurrence, cause etc.
8. Who authored Book(s) entitled (i) Fungi and Diseases in Plants and (ii) Fungi and Plant Diseases? Name the Publishers and year of Publication of the above Books.
9. Briefly describe landmarks in the History of Phytopathology?
10. Explain the impact and causes of Irish famine?

Suggested Readings

Schumann, G. L., & D'Arcy, C. J. (2010). *Essential Plant Pathology* (No. PA 581.2 S38 2013.). St. Paul, MN: APS Press.

Agrios, G. N. (2005). *Plant Pathology*. Elsevier.

Mehrotra, R. S. (2003). *Plant Pathology*. Tata Mcgraw-Hill education.

Aneja, K. R. (2007). *Experiments in microbiology, Plant Pathology and Biotechnology*. New Age International.

Lucas, J. A. (2009). *Plant Pathology and Plant Pathogens*. John Wiley & Sons.

Strange, R. N., & Scott, P. R. (2005). Plant disease: a threat to global food security. *Annual review of phytopathology*, *43*(1), 83-116.

Moreno, P., Ambrós, S., Albiach-Martí, M. R., Guerri, J., & Peña, L. (2008). Citrus tristeza virus: a pathogen that changed the course of the citrus industry. Molecular plant pathology, 9(2), 251–268. https://doi.org/10.1111/j.1364-3703.2007.00455.x

Sherwood, J. L., German, T. L., Moyer, J. W., & Ullman, D. E. (2009). Tomato spotted wilt virus. *Tomato spotted wilt virus.*

Johnson, K. B. (2000). Fire blight of apple and pear. *The Plant Health Instructor*, *2015*.

Jones, L. R. (1900). A soft rot of carrot and other vegetables caused by Bacillus carotovorus Jones. *Vt. Agric. Exp. Stn. Rep*, *13*, 299-332.

Schumann, G.L. (1991). Plant Disease: Their Biology and Social Impacts, APS Press, Minn.

Chapter - 2

History of Plant Pathology

Every ecologist knows the basic history of evolutionary theory. Much less well known is the struggle to establish the germ theory of disease, with many defenders and skeptics for well over a century before it gained acceptance. Both mycology and phytopathology made notable advances during the 1700s-with small steps toward a germ theory-though these advances were not widely known or appreciated. Researchers during the 1800s built upon the work of their predecessors. Not until the late 1800s were there evidences of a virus as a cause of plant disease. Nature is one, but scientists partition it into sciences they can master.

Although phytopathology and animal parasitology developed as separate sciences, their concerns overlap in cases of fungal diseases of animals, including humans, and nematode diseases of plants; and both phytopathology and parasitology overlap with bacteriology and virology concerning bacterial and viral diseases. The history of nematode parasitism of plants was outside the scope of histories of phytopathology, but nematodes are discussed in textbooks on phytopathology. With many simultaneous developments occurring, it seems best to discuss studies of fungi first, nematodes and bacteria second and third, and viruses last.

The history of phytopathology divides itself into eras, and these again into periods.

1. The Ancient Era (ancient to 5th century)
2. The Dark or Middle or Medieval Era (5th – 16th century)
3. The Pre-modern or Autogenetic Era (17th century to 1853 AD)
4. The Modern or Pathogenetic Era (1853 AD – 1906 AD)
5. The Present Era (1906 AD onwards)

1. The Ancient Era (ancient to 5th century)

- 3700 BC - *Rigveda* classification of plant diseases and germ theory of disease are mentioned and they were aware that the diseases were caused by microorganisms.
- 700 BC - Roman King Numa Pompilius: Initiated the celebration of annual festival of "Robigalia" to ward-off rust disease. The Robigalia was a festival in ancient Roman religion held April 25, initiated by the King Numa Pompilius named for the god Robigus and goddess Robigalia.
- 600 BC - Susruta, the great Indian pioneer in medicine and surgery mentioned a phanerogamic semiparasite *Loranthus longiflorus* as a cause of disease.
- 400 BC - A phanerogamic semiparasite *Loranthus longiflorus* Disr. was mentioned by Susruta.
- 384 BC - Aristotle reported that the life arose spontaneously from the decomposing organic matter.
- 370 BC - Theophrastus the Greek Philosopher and Great Botanist mentioned and speculated about diseases of cereals, legumes, and trees in his books entitled *Enquiry into plants* (*Historia Plantarum*), *Reasons of vegetable growth* and *The Nature of Plants*. He is also known as the "Father of Botany".

Plant disease symptoms are also mentioned in old testaments like Bible, Shakespear's poems and dramas and other Christian literature. Rust, smut, mildew and blights are very often quoted in Bible.

2. The Dark or Middle or Medieval Era (5th – 16th century)

- 1000 AD - *Vriksha Ayurveda* written in 11th century by the saint Surapal is the Indian book, which gave first information on plant diseases.
- 1267 AD - Roger Bacon invented the simple lens.
- 1585 AD - William Shakesphere describes about plant diseases in his poems.
- 1600 AD - Dodder (*Cuscuta reflexa* Roxb.), as a parasite has been mentioned in *Bhavaprkashanighantu.*

3. The Pre-modern or Autogenetic Era (17th century to 1853 AD)

- 1605 - Jahangir in his memoirs described a disorder of marigold, which could be ascribed today to species of *Alternaria, Botrytis,* or *Sclerotinia.*
- 1665 - Robert Hooke improved the compound microscope and saw the sections of cork and other plant tissues. He figured the teliospores of a *Phragmidium* taken from yellow spots on leaves.
- 1675 - Antonie van Leeuwenhoek Dutch worker developed the compound microscope,

discovered bacteria, yeasts and many other microorganisms, and opened up a new science of microbiology. He is called as Father of Bacteriology.

- 1705 - De Tournefort divided plant diseases into two classes; the first is due to the internal causes and the second to external causes.
- 1729 - P.A. Micheli Italian botanist proposed fungi comes from spores. He published a book "*Nova Plantarum Genera*" in which he gave descriptions about 1900 species in Latin out of which 900 were fungi. Thus, he took one of the first steps towards the overthrow of the theory of spontaneous generation. He is called as Father of Mycology.
- 1753 - Carolus Linnaeus mycology entered a dominantly taxonomic period. Although Linnaeus worked little with fungi, he made an important contribution to mycology by including them in his Latin binomial system.
- 1755 - Mathieu Tillet French botanist published a paper on bunt or stinking smut of wheat and discovered bunt is a disease of wheat. Experimental proof that wheat bunt is contagious and it can be partly prevented by seed treatments. Great Grand Father of Phytopathology.
- 1801 - C.H. Persoon worked on classification and nomenclature of fungi and discovered the life cycle of bunt fungus. He published a book "*Synopsis Methodica Fungorum*" for nomenclature of Ustilaginales, Uredinales and Gasteromycetes. He also published Mycologica Europica in 1822. He gave the name to rust pathogen of wheat as *Puccinia graminis.* He thought that some fungi arose spontaneously and that some grew from spores. He regarded smut fungi as products of the diseased plants.
- 1803 - Joseph Holt, a farmer, gave an accurate description of a rust-infected wheat crop at Australia.
- 1807 - I.B. Prevost French scientist showed bunt of wheat is a fungus and showed evidence that a disease is caused by a microorganism. He showed that the solution containing copper sulphate prevented the germination of bunt spores and can be used for control of bunt diseases.
- 1815-1885 - Tulasne brothers Louis Rene Tulasne called The Reconstructor of Mycology Tulasne brothers (L.R. Tulasne and Charles Tulasne) gave the idea of Polymorphism in fungi made illustrated diagrams of rust, smut and ascomycetes. They published nicely illustrated book called "*Selecta Fungorum Carpologia*"
- 1821 - E.M. Fries published three volumes of "*Systema Mycologicum*" for nomenclature of hymenomycetes. Fries regarded the rust and the smut fungi as products of the diseased plants. He is known as Linnaeus of Mycology, Father of Mycology and Founder of Modern Fungal Taxonomy/ Mushroom Taxonomy. Persoon and fries first time introduced binomial system of nomenclature to classify the fungal organisms.
- 1821 - Robertson in England he stated that sulphur is effective against peach mildew.

4. The Modern or Pathogenetic Era (1853 AD – 1906 AD)

» 1845 - P.A. Saccardo collected and fetched together the dispersed knowledge on systematic mycology under his massive 25-volume work, the *Sylloge Fungorum Omnium Hucusque Cognitorum* (1882 AD–1925 AD).

» 1858 - J.G. Kuhn published first textbook in Plant Pathology "*The Diseases of Cultivated Crops, their Causes and their Control*". He was one of the founders of Plant Pathology.

» 1861 - Heinrich Anton de Bary worked out the life cycle of potato late blight and first to prove experimentally *Phytophthora infestans* (*Peronospora infestans*) is the cause of potato late blight. He proved that fungi are causes but not the results of diseases. He is the Father of Modern Plant Pathology. He reported autocious and heteroecious nature of wheat stem rust, life cycle of Uredinales (rust) and Ustilaginales (smuts) Also gave a detailed account on life cycle of downy mildew genera. He did research on slime molds and he coined terms, symbiosis, mycetozoa, teleutospores, haustorium (the lichens consist of a fungus and an algae). He reported the role of enzymes and toxins in tissue disintegration caused by *Sclerotinia sclerotiorum.* He is also called as Father of Experimental Plant Pathology and Father of Modern Mycology. He trained M.S. Woronin of Russia, Brefeld of Germany, P.M.A. Millardet of France, H.M. Ward of England, W.G. Farlow of USA and Fischer of Switzerland.

» 1874 - Robert Hartig published a book entitled, "Important Diseases of Forest Trees". Father of Forest Pathology.

» 1875 - Farlow started independent course of plant pathology at Harvard University

» 1875-1912 - Oscar Brefeld discovered the methods of artificial culture of microorganisms. He also illustrated the complete life cycles of cereal smut fungi and diseases caused by them.

» 1876 - Louis Pasteur disproved the theory of spontaneous generation of diseases and propose germ theory in relation to the diseases of man and animal should be credited to advocate the theory of Pasteur in fungal pathology. Louis Pasteur is a Father of Modern Bacteriology and Founder of Microbiology

» 1876 - Robert Koch belongs to Germany and described the theory called "*Koch postulates.*" He established the principles of pure culture technique. He is Father of Modern Bacteriological Techniques, Father of Microbial Techniques, Founder of Modern Bacteriology.

» 1877 - M.S. Woronin discovered and named the Club root of Cabbage pathogen as *Plasmodiophora brassicae* and found out the life cycle of potato wart disease and worked on sunflower rust.

» 1881 - Harry Marshall Ward worked out the life cycle of coffee leaf rust. He is called as Father of Tropical Plant Pathology.

- 1881 - A. Barclay started studying the Indian rusts.
- 1882 - T.J. Burrill was an American Plant Pathologist who first time proved that fire blight of pear was caused by a bacterium called by him *Micrococcus amylovorus* (now known as *Erwinia amylovora*). He is a Founder of Phytobacteriology.
- 1885 - A. B. Frank defined and named mycorrhizal associations in plant roots.
- 1885- K.R. Kirtikar was the first Indian scientist who collected and identified the fungi in the country.
- 1886 - Adolf Mayer German agricultural chemist described a disease of tobacco called mosaikkranheit (tobacco mosaic). Adolf Mayer demonstrated the sap transmission of Tobacco Mosaic Virus disease. He led to the foundation of the field of virology.
- 1887 - Mason introduced the Burgundy mixture was in France.
- 1889 - D.D. Cunningham proposed the new genus *Mycodea* for the parasitic alga, now known as *Cephaleuros virescens* responsible for red rust of tea.
- 1892 - Dimitri Iosifovich Ivanowski is a Russian botanist demonstrated that the causal agent of tobacco mosaic could pass through bacterial filter with capable of permeating porcelain Chamberland filters.
- 1894 - Eriksson is a scientist described the phenomenon of physiologic races in cereal rust fungus, *Puccinia graminis*.
- 1898 - Nocard and Roux discovered Mycoplasma like Organism caused in animals.
- 1898 - M.W. Beijerinck is a Dutch microbiologist and founder of virology proved that the virus inciting tobacco mosaic is not a microorganism. He believed it to be *contagium vivum fluidum* (infectious living fluid). He was the first to use the term *virus*, which is the Latin word for poison. He is known as Father of Plant Virology.
- 1899 - W.A. Orton is considered as a pioneer worker in the development of disease-resistant varieties.
- 1903 - Bagchee started research in forest pathology at the Forest Research Institute, Dehradun in the Himalayas.
- 1904 - A.F. Blakeslee is a American Geneticist founded heterothallism in *Rhizopus*.
- 1904 - R.H. Biffen first to show that resistance to pathogens in plants can be inherited as a Mendelian character; pioneer in genetics of plant disease resistance.

5. The Present Era (1906 AD onwards)

- 1901-1920 - E.F. Smith gave the final proof of the fact that bacteria could be incitants of plant diseases. He also worked on the bacterial wilt of cucurbits and crown gall disease. He is also called as "Father of Phytobacteriology".

- » 1915 - Frederick Twort were the first to recognise viruses which infect bacteria.
- » 1917 - Felix d'Herelle called bacteriophages (eaters of bacteria).
- » 1917 - E.C. Stakman demonstrated physiologic forms in stem rust of wheat.
- » 1918 - E.J. Butler published book on *Fungi and Disease in Plants*; He is called as the Father of Modern Plant Pathology in India and Father of Indian Mycology; He worked at IARI for 20 years from 1901 to 1920. He joined as the first Director of Imperial Bureau of Mycology (Commonwealth Mycological Institute, CMI) now CAB International Mycological Institute in Kew, England in 1920. He began the journal *Review of Applied Mycology*. Published the monograph *Pythiaceous and allied fungi.* Books Fungi and Disease in Plants (1918), Fungi in India (with B.R. Bisby) and Plant Pathology (with S.G. Jones) 1949.
- » 1927 - Cragie and Dodge enabled genetic study of pathogenic fungi.
- » 1929 - Sir Alexander Fleming isolated the antibiotic, Penicillin from the fungus, *Penicillium notatum*.
- » 1929 - Mc Kinney observed that tobacco infected with a virus causing a mild mosaic disease was protected against a more severe form of that virus. Cross protection was used to establish virus relationships; as only related viruses would show the response.
- » 1932 - H.N. Hansen and R.E. Smith first to demonstrate the origin of physiologic races through heterokaryosis.
- » 1934 - W.H. Tisdale and I. Williams studied the organic fungicides by discovering alkyl dithiocarbamates.
- » 1935 - W.M. Stanley proved that viruses can be made as crystals. He got Nobel Prize in 1946.
- » 1938 - Norman Borlaug discussed the manifestation of the plant disease rust, a parasitic fungus that feeds on phytonutrients, in wheat, oat and barley crops across the US. His discovery on special plant breeding methods created plants resistant to rust. He developed multiline cultivars for wheat Father of Green Revolution.
- » 1942 - H.H. Flor developed gene-for-gene hypothesis in *Melampsora lini* causing flax rust.
- » 1945 - J.G. Horsfall explored the mechanism of fungicidal action.
- » 1948 - B.B. Mundkur started Indian Phytopathological Society with its journal Indian Phytopathology. He has written a book *'Fungi and Plant Diseases'* in 1949, which is the second, book in plant pathology in India.
- » 1948-67 - Patel. M.K. worked on more than 50 Plant pathogenic bacteria. Established

School of Plant bacteriology in Pune.

» 1952 - J. Lederberg coined the term plasmid.

» 1952 - S.A. Waksman won Nobel prize for the discovery of streptomycin.

» 1952 - Zinder and J. Lederberg discovered transduction in bacteria.

» 1953 - Pontecorvo et al. demonstrated parasexualism in fungi.

» 1957 - E.C. Stakman with J.G. Harrar wrote a book *Principles* of *Plant Pathology.*

» 1962 - H. Stolp and Starr discovered bdellovibrios (Gram negative, rod or curved (comma shaped), obligate aerobic bacterium parsitizing another. Gram negative bacterium) (*Bdellovibrio bacteriovorus*).

» 1963 - J.E. Van der Plank found out vertical and horizontal types of resistance in crop plants.

» 1966 - van Schmeling and Marshall Kulka first to find out systemic fungicides (oxathiin compounds - carboxin and oxycarboxin).

» 1966 - Kassanis discovered the satellite viruses.

» 1967 - Doi *et al.* and Ishiie *et al.* the Japanese scientists found that mycoplasma-like organisms (MLO) could be responsible for the disease of the yellows type. Doi observed that MLOs are constantly present in phloem while Ishiie observed MLOs temporarily disappeared when the plants are treated with tetracycline antibodies.

» 1970 - S.D. Garrett investigated the management of root diseases and he is the pioneer worker in the field of biological control.

» 1971 - T.O. Diener and Raymer discovered viroids, which only consist of nucleic acids. Smaller than viruses, caused potato spindle tuber disease (250-400 bases long of single-stranded circular molecule of infectious RNA).

» 1972 - Davies *et al* observed that a motile, helical wall-less microorganism associated with corn stunt diseases, which could be cultured and characterized and they named it as Spiroplasma.

» 1972 - G. Rangaswami wrote a book on 'Diseases of Crop Plants in India'.

» 1972 - I.M. Windsor and L.M. Black observed a new kind of phloem inhabiting bacterium causing clover club leaf disease.

» 1973 - Davis observed the Gram negative xylem limiting bacteria in grape plants affected with pierce's disease (*Xylella fastidiosa*). This bacterium was known earlier on as a rickettsia-like organism but is now referred to as a fastidious bacterium. It is not related to the *rickettsiae.*

» 1973 - Goheen et al. reported the association of RLO in Pierce disease of grapes.

- 1973 - Hopkins et al. reported the association of RLO in Phony disease of peach.
- 1974 - Chilton et al demonstrated that crown gall bacterium transforms plant cell to tumour cell by introducing into them a plasmid.
- 1975 - Kohler and Milstein isolated the first monoclonal antibodies from clones of cells.
- 1991 - De Wit et al cloned first fungal avirulence gene, *avr* 9, from *Cladosporium fulvum.*

Plant Nematology

- 1743 - John Turberville Needham was an English biologist and Roman Catholic priest. He reported wheat gall disease in association with plant parasitic nematode (*Anguina tritici*) in England
- 1857 - Casimir-Joseph Devaine studied the life cycle of *Anguina tritici.*
- 1889 - N.A. Cobb published an index of Australian fungi in 1893 AD. From 1891 AD to 1903 AD, he described and illustrated many plant diseases present in New South Wales. He endeavoured to determine the factors of importance to resistance to rust in wheat. According to Buhrer, Cobb became the father of American plant nematology. In fact, he gave nematology its name.
- 1892 - Atkinson association of root knot nematode with the fusarial wilt of cotton.
- 1901 - Hunger described the severity of bacterial wilt of tomato was facilitated by root knot nematode.
- 1958 - Hewit et al. demonstrated the role of nematodes as vectors of plant diseases.

Michigan State University was the first institution of higher learning in the United States to teach scientific agriculture dates back to 1855. Initially, studying plant pathogens and the diseases they produce was considered a subfield of botany or even an application of mycology. In 1883, the Royal Veterinary and Agricultural University in Copenhagen, Denmark, established one of the earliest jobs in the field of plant pathology. During that time period, plant pathology was taught at a variety of institutions around the world, though typically within botanical or mycological departments. Emil Rostrup was the first instructor, and he was made professor in 1902. Before joining the Copenhagen plant pathology department, he researched plants and plant diseases. In the early years of the 20th century, educational programmes in plant pathology began.

Contributions of Plant Pathologists in India

- J.F. Dastur and N. Prasad — Late blight of Potato
- G.S. Kulkarni — Downy mildew of sorghum

- » K.C. Mehta — Father of Indian Cereal Rust
- » J.C. Luthra — Solar treatment of wheat seeds
- » M.J. Thirumalachar — Antibiotics in disease control
- » R.S. Singh — *Pythium* and Book on Plant Diseases
- » Y. L. Nene — Fungicides in Plant Disease Control (Book)
- » V.V. Chenulu, S.P. Raychudhuri, Kapoor and A. Varma - Virus diseases
- » S. Nagarajan — Plant Disease Epidemiology
- » S.B. Mathur — Seed Pathology
- » P. Vidhyasekaran — Physiological Plant Pathology and Seed Pathology
- » A. Mahadevan — Biochemical changes in diseased plants
- » M. Mitra — Karnal bunt of Wheat at Haryana
- » T.S. Ramakrishnan — Diseases of Millets
- » T.S. Sadasivan — Concept of vivotoxins and mechanism of wilt in cotton
- » Y.L. Nene — Khaira disease of rice
- » Luthra and Sattar — Loose smut of wheat
- » S.N. Das Gupta — Black tip of mango
- » S. Nagarajan and H. Singh — Indian Stem Rust Rules
- » S. Nagarajan — Puccinia path in India
- » C.D. Mayee — Phytopathometry

In 1857, the first Indian colleges were founded in Calcutta, Bombay, and Madras, with a focus on mycology. Plant Pathology was first taught as a major in 1930 at Universities of Madras (1857), Allahabad (1857), and Lucknow universities est. 1921). Madras University was the first university in India to accept Plant Pathology as an academic discipline. In 1945, Agra University established a postgraduate Plant Pathology programme at the Government Agricultural College in Kanpur. After the founding of Agricultural Universities in the 1960s, Plant Pathology instruction has become an integral component of graduate and post-graduate programmes. After the establishment of Agricultural Universities in different states of India in 1960, the subject was accorded its due importance, and the teaching of its supporting courses, namely mycology, bacteriology, virology, and nematology, was incorporated into both undergraduate and graduate programmes of Agriculture. Currently, the majority of Plant Pathology-related courses have been rewritten, and Molecular Plant Pathology has been included to keep up with scientific progress.

Practice Questions

1. .... is recognized as father of Indian Phytopathology.
2. .... was a festival in ancient Roman religion to ward-off rust disease
3. .... Indian book authored by Surapal which gave the first information on plant diseases
4. ....wrote a book "Synopsis Methodica Fungorum" for nomenclature of Ustilaginales, Uredinales and Gasteromycetes
5. .... called as Father of Experimental Plant Pathology and Father of Modern Mycology
6. ....father of plant nematology
7. ....coined the term virus
8. Write the contributions of Anton de Bary?
9. Elaborate the contributions of E.J. Butler in Indian Phytopathology?
10. Briefly describe landmarks in the History of Phytopathology?
11. Explain the historical developments happened in ancient era in plant pathology?
12. Describe the developmental events and contributions on fungicide development?
13. What is Khaira disease of paddy? Who discovered this disease for the first time in India? What are its causes? Also, suggest the control measures.

Suggested Readings

Ainsworth, G. C. (1981). *Introduction to the history of plant pathology*. Cambridge University Press.

Raychaudhuri, S. P., Verma, J. P., Nariani, T. K., & Sen, B. (1972). The history of plant pathology in India. *Annual Review of Phytopathology*, *10*(1), 21-36.

Orlob, G. B. (1971). History of plant pathology in the middle ages. *Annual review of Phytopathology*, *9*(1), 7-20.

Lucas, G. B., Campbell, C. L., & Lucas, L. T. (1992). History of plant pathology. In *Introduction to Plant Diseases* (pp. 15-19). Springer, Boston, MA.

Akai, S. (1974). History of plant pathology in Japan. *Annual Review of Phytopathology*, *12*(1), 13-26.

Singh, R. S. (2017). *Introduction to principles of plant pathology*. Oxford and IBH Publishing.

Ravichandra, N. G. (2013). *Fundamentals of plant pathology*. PHI Learning Pvt. Ltd..

Borkar, S. G. (2017). *History of Plant Pathology*. CRC Press.

Ainsworth, G. C. (1969). History of plant pathology in Great Britain. *Annual review of Phytopathology*, *7*(1), 13-30.

Nolla, J. A. B., & Valiela, M. V. F. (1976). Contributions to the history of plant pathology in South America, Central America, and Mexico. *Annual Review of Phytopathology, 14*(1), 11-29.

Horsfall, J. G. (Ed.). (2012). *Plant Pathology V1: The Diseased Plant*. Elsevier.

Dasgupta, M. K. (1988). *Principles of plant pathology*. Allied Publishers.

Tronsmo, A. M., Collinge, D. B., Djurle, A., Munk, L., Yuen, J., & Tronsmo, A. (2020). *Plant pathology and plant diseases*. CABI.

Sharma, P. D. (2013). *Plant pathology*. Deep and Deep Publications.

Turner, R. S. (2005). After the famine: Plant pathology, Phytophthora infestans, and the late blight of potatoes, 1845–1960. *Historical Studies in the Physical and Biological Sciences, 35*(2), 341-370.

Chapter - 3

Classification of Plant Diseases

A successful infection will lead to the development of visible symptoms of disease or signs of the pathogen itself. Symptoms and signs may be found on all parts of a plant (i.e. roots, stems, leaves or flowers) and on the seeds and fruits that the plant produces. Plant disease are classified on the basis of various parameters, based on appearance of of symptoms, following terms are used to define it.

Terms and Concepts in Plant Pathology

Pest: a destructive insect or other animals that attacks crops, food, livestock, etc.

Pathogen: An entity, mostly a microorganism that can cause disease. In a literal sense a pathogen is any agent that causes *pathos* (ailment, suffering) or damage. However, the term is generally used to denote living organisms (Fungi, bacteria, MLO's, nematodes etc.,) and viruses but not nutritional deficiencies.

Disease	Pathogen
A pathogen is the causal agent of a disease and is usually a microorganism, virus or nematode. The pathogen has a scientific binomial (essentially Latin) name that is written in italics, or if it is a virus, an abbreviation. The pathogen's name should not be used as a synonym for the disease's name.	A disease is caused by a pathogen. The disease generally has a trivial name. Sometimes it is part of the scientific name, but not written in italics or with capital letters (unless it contains the name of a person or a place).

For example:	Examples include:
• The fungus *Puccinia graminis* causes the rust disease in cereals. • The bacterium *Erwinia amylovora* causes fireblight in fruit trees.	• Rust in cereals is caused by the fungus *Puccinia graminis*. • Fireblight in fruit trees is caused by *Erwinia amylovora*.

Parasite: Organisms which derive the materials they need for growth from living plants (*host or suscept*) are called parasites. All the parasites are pathogens, but not all the pathogens are parasites.

Parasite	Non-parasite
Unculturable Possess absorption and anchoring organ E.g., • Downy mildew chromista, • Powdery mildew fungi, • Rust and Smut fungi, Phenerogamic parasites, • *Candidatus* Phytoplasma, • Virus and Viroid	Culturable May be possess absorption and anchoring organ E.g., • Leaf spot fungi and bacteria • Wilt, root rot fungi and bacteria • Tumour inducing fungi and bacteria

The damage and losses

A classification of losses describes the complexity and inter dependence of loss at all levels of society. He also proposed three useful concepts for describing the dynamics of crop destruction. They are;

a. **Injury**: Any observable deviation from the normal crop; injury may lead to damage.
b. **Damage**: Any decrease in quantity and quality of a product; damage may lead to loss.
c. **Loss**: Any decrease in economic returns from reduced yields and cost of agricultural activities designed to reduce damage.

Appearance of disease symptoms

Syndrome: The set of varying symptoms characterizing a disease are collectively called a syndrome.

Disorder: Non-infectious plant diseases due to abiotic causes such as adverse soil and environmental conditions are termed disorders. The common characteristic of non-infectious diseases of plants is that they are caused by the lack or excess of something (temperature, soil moisture, soil nutrients, light, air and soil pollutants, air humidity, soil structure and pH) that supports life. Non-infectious plant diseases occur in the absence of pathogens, and cannot, therefore, be transmitted from diseased to healthy plants.

Sign: The pathogen or its parts or products visible on a host plant. This could be in the form of fungal spores or mycelia, different resting structures of the pathogen (such as sclerotia), nematode cysts or galls and bacterial slime.

- It is any objective evidence of disease.
- Visible structure of the pathogen produced in or on the diseased tissues

These include:

1. Mycelium or mold growth, under some conditions, readily visible to the naked eye.
2. Conks and mushrooms - the familiar structures of some fungi that are formed by some pathogenic fungi.
3. Fruiting bodies - reproductive structures of some fungi that are embedded in diseased tissue, often requiring a hand lens to see.
4. Sclerotia – hard resistant structures of some fungi.
5. Rust pustules.
6. Bacterial ooze and specific odours associated with tissue macerations by certain pathogens.

Symptoms: External or internal reactions or alterations of a plant as a result of disease.

- It is any subjective evidence of disease.
- Indicating physical condition of a disease/ visible effect of pathogen.
- These symptoms are collectively called as syndrome.
- Morphological and physiological reactions or alterations as result of disease/ disorder/ deficiency/ infestation.
- **Morphological symptoms:** Symptoms appear externally on the whole or part of the the diseased plant.
- **Physiological (or) pathological symptoms:** Symptoms appear in the tissues of affected plants.
- **Cytological symptoms:** changes in the individual cells.
- **Histological symptoms:** changes in the tissue structure.

Disruption of plant functions resulting in disease symptoms:

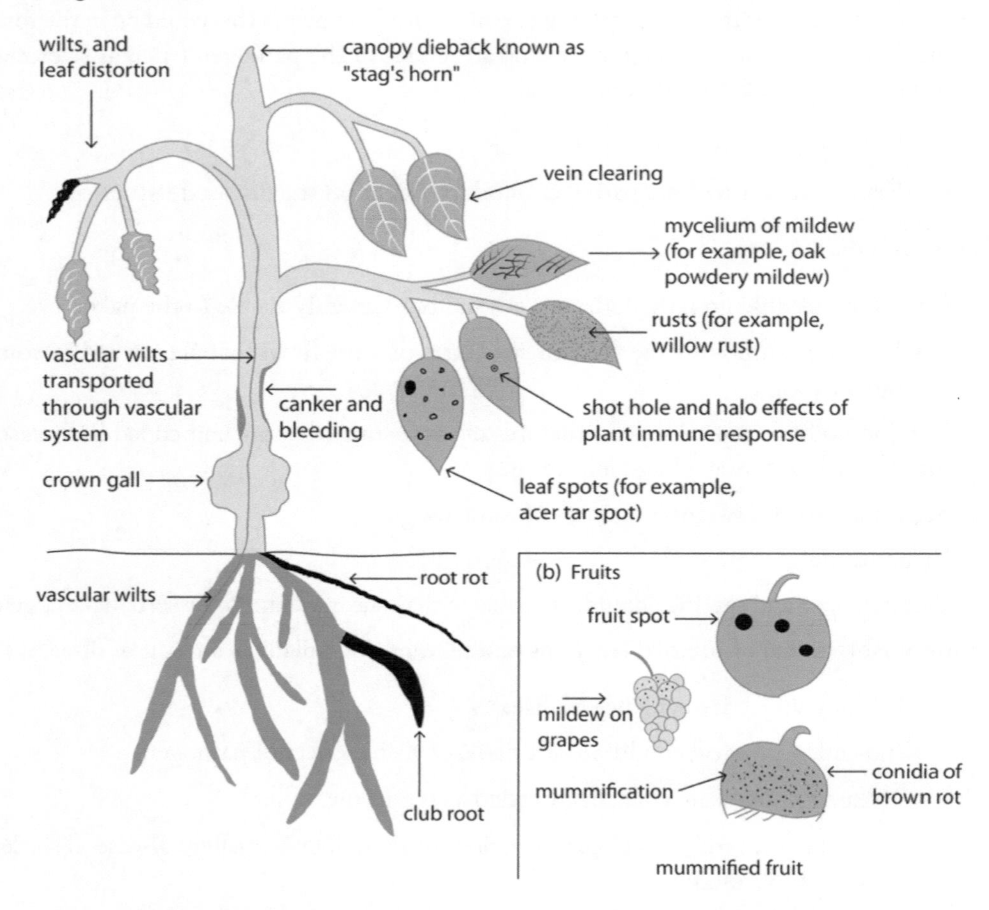

Infected plant organs showing a range of potential disease threats

The **disruption** of:

1. Plant Growth:
   - » Meristematic Activity
   - » Stunting
   - » Itiolation
   - » Epinasty
   - » Abscission

   » Storage Organs
   » Yield Loss
2. Absorption of Nutrients and Water:
   » Root Rots
   » Foot Rots
3. Translocation
4. Secondary Metabolism: Pigmentation Change
5. Normal Plant Functions:
   » Necrosis
   » Hydrosis
   » Yellowing
   » Plant Exudation
6. Water Transport: Vascular Wilts
7. Storage Organs
8. Reproduction

Major category of symptoms

1. **Colour change** - It is the most prevalent symptom caused by the disease effect on the host plant. The green pigment may completely disappear and be replaced with a yellow pigment known as yellowing or chlorosis. Etiolation describes chlorosis brought on by a lack of light. Sometimes, the diseased area turns colourless or devoid of colour. It is further referred as Chlorosis/ Yellowing, Albinism, Chromosis. E.g., Mosaic and Yellows.

   a. Chrorosis: devoid of chlorophyll (Yellowing), it can develop into necrosis at later stages.
   b. Chromosis: reduction of chlorophyll (Red or Orange or Purple discoloration)
   c. Albinism: devoid of any pigment (Colorless/ Hyaline)

2. Necrosis – Appearance of black or brown spots, sometimes whitish, on leaves or other plant parts, or as dry rots, internally degeneration of protoplast and death of tissues. Holonecrotic refers to plant cells or tissues that are completely dead, while plesionecrotic is used to describe nearly dead tissues such as the surrounding zone between necrotic and the healthy tissue of a fungal leaf spot.

   a. Restricted - E.g., Damping off, Spot, Speck, Blotch, Streak, Stripe

b. Extensive - E.g., Blight, Scald, Blast, Scorch, Anthracnose

c. Woody tissues - E.g., Cankers, Gummosis and Guttation

d. Storage organs - E.g., Rots, Leak, Mummification

3. **Hypertrophy** (Excessive enlargement of cells/ Increase in size)

Hyperplasia (Mass multiplication of cells/ Increase in number)

E.g., Galls, Tumours, Club root, Curls and Witches broom.

4. **Hypoplasia** – Reduced development

E.g., Shoe-string, Fern leaf, Little leaf, Dwarfing, and Stunting.

1. Type of infection

a. **Localized diseases:** These diseases are limited to a definite area of an organ or part(s) of a plant. E.g., leaf spots and anthracnose caused by fungi.

b. **Systemic diseases:** In these diseases, the pathogen spreads from a single infection point to infect all or most of the host tissues. E.g., Downy mildews caused by oomycete, mosaics, and leaf curls caused by viruses.

2. Type of perpetuation and spread

a. **Soil-borne diseases:** Damping off caused by fungi *Pythium* spp. and root rot caused by *Rhizoctonia* spp.

b. **Seed-borne diseases:** Loose smut of wheat caused by *Ustilago nuda tritici* (internally seed-borne) and Blast of rice caused by *Magnoporthe grisea* (externally seed-borne).

c. **Air-borne diseases:** Early leaf spot and late leaf spot of groundnut caused by *Cercospora arachidicola* and *Phaeoisariopsis personata* respectively.

d. **Vector-borne diseases: (Most of the virus):** *Cowpea aphid-borne mosaic virus.*

3. Extent of occurrence and geographic distribution

a. **Endemic diseases:** It is also known as enphytotic disease. When a disease is more or less constantly occurring year after year in a moderate to severe form in a confined locality then it is called as an endemic disease.

E.g.,

Wart disease of potato caused by *Synchytrium endobioticum* in Darjeeling.

Club root of cabbage caused by *Plasmodiophora brassicae* in Nilgiris

Sorghum rust caused by *Puccinia purpurea* is endemic in India, and

Black wart of potato (*Synchytrium endobioticum*) in Himachal.

b. **Epidemic or epiphytotic diseases:** An epidemic or epiphytotic refers to sudden outbreak of a disease periodically over a widespread area in a devastatingly severe form causing extensive losses or complete destruction.

E.g.,

Late blight of potato (*Phytophthora infestans*)

Wheat stem rust (*Puccinia graminis tritici*)

Powdery mildew of wheat (*Erysiphe graminis* var. *tritici*)

Powdery mildew of grapevine (*Uncinula necator*)

Sugarcane red rot *(Physalospora tucumanensis)*,

Downy mildew of grapevine (*Plasmopara viticola*) and

Rice blast *(Pyricularia oryzae)*.

c. **Sporadic diseases:** Sporadic diseases are those, which occur at irregular intervals over limited areas or locations.

E.g.,

Wheat loose smut (*Ustilago nuda tritici*) and

Udbatta disease of rice (*Ustilaginoidea virens*).

d. **Pandemic diseases:** A disease is said to be pandemic when it is prevalent throughout the country, continent, or world, involving mass mortality.

E.g.,

Late blight of potato (*Phytophthora infestans*), and

Wheat stem rust (*Puccinia graminis* var. *tritici*).

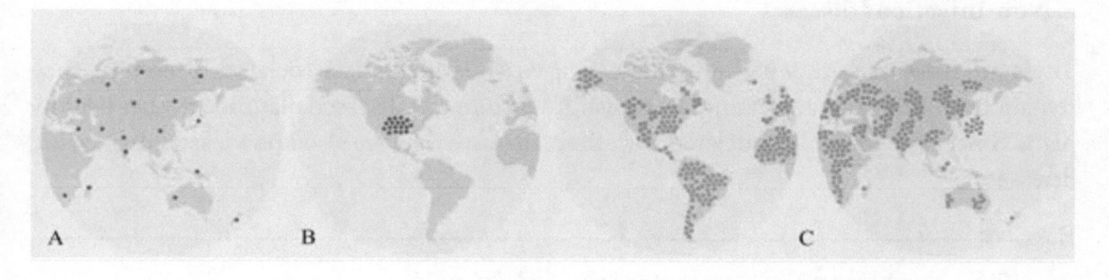

A. Endemic, B. Epidemic, C. Pandemic

4. Type of inoculum

- **a. Exogenous**: primary inoculum from outside sources. E.g., Wheat stem rust
- **b. Endogenous**: primary inoculum arrived within the area. E.g., Rice blast, Rice bacterial blight

5. Multiplication of inoculum

Based on the multiplication of inoculum, diseases are classified as

- a. In **simple interest disease** (monocyclic epidemics) the disease increase is like simple interest in money. Here inoculum comes from a reservoir and hence amount of inoculum for a given season's crop is fixed. So there is no repetition of the disease cycle within the crop season. Hence the disease spread will be slow. E.g., Soil pathogens like *Pythium* sp., *Rhizoctonia* sp. and *Sclerotium* sp.
- b. In **compound interest disease** (polycyclic epidemics) the disease increase is like compound interest in money. Inoculum gets multiplied several times (every 7 to 15 days for wheat rust) during crop growth in a season. So the disease spread is fast. E.g., Wheat stem rust, rice blast, powdery mildew diseases.
- **c. Bimodal polycyclic disease** single inoculum affects two growth stages of crop. E.g., blossom and fruit infection pear by fire blight disease, nursery and tillering stage infection of ragi blast.
- **d. Polyetic disease** (multiyear disease) can be caused by monocyclic and polycyclic pathogens. In some pathogens take several years to produce the inoculum. The amount of infection does not increase greatly within a year. Inoculum may increase steadily from year to year and cause severe outbreak. E.g., Apple powdery mildew is an example of a polyetic epidemic caused by a polycyclic pathogen, Dutch Elm disease and *Citrus tristeza virus* is a polyetic epidemic caused by a monocyclic pathogen

6. Based on causes

a. Non-infectious diseases

These are diseases with which no animate or virus pathogen is associated. Therefore, they remain non-infectious and cannot be transmitted from one diseased plant to another healthy plant. Since no parasite is associated with these diseases they are also known as non-parasitic diseases.

E.g.,

Tip rot or Necrosis of mango due to boron deficiency.

Black heart of potato due to unfavourable oxygen relations in stores and in field

Khaira disease of paddy due to non-availability of zinc to the plant.

Black tip of mango is due to sulphur toxicity.

Hollow and black heart of potato is due to excessive accumulations of CO_2 in storage

Bitter pit of apple is due to Ca deficiency.

b. Infectious diseases

These are diseases, which are incited by foreign organisms under suitable environments. A specific pathogen is responsible for the diseases. These diseases are always infectious sometimes contagious and are transmitted from diseased to healthy plants in the field and from one place to another through various agencies. When an animate cause is isolated or identified, its pathogenicity must be proved to confirm that it is the actual cause of the disease. There are many chances that more than one organism will be found in or on the diseased organ. Robert Koch (1882), a German scientist proposed a set of rules to demonstrate the association of a microbe with the disease in a host. These rules, known as Koch's postulates.

7. Based on causal agents

a. Parasites

i. Biotic agents: These are also called as animate or living organisms.

Prokaryotes

» Bacteria E.g., Citrus canker.

» Fastidious Vascular Bacteria E.g., Citrus greening, Pierce's disease of grapevine

» *Candidatus* Phytoplasma (MLO/ Phytoplasma) E.g., Little leaf of brinjal

» Spiroplasma E.g., Corn stunt, Citrus stubborn

Eukaryotes (organisms with membrane bound true nuclei)

» Fungi E.g., Wilt of banana

» Protozoa E.g., Hart rot of coconut

» Algae E.g., Red rust of mango

» Parasitic flowering plants or phanerogamic parasites E.g., Cuscuta

» Animals E.g., Nematodes

» Mesobiotic agents: They include viruses and viroids. They are infectious agents. They can be crystallized and are considered non-living. But their multiplication in the living plants ensures that they are living.

Viruses E.g., Tobacco mosaic

Viroids E.g., spindle tuber of potato

b. Non-parasites or Abiotic agents: These are also called non-infectious or physiological disorders.

- » Too low or too high temperature
- » Chilling injury
- » Freezing or frost injury
- » High temperature and dry winds
- » Lack or excess of soil moisture
- » Lack or excess of light
- » Lack of oxygen
- » Air pollution
- » Mineral deficiencies or toxicities
- » Soil acidity or alkalinity
- » Toxicity of pesticides / Latrogenic disease
- » Improper agricultural practices.

8. Based on dominance

a. **Pathogen-Dominant Diseases**: The pathogen is dominant over the host, but the relationship is transitory because the resistance of the host is less initially than it becomes eventually. Such pathogens are tissue nonspecific and attack young, immature root tissues or senescent tissues of mature plant roots. They seldom damage a rapidly growing, maturing root, so the period of disease development is short. Sometimes such pathogens are macerative, sometimes they are toxicogenic, or sometimes both can occur. The pathogenesis is due primarily to the virulence of the pathogen. Physiological specialization is relatively uncommon.

   Important pathogens are *Macrophomina*, *Phytophthora*, *Pythium*, *Rhizoctonia*, *Sclerotium*, etc.

b. **Host-Dominant Diseases**: In host-dominant diseases the host is dominant and the pathogen is successful only when factors favour the pathogen over the host. The resistance of the host is strong enough to keep the pathogen from advancing too rapidly against the host defenses during the vegetative growth phase and the host thereby prolongs the relationship. Damage is most severe in plants in the reproductive and senescent phases. In this group some pathogens are tissue non-specific but most are tissue specific. The pathogen may be macerative, toxicogenic or both.

Important pathogens are the species of *Armillaria*, *Polyporus*, *Poria*, *Helminthosporium*, *Fusarium*, etc.

Koch's postulates

Robert Koch (1843-1910) was the first to prove in 1876 that the animal disease, anthrax is incited by a bacterium, *Bacillus anthracis*.

To confirm scientifically whether a specific organism has caused a disease, it is necessary to perform Koch's postulates. He put forth the famous **'Koch's postulates'** for proving that a particular organism is the cause for a particular disease.

1. The suspected causal organism must be found associated with the disease in the plants.
2. It must be isolated and grown in pure culture on nutrient media.
3. When a healthy plant is inoculated with it, the original disease should be reproduced.
4. The same organism must be re-isolated in pure culture from experimentally infected plant and its characteristics must be exactly like that of original culture.

Practice Questions

1. ....is an endemic disease in darjeling
2. ....type of plant disease classification is highly useful for managing the disease
3. Koch postulate was proposed by............
4. ....disease is said to be pandemic when it is prevalent throughout the country, continent, or world, involving mass mortality.
5. Non-infectious plant diseases due to abiotic causes such as adverse soil and environmental conditions are termed as ....
6. Classify plant diseases on the basis of causes.
7. Explain the concept and steps in Koch postulates
8. Explain plant diseases on the basis of inoculum multiplication
9. Write the differences between disease, deficiency and disorder
10. Describe in detail about the occurance of diseases based on geographical regions

Suggested Readings

Manners, J. G. (1993). *Principles of plant pathology*. Cambridge University Press.

Surico, G. (2013). The concepts of plant pathogenicity, virulence/avirulence and effector proteins by a teacher of plant pathology. *Phytopathologia Mediterranea*, 399-417.

Bos, L., & Parlevliet, J. E. (1995). Concepts and terminology on plant/pest relationships: toward consensus in plant pathology and crop protection. *Annual review of Phytopathology*, *33*(1), 69-102.

Strange, R. N. (2003). *Introduction to plant pathology*. John Wiley & Sons.

Kranz, J. (1974). Comparison of epidemics. *Annual Review of Phytopathology*, *12*(1), 355-374.

Gilligan, C. A., Truscott, J. E., & Stacey, A. J. (2007). Impact of scale on the effectiveness of disease control strategies for epidemics with cryptic infection in a dynamical landscape: an example for a crop disease. *Journal of the Royal Society Interface*, *4*(16), 925-934.

Arneson, P. A. (2001). Plant disease epidemiology. *The Plant Health Instructor*.

Zadoks, J. C., & Schein, R. D. (1979). Epidemiology and plant disease management. *Epidemiology and plant disease management*.

Dhingra, O. D., & Sinclair, J. B. (2017). *Basic plant pathology methods*. CRC press.

Chapter - 4

Eukaryotic Fungi and Related Pathogens: Characters and Structures

The organisms we commonly refer to as fungi belong to three different kingdoms: the kingdom Fungi (Eumycota - true fungi), the kingdom Chromista, and the kingdom Protozoa.

Fungi: Fungi (properly pronounced 'fun-gee' with a hard gas is 'geyser') are eukaryotic, spore-bearing achlorophyllous, heterotrophic organisms that generally reproduce sexually and asexually and whose filamentous, branched somatic structures are typically surrounded by cell walls containing chitin or cellulose or both with many organic molecules and exhibiting absorptive nutrition.

The largest organism on Earth is thought to be a fungus, whose mycelium covers some 37 acres, is estimated to be about 700 years old and to weigh over 100 tonnes. However, it is not easy to prove whether this mycelium remains continuous or whether it has fragmented into several smaller mycelia

According to Hawksworth (1991), there are about 4,300 valid genera, many more that are synonyms, and about 70,000 species living as parasites or saprophytes on other organisms or their residues. More than 8,000 species cause plants disease. Fungi are divided into three kingdoms and eleven phyla.

» Oomycota, Zygomycota and Chytridiomycota are now generally accepted as separate phyla. The mycelium of these three phyla has many nuclei which are not marked off by crosswells (or non-septate mycelium) except where reproductive structures arise, a condition known as coenocytic. Asexual reproduction is by means of spores borne in sacs called sporangia.

Kingdom: **Protozoa** (Protists)	Kingdom: **Chromista** (Stramenopila)	Kingdom: **Fungi**
Phylum: Plasmodiophoromycota (Endoparasitic slime molds) Phylum: Dictyosteliomycota (Dictyostelid slime molds) Phylum: Acrasiomycota (Acrasid slime molds) Phylum: Myxomycota (True slime molds)	Phylum: Oomycota Phylum: Hyphochytriomycota Phylum: Labyrinthulomycota (slime molds)	Phylum: Chytridiomycota Phylum: Zygomycota Phylum: Ascomycota Phylum: Basidiomycota

» The Zygomycota have sexual spores called zygospores which are formed by the union of two similar sex cells or gametes; the Oomycota have sexual spores called oospores formed from dissimilar gametes; the Chytriodiomycota have neither type of sexual spore; the Ascomycota have septate mycelium and sexual spores in asci; the Basidiomycota have septate mycelium, frequently with clamp connections, and sexual spores.

» The protozoa have thalli as a motile mass of protoplasm (a plasmodium or myxamoeba – no mycelium) which is transformed into a mass of small, aseptate resting spores that on germination form motile cells with or without flagella. The protozoa include protists with amoeboid thalli and their status as fungi often has been questioned. The thalli of the protozoa are naked, amoeboid, plasmotic masses without cell walls and are termed plasmodia or pseudoplasmodia. They are also able to move by the formation of pseudopodia and by plasma-streaming. The Plasmodiophoromycetes is the only class which includes parasites of vascular plants. The best known species is *Plasmodiophora brassicae*, which causes "club root" of cabbage.

Somatic structures

1. **Thallus/ Stoma:** It is commonly called as vegetative body or fungal body. A thallus (plural: thalli) is a simple, entire body of the fungus devoid of chlorophyll with no differentiation into stem, roots, and leaves lacking a vascular system.
2. **Hypha** (hypha=web) (plural: hyphae): Hypha is a thin, transparent, tubular filament filled with protoplasm. It is the unit of a filamentous thallus and grows by apical elongation. Hyphae are usually with a diameter of 2-10 μm.
3. **Mycelium** (plural: mycelia): (mass of hyphae) A network of hyphae (aggregation of hyphae) constituting the filamentous thallus of a fungus. It may be colourless i.e., hyaline or coloured due to presence of pigments in cell wall. The mycelium may be ectophytic or endophytic.

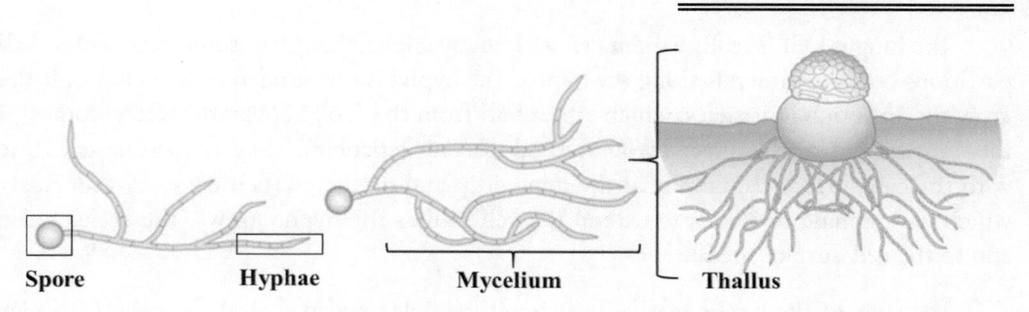

Fungi take a very different approach to the cellular animals and plants, in that they build bodies from hyphal filaments. It is a long, tubular, thread like filamentous and branched structures of the eucarpic thallus. It comprises of an outer cell wall and a cavity lined with protoplasm. The cytoplasmic streaming in the hyphae is unidirectional towards the tip. Hypha is usually uniform in thickness and its diameter is always more than 1 micron (1μ or micron), usually 1-2 μm in diameter. The length of the hypha may be very few micrometers in some fungi, but in others it may be several meter length.

The hyphae often have cross-walls, with pores connecting the cytoplasm on either side, and so can in this case be considered as chains of cellular units. However, although we sometimes refer to fungal cells, with the exception of the cellular yeasts, it is not always easy to say what exactly a fungal 'cell' is: is it a hypha or one unit in a hypha partitioned by cross-walls.

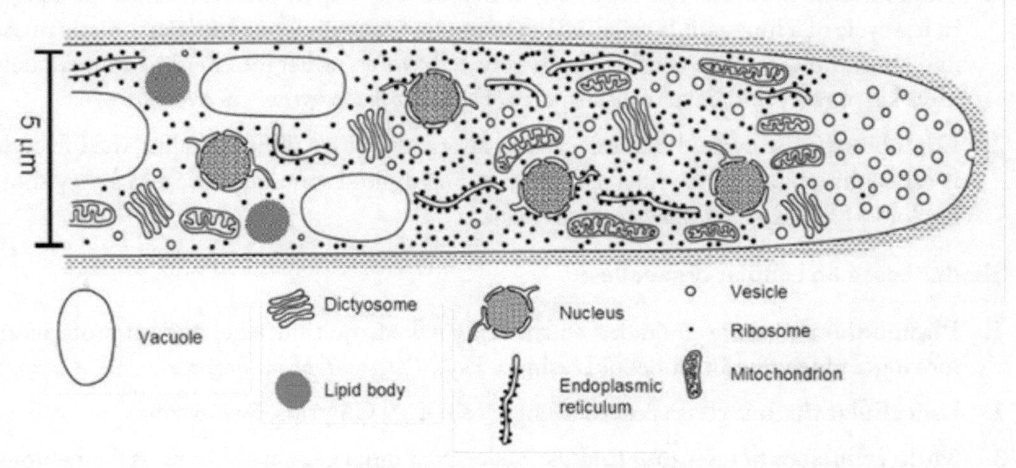

Mycelium components (Fungal cell)

The fungus 'cell' is really a giant cell with many nuclei, though in some fungi cross-wall partitions occur at intervals along the hypha. The hypha is surrounded by a hyphal wall. the growing tip is rich in vesicles, which either bud from the Golgi apparatus (dictyosomes) if these are present, or else directly from the endoplasmic reticulum. Many of these vesicles fuse with the cell-surface membrane at the hyphal tip and discharge their cargo by exocytosis, which will include materials to extend the cell wall as the hypha grows. The vesicles also add to the cell-surface membrane.

The apex of the hypha is usually rich with vesicles, called Apical Vesicular Complex (AVC) and dense vacuolar bodies namely Spitzenkorpers. The fungal growth takes place only at the thin and plastic apex of hyphae. The zone below hyphal tip is incapable of elongation but helps in the transportation of polysaccharides formed by the secretary action of Golgi apparatus to the growing tip. The Spitzenkorpers helps in the growth by synthesizing cytoplasm. The lateral branches are formed behind the apex in acropetal order. Each of these daughter branches behave like the parent hyphae and form secondary and tertiary branches behind their growing tip. The rate of mycelial growth is a character unique to each organism. Typically, the lower fungi have faster rate of mycelial growth in vitro than higher fungi. Hyphae may be lying in between the cells (intercellular) or lying within the cells (intracellular) in host tissues.

Various life cycle patterns displayed by fungi:

1. **Haplobiontic life cycle**: If there is only one free living thallus, which is haploid or diploid in life cycle of a fungus, it is called as haplobiontic life cycle. (long haploid somatic phase and short diploid phase confined to zygote cell, which undergoes meiosis immediately after karyogamy and develop ascospores) E.g., *Schizosaccharomyces octosporus.*
2. **Diplobiontic life cycle:** If haploid thallus alternates with a diploid thallus, the life cycle is called diplobiontic life cycle, which has a long diploid somatic phase and a very short haploid phase. E.g., *Saccharomyces ludwigii.*

Thallus based on cellular organelles:

1. **Plasmodium** (plasma = moulded body): It is a naked, multinucleate mass of protoplasm moving and feeding in amoeboid fashion. E.g., *Plasmodiophora brassicae.*
2. **Unicellular thallus:** consisting of a single cell. E.g., Chytrids, *Synchytrium*
3. **Multi cellular or filamentous thallus**: Majority of fungi i.e., a true fungus is filamentous, consisting of a number of branched, thread like filaments called hyphae. E.g., Many fungi, *Alternaria.*

Thallus based on nature of activity:

1. **Holocarpic/ Unicellular/ Non-mycelial type (holos=whole+ karpos=fruit)**: If the thallus is entirely converted into one or more reproductive structures, such thallus is called holocarpic thallus. E.g., *Synchytrium* and *Olpidium*

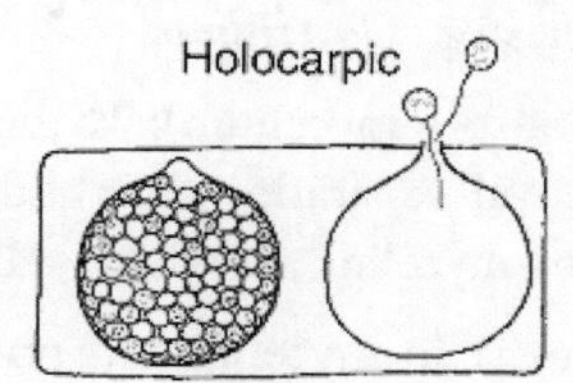

2. **Eucarpic/ Filamentous/ Mycelial type (Eu=good+ karpos=fruit)**: If the thallus is differentiated into a vegetative part which absorbs nutrients and a reproductive part which forms reproductive structures, such thallus is called eucarpic thallus. E.g., *Pythium* (Oomycota), Basidiomycota, Ascomycota and Mucoromycotina.

Eucarpic thalli may have one or several sporangia and are then termed monocentric or polycentric, respectively. In some species, there are both monocentric and polycentric thalli, so that these terms have descriptive rather than taxonomic significance.

A further distinction has been made, especially in monocentric forms, between those in which only the rhizoids are inside the host cell whilst the sporangium is external (epibiotic), in contrast with the endobiotic condition in which the entire thallus is inside the host cell. Examples of eucarpic fungi: All *Rhizophydium* spp.

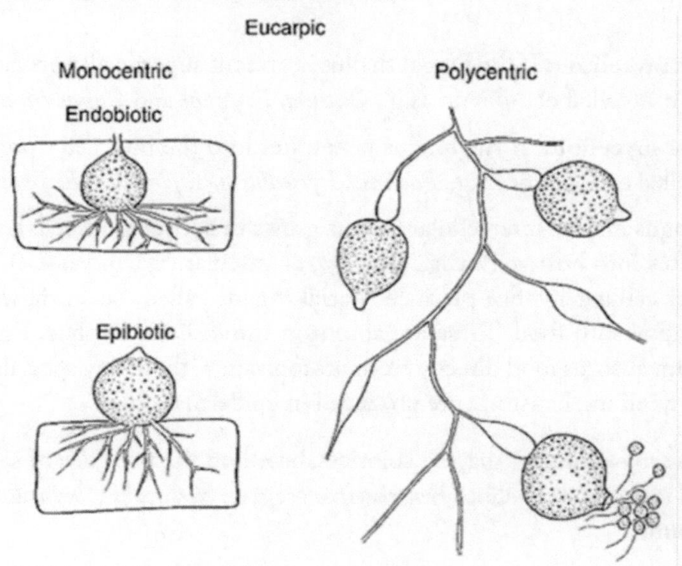

3. **Rhizomycelium** (Rhiza= root + mycelium): Root like absorptive mycelium. There are very few groups of fungi which do not have the filamentous type of body or instead they remain unicellular and may be round, oval or spherical in shape. They are mostly naked and uni or bi or multinucleate. If at all any hyphal growth is seen they will perhaps be rudimentary which is called rhizomycelium. Its function is to absorb nutrition. It may or may not penetrate the host. E.g., *Chytridium*
4. **Pseudomycelium** (Pseudo= false + mycelium): Thallus with serious of cells adhering end to end due to incomplete wall separation after budding are called pesudomycelium. E.g., Yeast. It differs from true mycelium in which cells have true septum.
5. **Dimorphism**: Some fungi exist in mycelial and non-mycelial form in response to temperature and nutrient requirements. E.g., *Taphrina* and smut fungi (Ustilaginales; Urocystales) from mycelial thallus in the host plant and yeast like holocarpic thallus in culture.

Types of mycelium

1. **Intercellular mycelium:** If the mycelium occurs in between the cells of the host, it is called as intercellular mycelium. E.g., White rust or white blisters of crucifers - *Albugo candida.*
2. **Intracellular mycelium:** If the mycelium occurs or present inside the host cell, it is called as intracellular mycelium. E.g., Wilt of linseed - *Fusarium oxysporum* f. sp. *lini.*
3. **Inter and intracellular mycelium**: If the mycelium occurs both in between and inside the host cell it is called as inter-intra cellular mycelium. E.g., Wilt of pigeonpea - *Fusarium udum*
4. **Ectophytic mycelium:** If the fungal thallus is present superficially on the surface of the host plant, it is called ectophytic. E.g., *Oidium, Erysiphe* and *Capnodium.*
5. **Endophytic mycelium**: If the fungus penetrates into the host cell / present inside the host, it is called endophytic. E.g., *Puccinia, Pyricularia, Taphrina, Rhizopus* and *Leveillula.*

Endophytic fungus may be intercellular (hypha grows in between the cells), or intra cellular (hypha penetrates into host cell). E.g., *Ustilago*, or vascular (xylem vessels) E.g., *Fusarium oxysporum.* Inter cellular hyphae produce special organs called haustoria which penetrate the host cell and absorb food. These are absent in intracellular hyphae. Endophytic intra cellular mycelium absorbs food directly from protoplasm without any specialized structures. In ectophytic mycelium, haustoria are produced in epidermal cells.

1. **Hyaline mycelium:** It may be colorless, based on the presence of septum, they are divided into septate hyaline (*Fusarium*) or aseptate hyaline (*Pythium* and *Phytophthora*) mycelium.

2. **Colored mycelium:** It may be variously pigmented. Based on the presence of septum, they are divided into septate colored (*Alternaria, Curvularia* and *Helminthosporium*) or aseptate colored (*Mucorales*) mycelium.
3. **Monokaryotic mycelium** (Haploid): mycelium contains single nucleus per cell. They usually form the haplophase of the life cycle of fungi (haplomycelium). E.g., Protozoa, Zygomycota and Ascomycota.
4. **Dikaryotic mycelium:** Mycelium has two sexually different compatible nuclei per cell. It denotes the dikaryophase of life cycle. It was maintained by clamp connection. E.g., Smut fungi.
5. **Diplo mycelium:** Mycelium with diploid nucleus. E.g., *Allomyces.*
6. **Homokaryotic mycelium**: Mycelium contains genetically identical nuclei may be by means of parasexuality. E.g., *Aspergillus nidulans.*
7. **Heterokaryotic mycelium**: Mycelium contains nuclei of different genetic constituents. Heterokaryosis is a condition in which genetically different nuclei are associated in same mycelium as result of mutation or anastomosis of hyphae. E.g., *Mucor* sp. and Smut fungi.

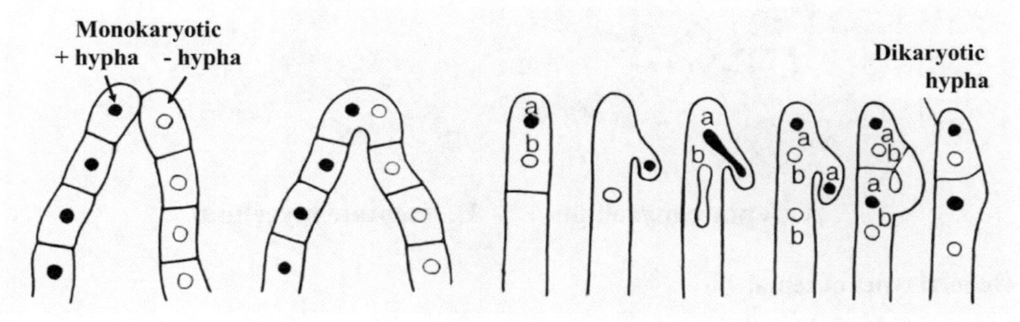

Septation in mycelium: (septum=hedge/partition) (plural: septa): Some fungal hyphae are provided with partitions or cross walls that divide the fungus into a number of compartments /cells. These cross walls are called septa.

1. **Aseptate hypha/ coenocytic hypha:** (Koinos=common, kytos=hollow vessel) A hypha without septa is called aseptate /non-septate/ coenocytic hypha wherein the nuclei are embedded in cytoplasm. E.g., Lower fungi like Oomycetes and Zygomycetes.
2. **Septate hypha**: A hypha with septa or cross walls is called septate hypha. E.g., common in higher fungi like Ascomycotina, Basidiomycotina and Deuteromycotina

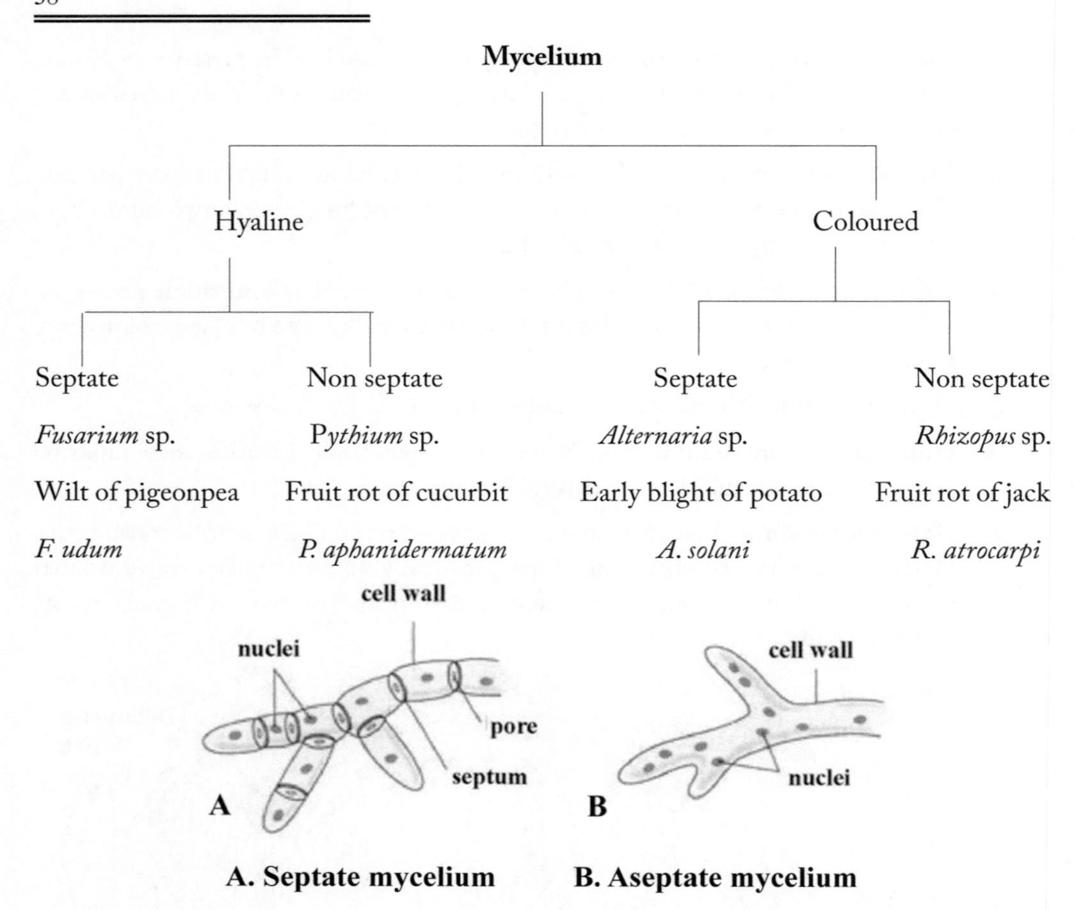

A. Septate mycelium **B. Aseptate mycelium**

General types of septa:

1. Based on formation:

a. **Primary septa:** These are formed in direct association with nuclear division (mitotic or meiotic) and are laid down between daughter nuclei separating the nuclei /cells. E.g., Higher fungi like Ascomycotina and Basidiomycotina.

b. **Adventitious septa:** These are formed independent of nuclear division and these are produced to delimit the reproductive structures. E.g., lower fungi like Oomycetes and Zygomycetes in which septa are produced below gametangia (sex organs) which separate them from rest of the cells.

2. Based on construction:

a. **Simple septa:** It is most common plate like, with or without perforation is.

b. **Complex septa**: A septum with a central pore surrounded by a barrel shaped swelling of the septal wall and covered on both sides by a perforated membrane termed the septal pore cap or parenthosome. E.g., Dolipore septum in Basidiomycotina.

3. Based on perforation:

a. **Complete/ solid septa:** A septum is a solid plate without any pore or perforations. E.g., Oomycota

b. **Incomplete septa:** A septum with a central pore was further classified.

1. **Perforated septa**: It is a simple septum perforated by a number of small bodies E.g., *Geotrichum*

2. **Bordered pit type septa**: Septal pore is surrounded by an overarching bifurcation of the septal margin looking like a typical bordered pot of traceid of *Pinus.* E.g., *Trichomycetes.*

3. **Ascomycetean type septa:** The septum has large central pore of 0.05-0.5 μm diameter. This type of septum commonly found in *Ascomycota, Uredinales* and *Ustilaginales.*

4. **Dolipore septa**: Dolipore septa are the complex type septa with barrel shaped central pore and hemispherical cap (parenthosome/ pore cap). They are found in both monokaryotic and dikaryotic mycelia of Agaricomycotina. The diameter of the septal pore varies from 0.1 to 0.2 μm.

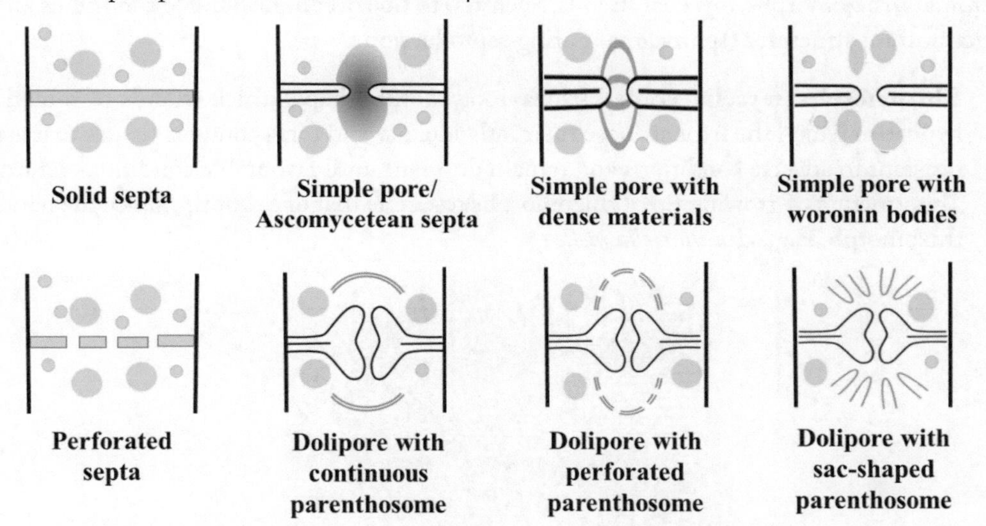

Fungal tissues: Plectenchyma: (plekein=to weave +enchyma=infusion)

Fungal tissues are called plectenchyma i.e., mycelium becomes organized into loosely or compactly woven tissue. These tissues compose various types of vegetative and reproductive structures.

Types of plectenchyma (fungal tissue):

a. **Prosenchyma:** It is a loosely woven tissue. The component hyphae retain their individuality which can be easily distinguishable as hyphae and lie parallel to one another. E.g., Trauma in *Agaricus.*

b. **Pseudoparenchyma**: It is compactly woven tissue. It consists of closely packed cells which are isodiametric or oval in shape resembling parenchymatous cells of plants and hence the name. The component hyphae loose their individuality and are not distinguishable as hyphae. E.g., Sclerotial bodies of *Sclerotium* and rhizomorph of *Armillariella.*

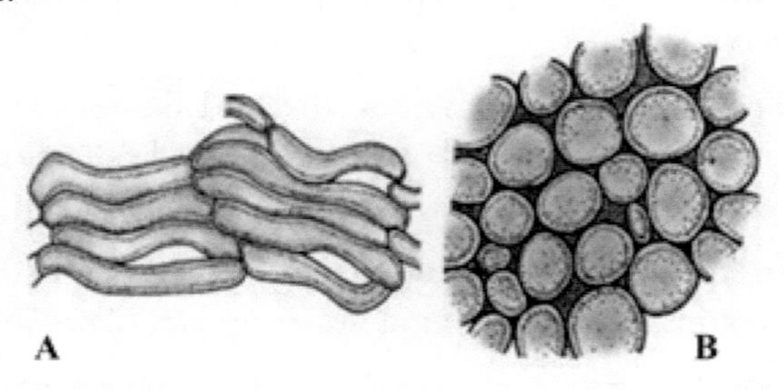

A. Prosenchyma B. Pseudoparanchyma

Usually the hyphal aggregations of fungal tissues make various somatic structures (stromata, rhizomorphs, mycelial strands, sclerotia) to tide over unfavourable conditions and fructification structures (sporocarps) during reproduction.

1. **Rhizomorphs (mycelial cords):** (rhiza=root, morph=shape) Thick strands of somatic hyphae in which the hyphae loose their individuality and form complex tissues that are resistant to adverse conditions and remain dormant until favourable conditions return. The structure of growing tip of rhizomorphs resemble that of a root tip, hence the name rhizomorph. E.g., *Armillariella mellea.*

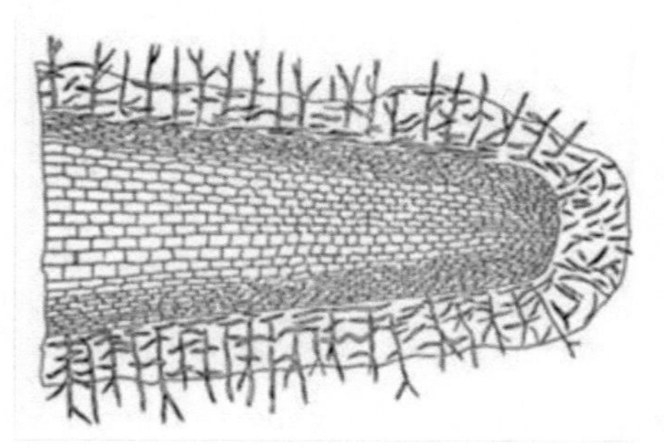

2. **Mycelial strand**: It is a parallel or interwoven prosenchymatous tissue of undifferentiated hyphae without well-defined apical meristem. E.g., *Basidiomycota* and *Ascomycota*.

3. **Stroma**: (stroma=mattress) plural: stromata. It is a compact somatic structure looks like a mattress or a cushion on which or in which fructifications (spores or fruiting bodies) are usually formed.

   a. **Substomatal stroma:** cushion like structure formed below epidermis in sub stomatal region from which sporophores are produced. E.g., Acerculi (*Colletotrichum*, *Pestalotia*); Sporodochia (*Fusarium*); Condiomata (*Cercospora personata*).

   b. **Perithecial stroma:** When reproductive bodies like perithecia of some fungi are embedded characteristically throughout periphery of stroma, such stroma are called perithecial stroma. E.g., *Claviceps, Xylaria.*

4. **Sclerotium:** (skleron=hard) (plural: sclerotia): It is a hard, round (looks like mustard seed)/ cylindrical or elongated (*Claviceps*) dark coloured (black or brown) resting body formed due to aggregation of mycelium, the component hyphae loose their individuality, resistant to unfavourable conditions by sysnthesis melanin and remain dormant for a longer period of time and germinate on the return of favourable conditions. E.g., *Sclerotium, Rhizoctonia.*

   » The size of the sclerotium varies from a few cells in some genera to as large as that of the size of man's head. E.g., *Polyporus mylittae*, in different genera.

   » It is usually dark on the outside and dull or white inside and madeup of prosenchyma and pseudoparenchymatous tissues.

   » It remains dormant and germinates on favourable conditions in the following ways

a. **Myceliogenous germination**: The sclerotia germinate directly into mycelium. E.g., *Sclerotium rolfsii* and *Rhizoctonia solani*

b. **Carpogenous germination**: The sclerotia germinate into sporocarp bearing stalks. The sclerotia of *Claviceps fusiformis* germinate and put forth small or long stalked stromata with a globose sporogenous head, which consists of flask shaped perithecium.

   » The sclerotia of *Sclerotinia sclerotiorum* germinate as stalked cups or apothecia which bearing asci and ascospore.

c. **Sporogenous germination**: the sclerotia give rise to spore. E.g., *Botrytis cinerea*

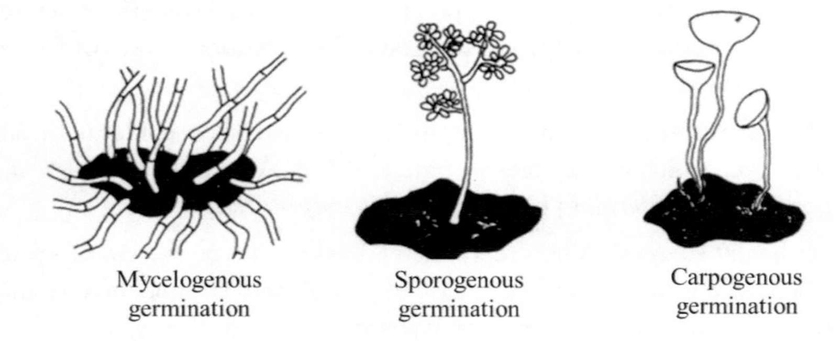

Modification of mycelium/ specialized somatic structures

Purpose of modifications is as follows

1. To obtain nourishment i.e., for nutrition.
2. To resist or tolerate unfavourable conditions for their survival i.e., over wintering, over summering.
3. To undergo reproduction.

Special somatic structure of fungi and chromista

1. **Haustorium**: (hauster=drinker) plural: haustoria. It is an outgrowth of intercellular or superficial hyphae or somatic hyphae regarded as special absorbing organ produced on certain hyphae by parasitic fungi for obtaining nourishment by piercing into living cells of host. They may be knob like (*Erysiphe*); globose (*Albugo*), simple (*Phytophthora*), branched (*Sclerospora*), elongated (*Uncinula*), finger like (*Peronospora*) and able to penetrate only in the cell wall not in the plasma membrane.

   Mostly present in obligate parasite and inappropriate

   E.g., Rust fungi, powdery mildew fungi, downy mildew parasites and phenarogamic parasites

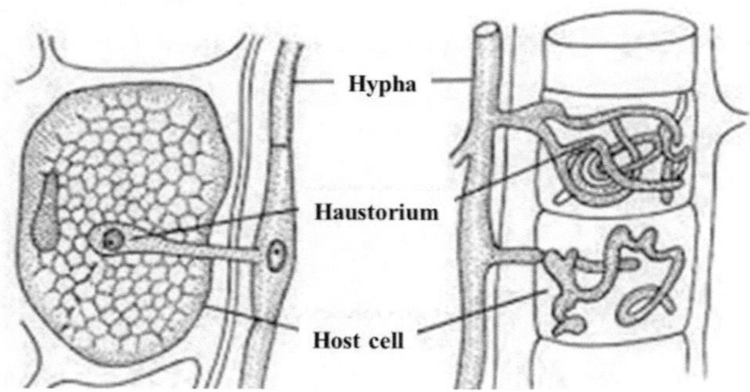

2. **Rhizoids**: (rhiza=root, oeides=like) it functions as both anchoring and absorbing organ. These are slender root like branched structures found in the substratum produced by some fungi which are useful for anchoring the thallus to substratum and for obtaining nourishment from the substrate. E.g., *Rhizopus stolonifer.*

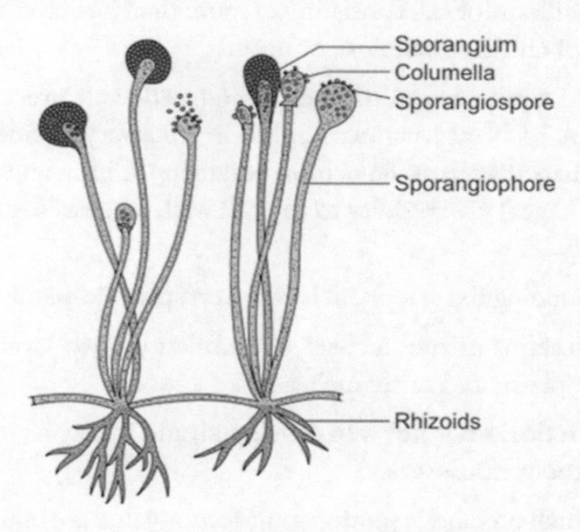

3. **Appresorium**: (apprimere=to press against) plural: appressoria. It is an anchoring organ of fungi. A flattened tip of hyphae or germ tube acting as pressing organ by attaching to the host surface and gives rise to a minute infection peg which usually grows and penetrates the epidermal cells of the host. E.g., *Puccinia, Colletotrichum, Erysiphe.*
   » Melonized appressoria: *Magnaporthe grisea*
   » Lobed appressoria: *Rhizoctonia solani*

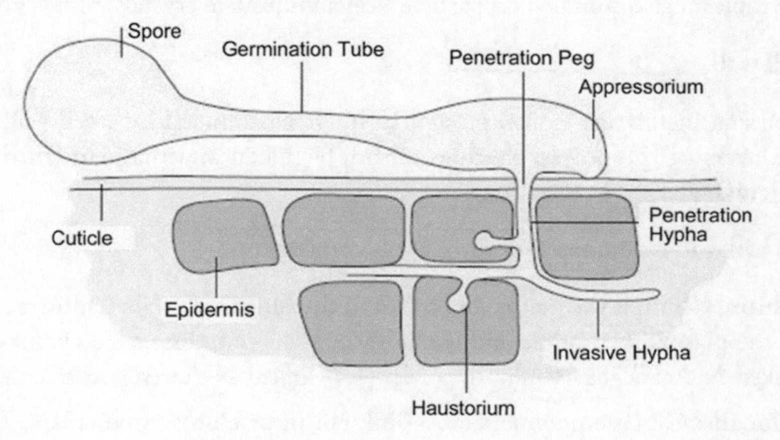

Fungal cell:

» Fungal cells are typically eukaryotic and lack chloroplasts.

» Cell is bounded by cell wall, which provides rigidity and shape to the cell is the outermost membrane of cell consisting of more than one layer with fibrous structure and made up of chitin or cellulose or both.

» Characteristic components of the hyphal cell wall are chitin (β-1,4-linked homopolymers of N-acetylglucosamine in microcrystalline state) and glucans (glucose polymers that cross-link chitin polymers). Chitin and β-glucans (via β-1,3- and β-1,6- linkages) give rigidity to the cell wall, whereas α-glucans are part of the cell wall matrix.

» The layer surrounding the cytoplasm is called cytoplasmic membrane or plasmalemma.

» Protoplasm contains a true nucleus surrounded by two layered membranes with nucleolus, cytoplasm and other inclusions.

» Endoplasmic reticulum is not well developed, and it may be rough with ribosomes or smooth without ribosomes.

» Vacuoles in which metabolic products are accumulated are bounded by a membrane called tonoplast.

» Ribosomes are protenaceous bodies scattered all over cytoplasm, play a role in protein synthesis.

» Mitochondria are the sites of respiratory activities.

» Lomasomes are the swollen membranous structures of plasmalemma.

» Cytoplasm also contains fat particles, calcium oxalate crystals, resins, glycogen.

Fungal cell wall

A single plasma membrane is also present in fungi, surrounded by a cell wall consisting of various layers of the polysaccharides chitin, β-glucan and mannan (in the form of mannoproteins).

The fungal cell wall is composed of three main components:

1. **Chitin:** Chitin is the main component of the fungal cell wall. Chitin is synthesized in the plasma membrane and is mainly consists of unbranched chains of β-(1,4)-linked-N-Acetylglucosamine or poly-β-(1,4)-linked-N-Acetylglucosamine (chitosan).
2. **Glucans:** Glucose polymers cross-link chitin or chitosan polymers. The glucose molecules like β-glucan are linked via β-(1,3)- or β-(1,6)- bonds while α-glucans are linked by α-(1,3)- or α-(1,4) bonds and protects the cell wall.

3. **Proteins:** In addition to structural proteins like glycosylated, the enzymes vital for the synthesis and breakdown of the cell wall are also present. Glycosylated contain mannose, and such proteins are also called mannoproteins or mannans.

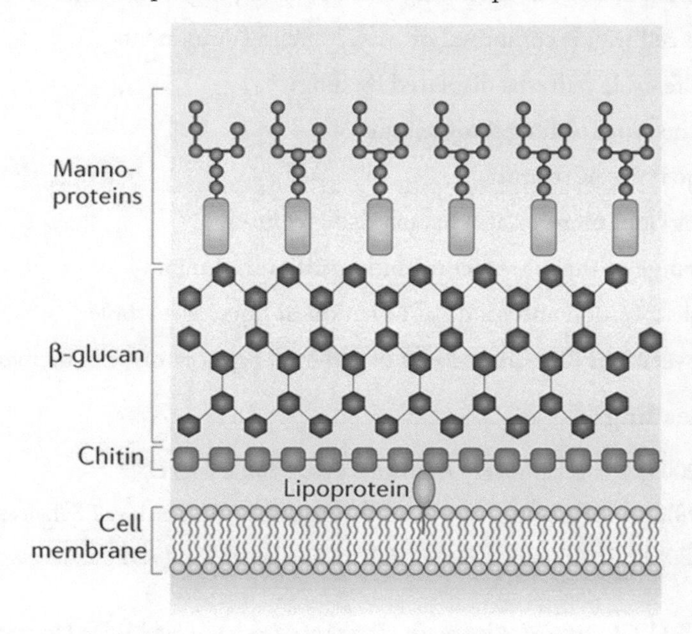

Fungal cell wall

Fungal nutrition:

» Fungi are heterotrophic with holophytic nutrition (absorptive type).

» The essential elements for fungi are, C, H, O, N, P, K, S, Zn, Fe, Mg, Mn, Mo, Cu and Ca.

» Reserve food material in the c ell may be either fat or carbohydrates.

» Fats may be present in the form of oil drops and carbohydrates in the form of glycogen or sugars.

» Starch is never present in the fungal cell.

Practice Questions

1. ……..commonly called as vegetative body or fungal body

2. ……. is a naked, multinucleate mass of protoplasm moving and feeding in amoeboid fashion

3. .... is an outgrowth of intercellular or superficial hyphae or somatic hyphae regarded as special absorbing organ by fungi.
4. .... functions as both anchoring and absorbing organ in fungi
5. The fungal cell wall is composed of .... main components
6. Describe life cycle patterns displayed by fungi
7. Illustrate the types of fungal mycelium
8. Explain the types of sclerotia
9. Define fungi and mention its characteristic features
10. Describe range of thallus structure and nutrition in fungi.
11. Describe distribution and mode of nutrition in fungi as a whole
12. What is mycelium? Give an account of different types of mycelia met within the fungi.

Suggested Readings

Agrios, G. N. (2005). Plant Pathology Academic Press. *San Diego, USA.*

Lucas, J.A. (1998). Plant Pathology and Plant Pathogens, 3rd ed. Blackwell Science.

Blackwell, M. (2011). The Fungi: 1, 2, 3... 5.1 million species?. *American journal of botany*, *98*(3), 426-438.

Ainsworth, G. C. (1976). *Introduction to the History of Mycology*. Cambridge University Press.

Money, N. P. (2016). Fungal diversity. In *The fungi* (pp. 1-36). Academic Press.

Carlile, M. J., Watkinson, S. C., & Gooday, G. W. (2001). *The fungi*. Gulf Professional Publishing.

Webster, J., & Weber, R. (2007). *Introduction to fungi*. Cambridge university press.

Watkinson, S. C., Boddy, L., & Money, N. (2015). *The fungi*. Academic Press.

Kavanagh, K. (Ed.). (2017). *Fungi: biology and applications*. John Wiley & Sons.

Deacon, J. W. (1997). *Modern mycology* (Vol. 3). Oxford: Blackwell Science.

Alexopoulos, C. J., Mims, C. W., & Blackwell, M. (1996). *Introductory mycology* (No. Ed. 4). John Wiley and Sons.

Chapter - 5

Asexual Reproduction in Fungi and Chromista

Reproduction: The life cycle of the majority of fungus includes both sexual (teleomorph stage) and asexual (anamorph stage) reproduction. In numerous fungi, the sexual stage might be uncommon or absent. Fungi reproduce predominantly by spores, which are reproductive units composed of one or a few cells. Spores may be produced by sexual reproduction or vegetative (asexual) reproduction, as in a type of budding in which the spore is genetically identical to the hypha from which it originated.

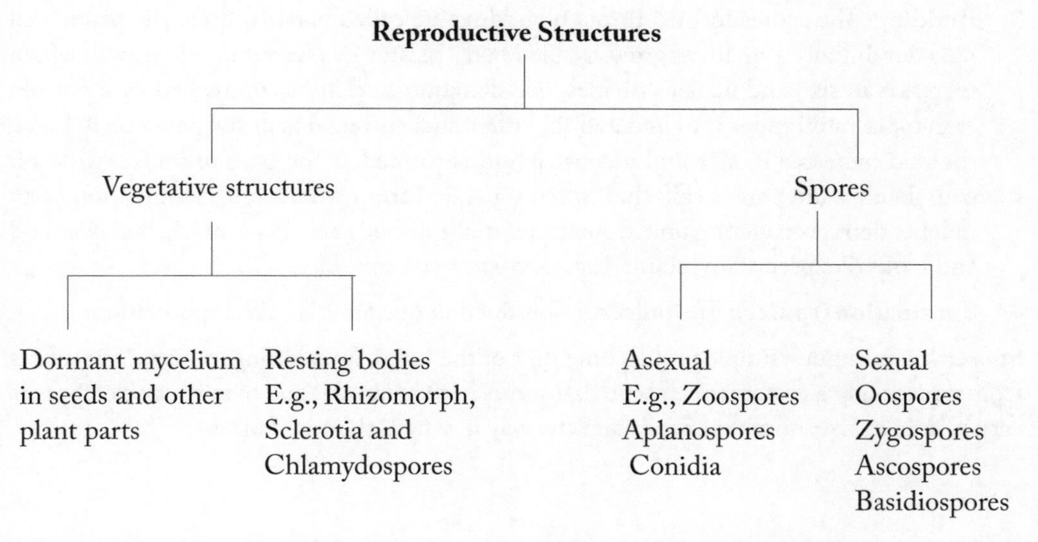

Asexual reproduction /non-sexual / vegetative / somatic reproduction/ anamorphic (without nuclear change): Asexual reproduction stage is also known as imperfect stage and technically called as anamorphic stage. There is no union of nuclei /sex cells/ sex organs. The most common type of asexual spore is a conidium (plural: conidia). Conidia may be formed by specialized hyphae or hyphal branches that cut off the terminal cell. A hypha that carries one or more conidia is a conidiophore.

» It is repeated several times during the life span of a fungus producing numerous asexual spores. Hence, it is more important for fungi than sexual reproduction.

» Asexual spores are formed after mitosis, hence also called mitospores.

Methods of asexual reproduction:

1. Fragmentation 2. Fission 3. Budding 4. Sporulation (production of spores)

1. **Fragmentation**: It is the most common method. Hypha of fungus breaks into small pieces, each broken piece is called a fragment, which function as a propagating unit and grows into a new mycelium. The spores produced by fragmentation are called **arthrospores** (arthron=joint) (spora=seed) or oidia. E.g., *Oidium, Geotrichum.* Sometimes, the contents of intercalary cells or terminal cells of hypha rounded off and surrounded by thick wall and formed as **chlamydospores** which are thick walled resistant spores produced either singly or in chains. E.g., *Fusarium oxysporum, Ustilago tritici.*
2. **Fission / Transverse fission:** The parent cell elongates, nucleus undergo mitotic division and forms two nuclei, then the contents divide into equal halves by the formation of a transverse septum and separates into two daughter cells. E.g., *Saccharomyces cerevisiae.*
3. **Budding:** The spores formed through budding are called **blastospores**. The parent cell puts out initially a small outgrowth called bud / blastos i.e., sprout or out growth which increases in size and nucleus divides, one daughter nucleus accompanied by a portion of cytoplasm migrates into bud and the other nucleus remains in the parent cell. Later, the bud increases in size and a constriction is formed at the base of bud, cutting off completely from parent cell. Bud, when separated from parent cell, can function as an independent propagating unit. Sometimes multiple buds are also seen i.e., bud over bud and looks like pseudomycelium. E.g., *Saccharomyces cerevisiae.*
4. **Sporulation (spores):** The process of production of spores is called sporulation.

Spore: It is a minute, simple propagating unit of the fungi, functioning as a seed but differs from it in lacking a preformed embryo that serves in the reproduction of same species. Spores vary in colour, size, number of cells and the way in which they are borne.

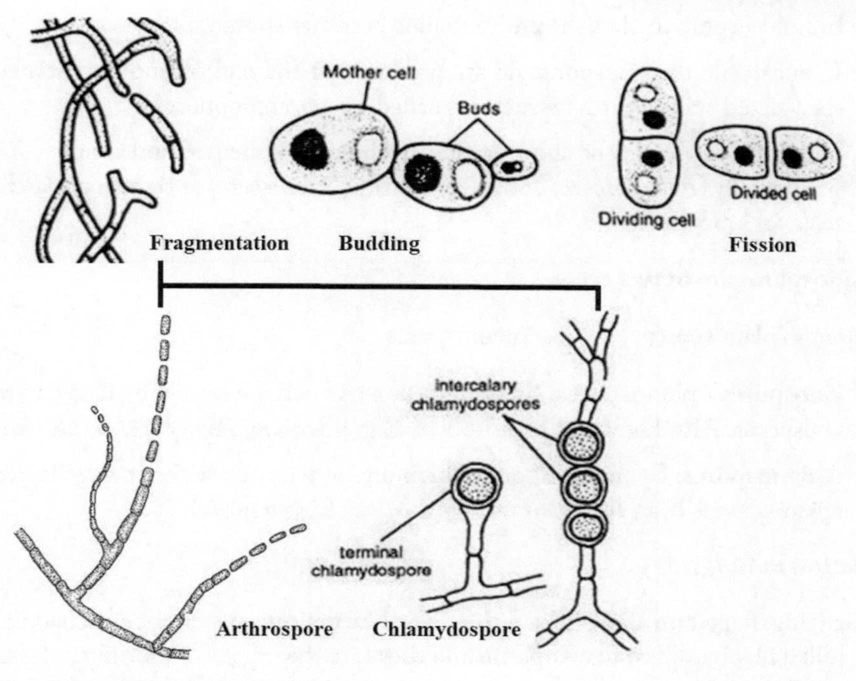

There are two main types of spores.

1. Endogenous (Sporangiospores)
2. Exogenous (Thallospores and Conidia)

Endogenous spores

Sporangiospores: Sporangiospores are spores that are produced in a sporangium (plural: sporangia). When the asexual spores are produced internally, within the sporangia, such spores are called sporangiospores. The sac like sporangia which produces sporangiospores is called sporangium. The special hypha bearing sporangium is called sporangiophore which may or may not be distinguishable from hypha. A small sporangium with or without columella containing a few or single spore is called as sporangiolum.

» The sporangia may be spherical, oval, cylindrical, variously lobed or pear shaped pigmented (*Rhizopus*) or hyaline (*Pythium, Phytophthora*); produced singly (*Pythium*) or in chains (*Albugo*) and intercalary (*Pythium*) or terminals (*Rhizopus*) of sporangiophore.

» Sporangium which is cylindrical in shape is called as merosporangium. E.g., *Saprolegnia* sp.

» Sporangium with columella is called as columellate sporangium. E.g., *Rhizopus stolonifer*.

- » In holocarphic thallus, the entire thallus becomes sporangium.
- » In eucarphic thallus, sporangia are produced at the end of undifferentiated or on specialized spore bearing structures called as sporangiophores.
- » Sporangiophore may be short or long, hyaline or pigmented and simple (*Albugo*) or sympodially (*Phytophthora*) or dichotomously (*Peronospora*) or right angle (*Plasmopara viticola*) branched.

Sporangiospores are of two types

a. Zoospores /planospores   b. Aplanospores

a. **Zoospores / planospores:** Sporangiospores which are motile by flagella are called zoospores. Also known as planospores. E.g., *Pythium, Phytophthora* (*Oomycota*).

b. **Aplanospores**: Sporangiospores which are non-motile without flagella are called aplanospores. E.g., *Rhizopus stolonifer, Mucor* (*Zygomycota*).

Flagellation in fungi:

Flagella: (sing. flagellum) Flagella are thin, hair like delicate structures attached to a basal granule called blepharoplast in cytoplasm and these are the organs of motility in lower fungi and aquatic fungi.

Types of flagella:

Based on flagella surface

a. Whiplash   b. Tinsel

a. **Whiplash:** A flagellum with long, thick, rigid basal portion and with a short, narrow, flexible, upper portion. It gives a whip like appearance to flagellum.

b. **Tinsel**: It is a feathery structure consisting of a long rachis with lateral hair like projections called mastigonemes or flimmers on all sides along its entire length. The number, position and nature of flagella play an important role in the classification of lower fungi.

Based on flagella numbers

a. **Uniflagellate zoospore:** A zoospore with a single flagellum, may be placed at anterior or posterior end of spore.

b. **Biflagellate zoospore:** A zoospore with two flagella, situated laterally or anteriorly on zoospore.

» One whiplash, one tinsel type flagella and equal in size. E.g., *Pythium aphanidermatum, Phytophthora infestans*,

» Both whiplash flagella, unequal in size (heterokont). E.g., *Plasmodiophora brassicae.*

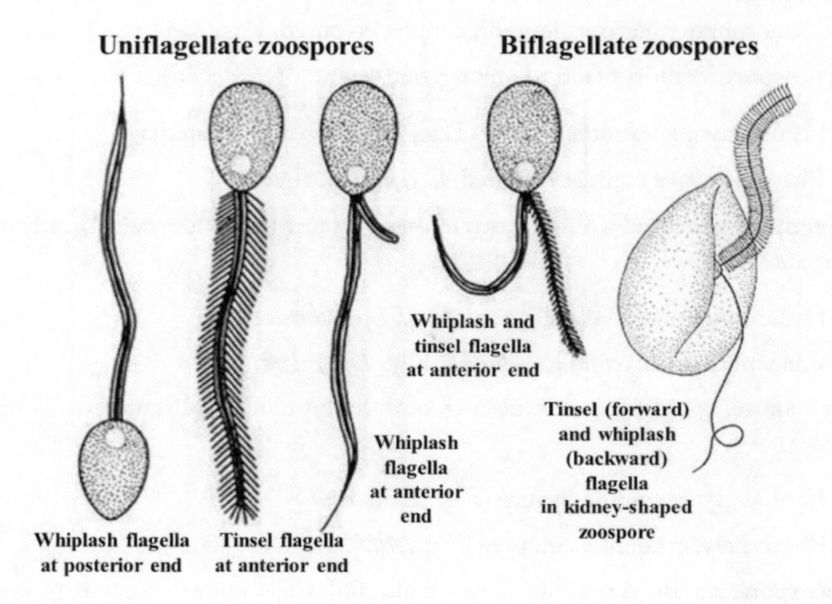

2. Exogenous

a. **Thallospores:** They arise directly from a pre-existing segment of the fungal thallus and detached from the parent hyphae and not by astrictions.

b. **Arthrospores**: It is specialized uninucleate cells, which function as spore. It is also called oidia. E.g., *Endomyces*, *Geotrichum*.

c. **Chlamydopores**: It is a thick-walled, non-deciduous, intercalary or terminal, asexual spore formed by the rounding of a cell or cells of the hypha. They do not detach from the parent hypha and remain viable even after the hypha decays, with reserve food materials and possess thick wall to withstand unfavourable conditions. The walls may often have covered with dark melanin pigment. E.g., *Fusarium*

d. **Conidia / Conidiospores**: (konis=dust; oides=like) Conidia are non-motile asexual spores which may arise directly from somatic hyphae or from specialized conidiogenous cells (a cell from which conidia are produced) or on conidiophore (hypha which bear conidia). Conidia are produced freely on conidiophore ie., at the tips or sides of conidiophore or may be produced in specialized asexual fruiting bodies *viz.*, pycnidium, acervulus, sporodochium and synnemata.

Saccardo described the conidia based on shape, septation and colour.

I. Amerospore: conidia non septate (single celled), spherical, ovoid to elongated, or short cylidric.

a. Hyalosporae (Hyalo =colourless): conidia hyaline. E.g., *Phoma*

b. Phaeosporae (phaeo=coloured): conidia coloured. E.g., *Sphaeropsis*

II. Didymospore: conidia ovoid to oblong, one septate (two celled)

a. Hyalodidymae: conidia hyaline. E.g., *Fusarium* micro conidia

b. Phaeodidymae: conidia coloured. E.g., *Botryodiplodia*

III. Phragmospore: conidia oblong, two to many septate (3 or more celled), only transverse septa present.

a. Hyalophragmae: conidia hyaline. E.g., *Pyricularia*

b. Phaeophragmae: conidia coloured. E.g., *Drechslera*

IV. Dictyospore: conidia ovoid to oblong, both longitudinal and transverse septa present (muriform).

a. Hyalodictyae: conidia hyaline. E.g., *Epicoccum*

b. Phaeodictyae: conidia coloured. E.g., *Alternaria*

V. Scolecospore: conidia thread like to worm like, filiform, septate or aseptate (one to several celled). E.g., *Cercospora*

VI. Helicospore (Allantospore): conidia spirally cylindrical, curved (allantoid), septate or aseptate. E.g., *Helicomyces*

VII. Staurospore: conidia stellate (star shaped), radially lobed, septate or aseptate (one to several celled). E.g., *Actinospora*

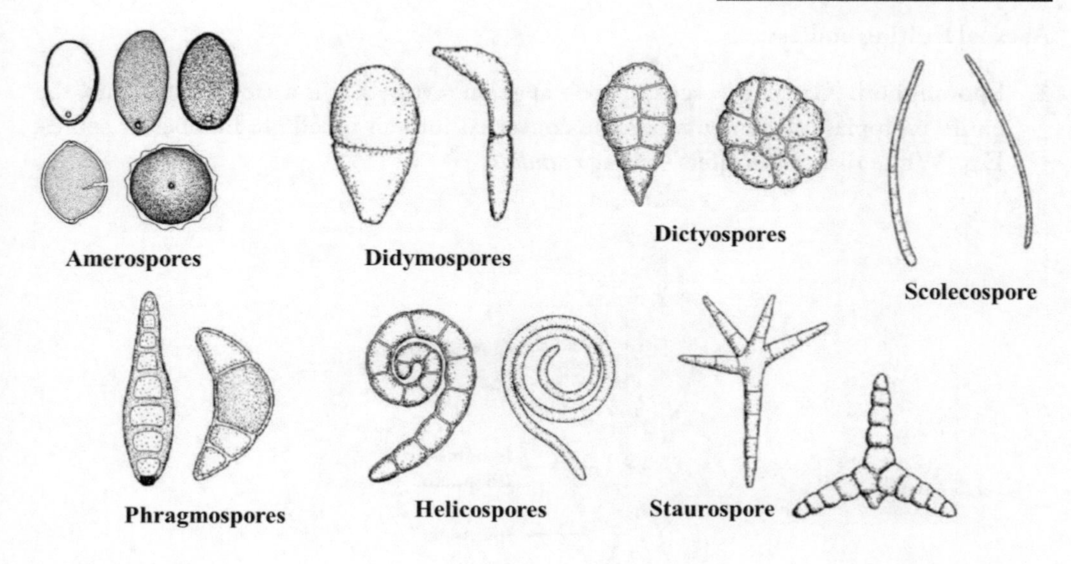

Color of conidia

a. **Hyalosporae**: Cell wall of conidia hyaline.

b. **Phaeosporae**: Cell wall of conidia coloured/ pigmented.

Classification of conidia based on colour and septation

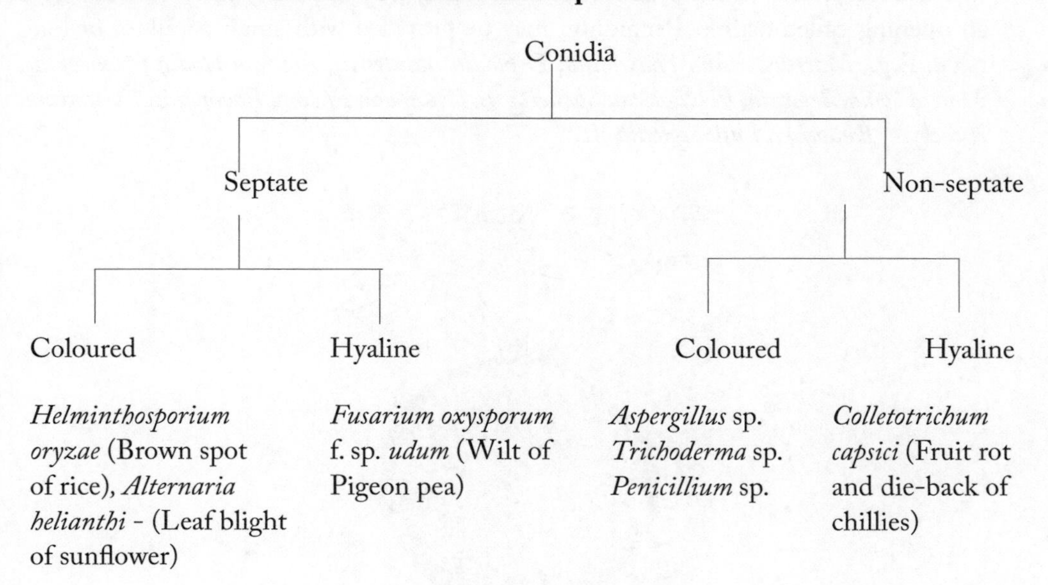

Asexual fruiting bodies:

1. **Sporangium:** (Gr. spora = seed, spore + angeion = vessel). It is a sac-like structure; the entire protoplasmic contents become converted into an indefinite number of spores. E.g., White rust of Crucifers - *Albugo candida.*

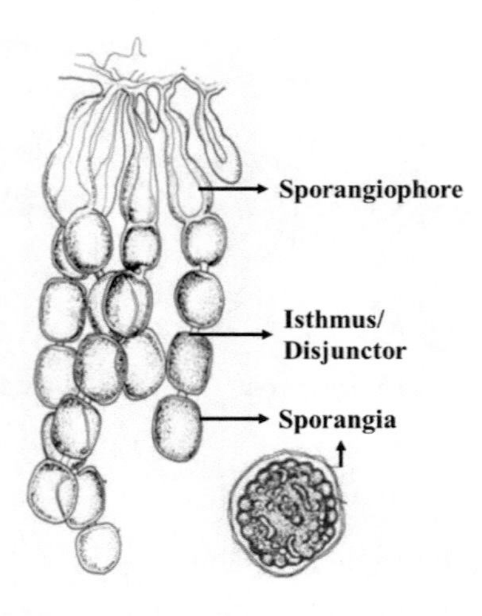

2. **Pycnidium: (**plural: pycnidia) It is a globose or flask shaped fruiting body lined in side with conidiophores which produce conidia. It may be completely closed or may have an opening called ostiole. Pycnidium may be provided with small papillum or long neck. E.g., *Macrophomina phaseolina, Diplodia natalensis, Botryodiplodia theobromae, Macrophoma, Septoria, Phyllosticta, Sphaeropsis, Cytospora, Phoma, Phomopsis, Fusicoccum, Ascochyta, Endothia, Lasiodiplodia* etc.

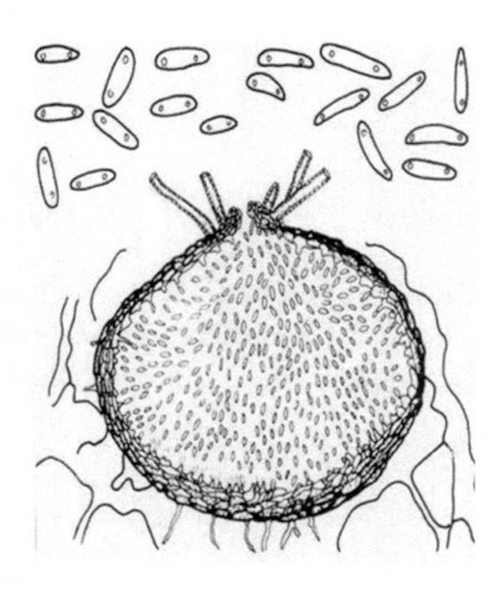

3. **Acervulus**: (plural: acervuli) A flat or saucer shaped fruiting body with a stromatic mat of hyphae producing conidia on short conidiophores. An acervulus lacks a definite wall structure and not having an ostiole or definite line of dehiscence. E.g., *Colletotrichum, Gloeosporium, Spaceloma, Melanconium, Cylindrosporium, Pestalotiopsis.*

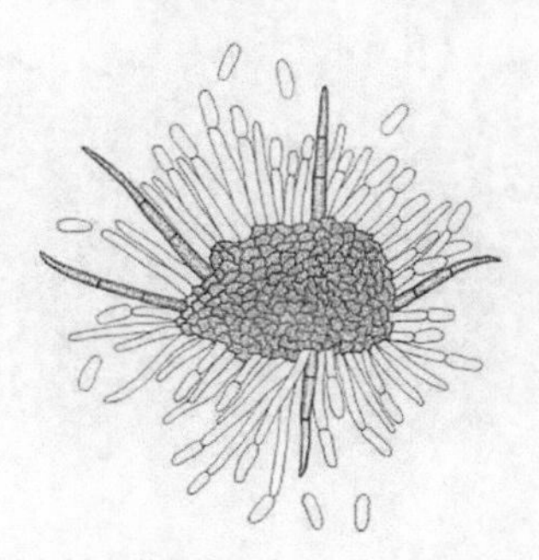

4. **Sporodochium:** (plural: sporodochia) A cushion shaped asexual fruiting body. Conidiophores arise from a central stroma and they are woven together on a mass of hyphae and produce conidia. E.g., *Fusarium, Epicoccum, Mycospherella, Tubercularia* etc.

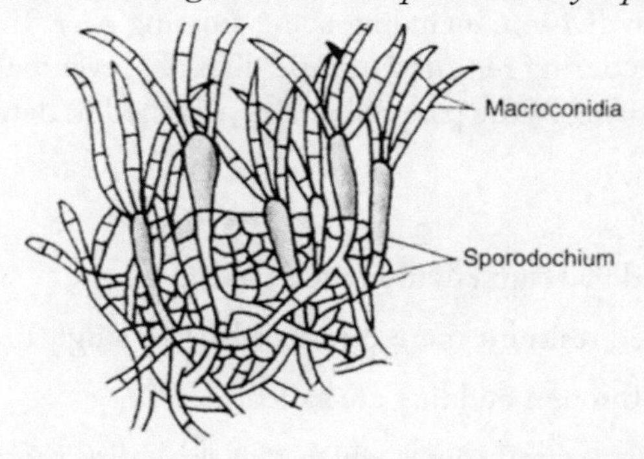

5. **Synnemata (Coremium):** (plural: synnema) A group of conidiophores often united at the base and free at the top. Conidia may be formed at its tip or along the length of synnema, resembling a long-handled feather duster. E.g., *Ceratocystis, Arthtrobotrys, Graphium.*

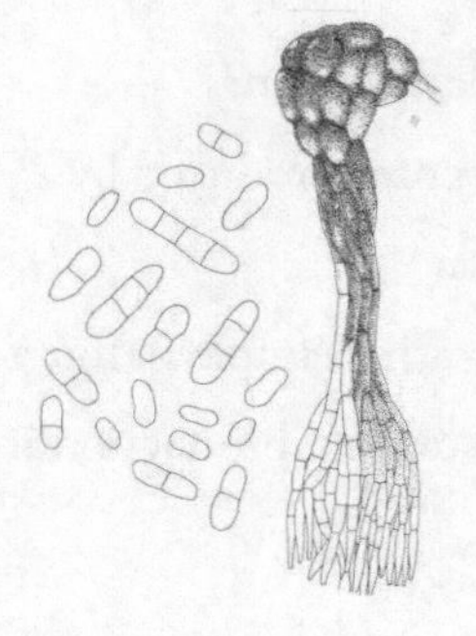

6. **Sorus** (Sori): (Gr- Sorus- Heap). The spore bearing hyphae are grouped into small to large masses or clusters E.g., *Puccinia, Hemileia* (Rust) and *Sporisorium* (Smut).

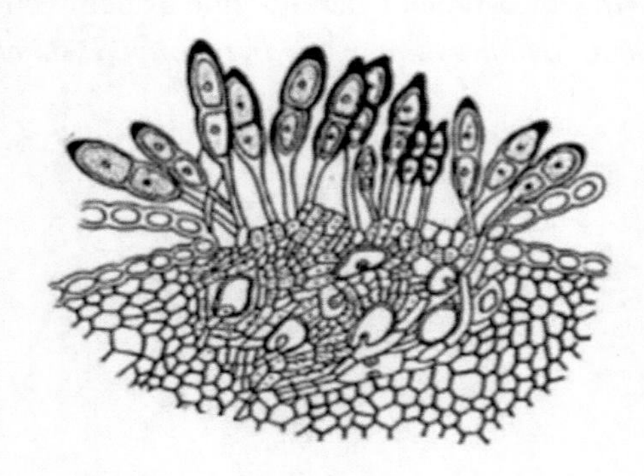

Somatic recombination: Some fungal species are known to undergo genetic recombination between individuals by parasexual events. This can occur when hyphae of two different individuals merge, resulting in genetically different nuclei in the cell. This is called a heterokaryon. The genetically distinct haploid nuclei may fuse, resulting in diploid nuclei that multiply by mitotic division, including some crossing over. This diploid homokaryon may shed chromosomes during repeated mitotic divisions, eventually producing a haploid nucleus that is a genetic mix of the parental strains, disussed in detail in parasexual cycle.

Practice Questions

1. The spores produced by fragmentation are called……..
2. ……….. thick walled resistant spores produced either singly or in chains by fusarium
3. The spores formed through budding are called……
4. ……. are non-motile asexual spores which may arise directly from somatic hyphae or from specialized conidiogenous cells
5. …….. flat or saucer shaped fruiting body with a stromatic mat of hyphae producing conidia on short conidiophores
6. Describe the asexual fruiting body of fungi
7. Explain the differences in zoospores produced by *Pythium* and *Phytophthora*
8. Describe somatic recombination
9. Illustrate the parts of acervuli with example pathogens
10. Describe various methods of asexual reproduction in fungi.

Suggested Readings

Descals, E., & Moralejo, E. (2001). Water and asexual reproduction in the ingoldian fungi. *Botanica complutensis*, *25*(1).

Gams, W., & Seifert, K. A. (2008). Anamorphic fungi. *eLS*.

Raghukumar, S. (2017). Fungi: Characteristics and classification. In *Fungi in Coastal and Oceanic Marine Ecosystems* **(pp. 1-15). Springer, Cham.**

Brandt, M. E., & Warnock, D. W. (2011). Taxonomy and classification of fungi. *Manual of clinical microbiology*, 1744-1755.

Elliott, C. G., & Maheshwari, R. (1994). Reproduction in fungi: genetical and physiological aspects. *Journal of Genetics*, *73*(1), 55.

Chapter - 6

Sexual Reproduction in Fungi and Chromista

Sexual reproduction involves union of two compatible nuclei or sex cells or sex organs or somatic cells or somatic hyphae for the formation of new individuals. Sexual stage is perfect stage and technically called as teleomorphic stage. Sexual cycle normally occurs once in the life span of the fungus. Sexual spores or sexual structures which contain sexual spores are thick walled, resistant to unfavourable conditions and are viable for longer period and thus these spores help the fungus to perpetuate from one season to another, hence these are called as resting spores. Sexual spores are definite in number.

Sex organs of fungi:

1. **Gametangia:** Sex organs of fungi are called gametangia containing gametes or gamete nuclei.
2. **Gametes**: Sex cells are called as gametes.
3. **Antheridium**: (plural: antheridia) Male gametangium is called as antheridium. Male gametangium is small and club shaped.
4. **Oogonium / Ascogonium**: (plural: oogonia/ascogonia): The female gametangium is called Oogonium (oomycetes) or ascogonium (ascomycotina). Female gametangium is large and globose shaped. Male gametes are called antherozoids or sperm or spermatozoids. Female gametes are called egg or oosphere.

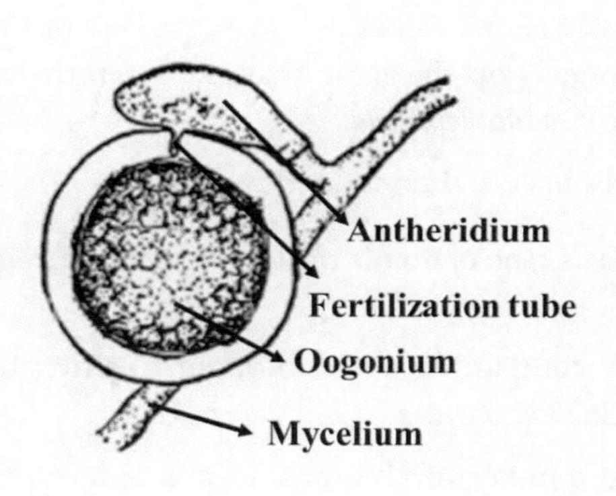

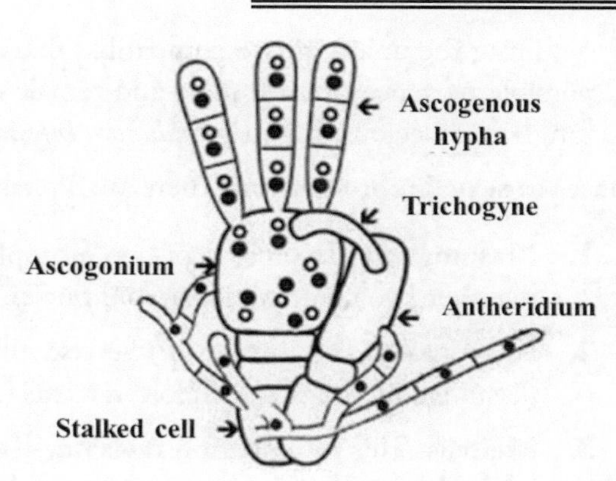

Nature of gametes

1. **Planogametes**: If gametes are motile, they are called planogametes.
2. **Isogametangia:** If gametangia are morphologically similar or identical i.e., indistinguishable as male and female, they are called as isogametangia.
3. **Isogametes**: If gametes are similar morphologically, they are called as isogametes.
4. **Heterogametangia:** If gametangia differ morphologically in size and structure, they are called as heterogametangia.
5. **Heterogametes**: If gametes differ morphologically, they are called heterogametes.
6. **+ or – signs:** In some sexually undifferentiated fungi, male and female are symbolically designated as '+' (male) and '–' (female).

Classification of fungi based on bisexualism

1. **Monoecious fungi / hermaphroditic fungi:** (mono=single, oikos=home) The fungi which produce distinguishable male and female sex organs on the same thallus, which may or may not be compatible are called monoecious/ hermaphroditic fungi. E.g., *Pythium aphanidermatum*.
2. **Dioecious fungi**: (di=two, oikos=home) The fungi which produce distinguishable male and female sexorgans on two different thalli ie., there will be separate male and female thalli. E.g., *Phytophthora infestans.*

Classification of fungi based on compatibility:

1. **Homothallic fungi:** Fungi in which both sexes occur on same thallus, which can reproduce sexually by itself without the aid of another thallus ie., self compatible / self fertile are called homothallic fungi. E.g., *Pythium aphanidermatum, Neurospora crassa.*
2. **Heterothallic fungi:** A fungal species consisting of self sterile (self incompatible) thallus

requiring the union of two compatible thalli for sexual reproduction, regardless of the possible presence of both male and female organs on the same thallus. Heterothallic fungi are dioecious. E.g., *Phytophthora infestans, Mucar muceda.*

Phases in sexual reproduction: There are 3 phases in sexual reproduction.

1. **Plasmogamy**: The fusion of two protoplasts (the contents of the two cells), brings together two compatible haploid nuclei.
2. **Karyogamy**: It is union of two sexually compatible nuclei brought together by plasmogamy to form a diploid nucleus (2n) i.e., zygote.
3. **Meiosis**: This is reduction division. The number of chromosomes is reduced to haploid (n) i.e., diploid nucleus results into haploid nucleus.

In fungi like parasites and lower fungi (Chromista, Chitridiomycetes and Zygomycotina) plasmogamy, karyogamy and meiosis occurs at regular intervals / sequence. i.e., karyogamy follows immediately after plasmogamy.

In higher fungi (Ascomycotina, Basidiomycotina), karyogamy is delayed, as a result the nuclei remain in pairs (dikaryotic phase- n+ n condition), which may be brief or prolonged.

Dikaryon: A pair of genetically different nuclei, lying side by side with out fusion for a considerable period of time is called dikaryon. A cell containing dikaryon is called dikaryotic cell. And the process is known as dikaryogamy.

Methods of sexual reproduction: five methods.

1. Planogametic copulation. 2. Gametangial contact 3. Gametangial copulation 4. Spermatisation 5. Somatogamy

1. **Planogametic copulation (gametogamy):** This involves the union of 2 naked gametes one or both of which are motile.
   - b. **Isogamy (Isogamous planogametic copulation)**: If both gametes are motile and similar. E.g., *Plasmodiophora brassicae.*
   - c. **Anisogamy (Anisogamous planogametic copulation)**: If both gametes are motile but dissimilar. E g. *Allomyces macrogynus*
   - d. **Heterogamy**: If gametes are dissimilar, one motile, another is non-motile. E.g., *Monoblepharis polymorpha.*

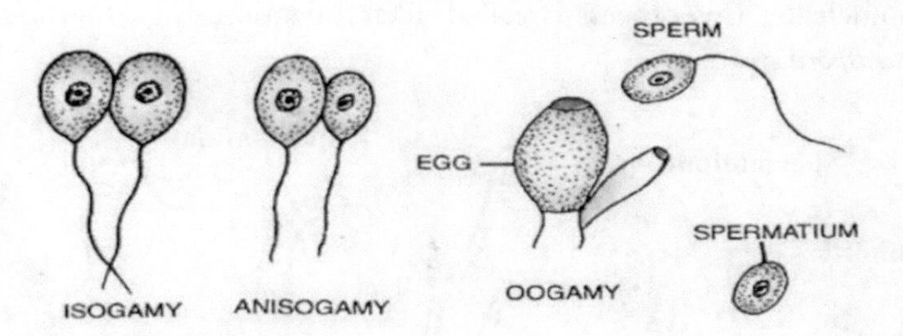

2. **Gametangial contact (gametangy / oogamy):** Male and female gametangia come in contact. At the place of contact, dissolution of wall occurs and a fertilization tube is formed. The contents of male gametangium migrate into female gametangium through a pore or fertilization tube developed at the point of contact.
   » The gametangia do not loose their identity. E.g., *Pythium aphanidermatum.*

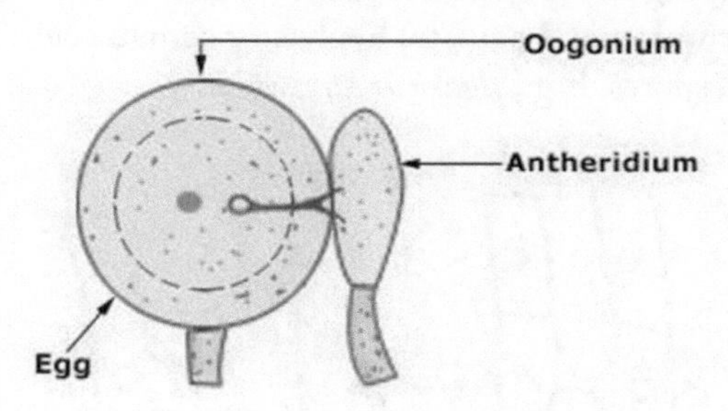

3. **Gametangial copulation (gametangiogamy):** The isogametangia come in contact, their intervening wall dissolves leading to fusion of entire contents of two contacting gametangia to form a single unit. Gametangia loose their identity. The protoplasts fuse and the unit increases in size. E.g., *Rhizopus stolonifer.*

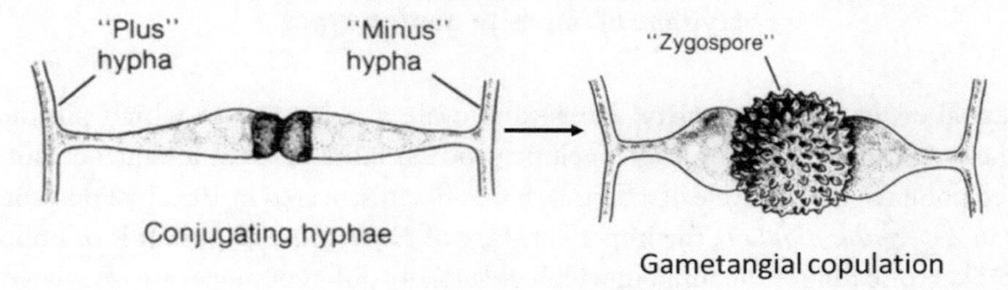

Gametangial copulation

4. **Spermatization**: Minute, uninucleate male cells called as spermatia which are produced on spermatiophores in a fruiting body (pycnium) are carried to female reproductive structures called receptive hyphae. Spermatia and receptive hyphae come in contact and contents of male spermatium migrate into female receptive hypha, thus making

the cell binucleate. This process is called dikaryotization. E.g., *Puccinia graminis* var. *tritici*, *Neurospora* sp.

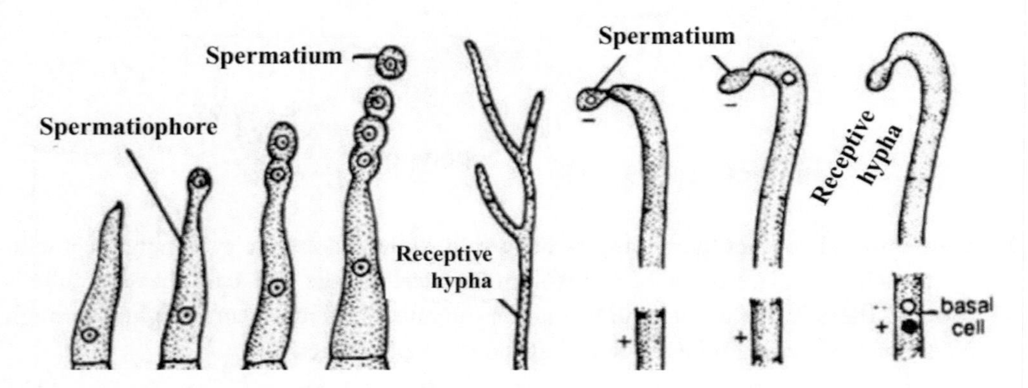

5. **Somatogamy:** Many higher fungi do not produce sex organs. In such cases somatogamy takes place. It is the union of 2 somatic hyphae or somatic cells representing opposite sexes to form sexual spores. E.g., *Agaricus campestris.*

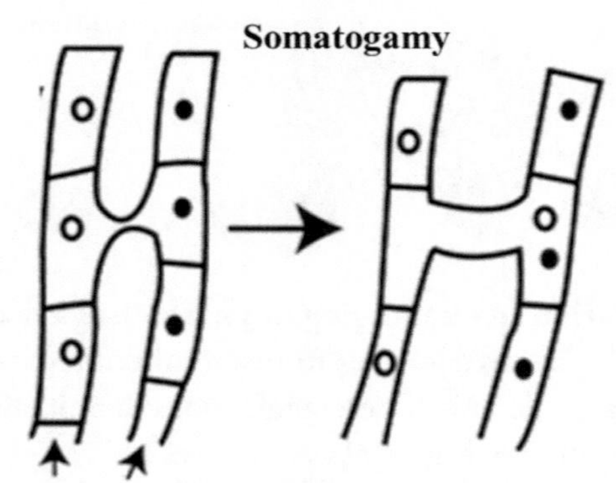

Parasexual cycle / Parasexuality: Parasexual cycle is a process in which plasmogamy, karyogamy and haploidisation (non-meiotic process) takes place in a sequence but not at specified points in the life cycle of a fungus. It was first discovered in 1952 by Pontecorvo and Roper in *Aspergillus nidulans*, the imperfect stage of *Emericella nidulans*. It is of importance in heterokaryotic fungi (The fungi in which genetically different nuclei are associated in the same protoplast or mycelium). This is one of the methods of producing variability of fungal pathogens. In majority of Deuteromycotina, the true sexual cycle is absent but derives many of the benefits of sexuality through this cycle.

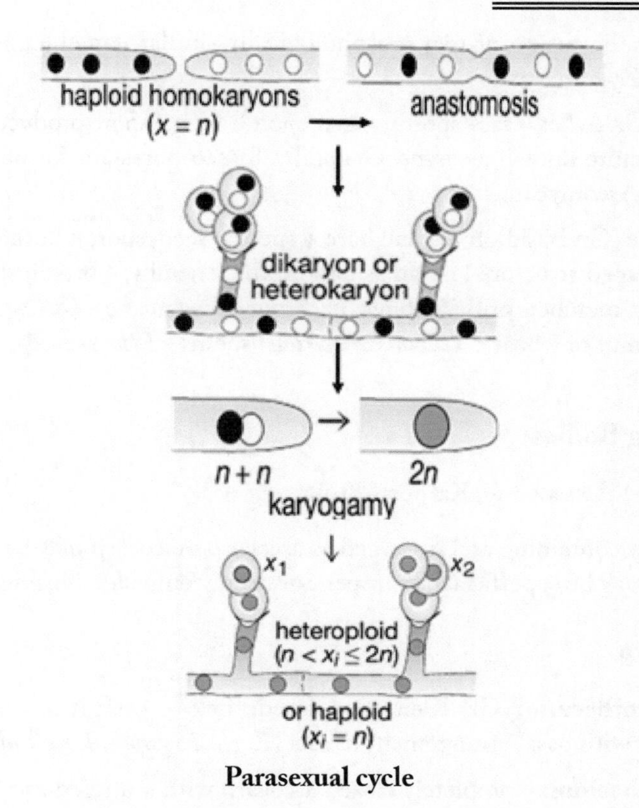

Parasexual cycle

The parasexual cycle parallels events in the sexual cycle, resulting in genetically unique haploid offspring but without a meiotic reduction. Hyphae of genetically unique homokaryotic parents grow towards each other by chemotaxis and fuse. Nuclei from each unique strain migrate within the fused hypha, which is now considered a heterokaryon. Haploid nuclei in the heterokaryon undergo karyogamy to create a heterozygous diploid nucleus. The diploid nucleus undergoes mitotic recombination to produce a recombined genotype. In growing hyphae, gradual loss of chromosomes due to repeated rounds of mitotic nondisjunction results in haploidization and unique genotypes in various sectors of mycelium.

Different types of sexual spores: Sexual spores are formed after meiosis, hence also called meiospores.

1. Oospores 2. Zygospores 3. Ascospores 4. Basidiospores

1. **Oospore** (Gr. Oon = egg + spora = seed / spore): A thick walled sexual resting spore produced by the union of two morphologically different gametangia. E.g., *Pythium, Phytophthora*, members of class Oomycetes.
2. **Zygospore** (Gr. zygos = yoke + spora: = seed, spore): A thick walled sexual resting spore

produced by the fusion of two morphologically similar gametangia. E.g., *Rhizopus*, members of Zygomycota.

3. **Ascospore** (Gr. Askos = sac+ sporos = seed, spore): Sexual spore produced in a specialized sac-like structure known as ascus. Generally, 8 ascospores are formed. E.g., *Erysiphe,* members of Ascomycota.
4. **Basidiospore** (Gr. Basidion = small base + spora = seed, spore): Sexual spore produced on a club-shaped structure known as basidium. Generally, 4 basidiospores are formed. E.g., *Puccinia,* members of Basidiomycota. Smut of sugarcane - *Ustilago scitaminae,* Bunt or stinking smut of wheat - *Tilletia caries*, mushrooms - *Pleurotus, Agaricus, Ganoderma, Armillaria* etc.,

Sexual Fruiting Bodies

1. Ascocarp: (Gr) Ascus - Sac; Karpos - Fruit)

The fruiting body containing asci is termed as ascocarp. Ascocarp may be formed singly or in groups. They may be superficial - erumpent or deeply embedded inside the substratum.

Types of ascocarp

a. **Chaesmothecium:** (Gr) Kleistos– Closed: theke – case; it is a completely closed ascocarp with basal arrangement of asci (E.g.,) *Erysiphe, Leveillula, Uncinula* etc.
b. **Cleistothecium:** Completely closed ascocarp with scattered asci. E.g., *Penicillium* and *Aspergillus*
c. **Apothecium**: (Apothecke - Store house) Saucer shaped open ascocarp. E.g., *Diplocarpon rosae, Sclerotinia* spp.
d. **Perithecium**: (Gr- Peri - around; theke – case) Flask shaped closed ascocarp with an opening called ostiole and ascus wall is single. E.g., *Claviceps*
e. **Ascostroma / Pseudothecium**: More or less flask shaped resembling perithecium whose asci are not arranged in hymenial layer and asci with bitunicate wall and disperse ascospores suddenly. E.g., *Venturia inaequalis* - Apple scab

2. Basidiocarp: Fruiting body of basidiomycotina bearing basidiospores.

a. **Mushroom** is a soft fruiting body without hymenial pore. E.g., *Pleurotus* and *Agaricus*
b. **Bracket** is a hard fruiting body with hymenial pores. E.g., *Ganoderma*

Practice Questions

1. Male gametangium is called as ……..
2. The fungi which produce distinguishable male and female sex organs on the same thallus is ………..
3. Sex organs of fungi are called gametangia containing…..
4. The union of two mycelial protoplasts are called as ………
5. Parasexual cycle was first discovered in 1952 by …….
6. Describe sex organs of fungi
7. Explain the differences between homothallic and heterothallic
8. Describe the phases in sexual reproduction
9. Illustrate the methods of sexual reproduction
10. Illustrate the parasexual cycle with steps involved

Suggested Readings

Wallen, R. M., & Perlin, M. H. (2018). An overview of the function and maintenance of sexual reproduction in dikaryotic fungi. *Frontiers in microbiology*, *9*, 503.

Ni, M., Feretzaki, M., Sun, S., Wang, X., & Heitman, J. (2011). Sex in fungi. *Annual review of genetics*, *45*, 405.

Heitman, J., Sun, S., & James, T. Y. (2013). Evolution of fungal sexual reproduction. *Mycologia*, *105*(1), 1-27.

Heitman, J. (2015). Evolution of sexual reproduction: A view from the fungal kingdom supports an evolutionary epoch with sex before sexes. *Fungal Biology Reviews*, *29*(3-4), 108-117.

Castrillo, L. A., Roberts, D. W., & Vandenberg, J. D. (2005). The fungal past, present, and future: germination, ramification, and reproduction. *Journal of invertebrate pathology*, *89*(1), 46-56.

Hawker, L. E. (2016). *The physiology of reproduction in fungi*. Cambridge University Press.

Chapter - 7

Taxonomy of Plant Pathogenic Fungi and Related Eukarya

Taxonomy is the study of naming and classification of organisms.

Nomenclature: naming of organisms.

Classification: Scientific categorization of the organisms in a hierarchical series of groups.

Classification and nomenclature are governed by the International Code of Botanical Nomenclature (ICBN). Naming of fungi as Binomial name (as in plants).

Two major kingdoms: *Plantae* and *Animalia*

1821 - E.M. Fries published *Systema Mycologicum* for naming fungi (Friesian System), based on morphology.

1860 - Ernst Haeckel proposed 3rd kingdom, *Protista* (Algae, bacteria, fungi and protozoa).

1938 - Herbert Copeland included 4th kingdom, *Monera* (bacteria and algae).

1969 - Whittaker recognized Fungi as separate kingdom.

1973 - G.C. Ainsworth proposed kingdom Mycota with two divisions viz., Myxomycota (slime moulds) and Eumycota (fungi and fungi like organisms).

Eumycota is further divided into five sub-divisions based on asexual and sexual reproduction.

- » Mastigomycotina (fungi like organisms).
- » Zygomycotina (bread moulds).
- » Ascomycotina (sac fungi).

» Basidiomycotina (cup fungi).

» Deuteromycotina (fungi imperfecti).

1978 - Whittaker and Margulies proposed five kingdoms as follows

» Monera (=prokaryotes).

» Protista (=algae and protozoa).

» Plantae (=plant).

» Mycetae (=fungi).

» Animalia (=animals) [Important beginning to establish monophyletic groups (groups with an ancestor and all its descendants)].

Carl Woese divided the kingdom *Monera* into two group's viz., *Eubacteria* and *Archaebacteria* based on molecular level comparisons of genes.

1990 - Carl Woese suggested new taxonomic category, Domain (Super kingdom) and divided all living organisms into 3 domains.

Based on their cell structure and ribosomal RNA (16S of prokaryotes and 18S of eukaryotes)

» ***Bacteria*** (includes bacteria)

» ***Archaea*** (includes archaeobacteria)

» ***Eukarya*** (includes eukaryotes)

In general, rRNA is highly conserved with the slow rate of evolutionary change. Hence, classification based on nucleic acid sequence analysis and cell ultra structure clearly shows the evolutionary relationship and helps to understand the close relatives of an organism.

This classification is known as Phylogenetic relationship, and this represented diagramatically by a phylogenetic tree or evolutionary tree or dendrograms or cladograms.

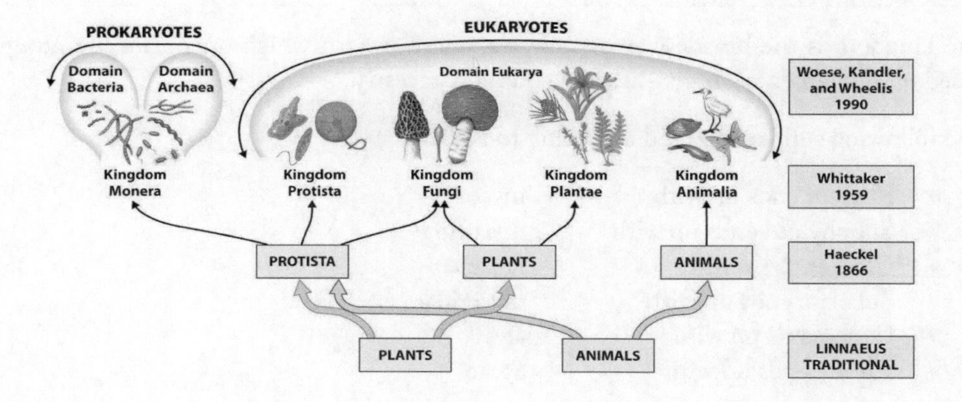

Differences between prokaryotes and eukaryotes

	Prokaryotes	**Eukaryotes**
Type of Cell	Always unicellular	Unicellular and multi-cellular
Cell size	Ranges in size from 0.2 μm – 2.0 μm in diameter	Size ranges from 10 μm – 100 μm in diameter
Cell wall	Usually present; chemically complex in nature	When present, chemically simple in nature
Nucleus	Absent. Instead, they have a nucleoid region in the cell	Present
Ribosomes	Present. Smaller in size and spherical in shape	Present. Comparatively larger in size and linear in shape
DNA arrangement	Circular	Linear
Mitochondria	Absent	Present
Cytoplasm	Present, but cell organelles absent	Present, cell organelles present
Endoplasmic reticulum	Absent	Present
Plasmids	Present	Very rarely found in eukaryotes
Ribosome	Small ribosomes	Large ribosomes
Lysosome	Lysosomes and centrosomes are absent	Lysosomes and centrosomes are present
Cell division	Through binary fission	Through mitosis
Flagella	The flagella are smaller in size	The flagella are larger in size
Reproduction	Asexual	Both asexual and sexual
Example	Bacteria and Archaea	Plant and Animal cell

The kingdom is the broadest taxonomic classification into which organisms are grouped based on fundamental similarities and common ancestry.

The following suffixes are used according to ICBN

» Phylum ends up with mycota
» Subphylum ends up with mycotina
» Class ends up with mycetes
» Subclass ends up with mycetidae
» Order ends up with ales
» Family ends up with aceae

» No standard ending for genera and species.

The species are sometimes further broken down into varieties/ sub-species/ *forma specialis* (f.sp.) or form species, physiological races and cultural races, variant, biotypes, etc.

Taxonomy of Fungi

The taxonomy was originally based on morphological characters, but molecular techniques now in use have resulted in several changes in classification, and further changes are expected. For classification and naming the different fungi, we follow the Catalogue of Life and Index Fungorum (Species fungorum).

» The fungi-like organisms belong to two distinct kingdoms. Historically, they have been categorized as fungi due to their shared physical characteristics (particularly their mycelial habit). Now they are classed as Chromista and Protozoa kingdoms. The oomycete pathogens belong to the Chromista kingdom and the Oomycota phylum.

» The Plasmodiophorida species are now classified in the Phylum Cercozoa, which is also part of the Chromista kingdom. This group of organisms is called heterokonts or stramenopiles because they possess two terminal flagella of uneven length throughout particular life cycles. Previously, Plasmodiophorida species were placed inside the kingdom Protozoa, to which slime moulds now belong, still it is under the review and debate. Hence, in this book, flagellate protozoa are placed in Protozoa kingdoms, like followed in the Indian schools and universities.

» Nematodes are animals and members of the kingdom Animalia's phylum Nematoda. They are included in this book because, in many ways, they resemble other disease-causing organisms, and nematology is frequently covered in plant pathology courses.

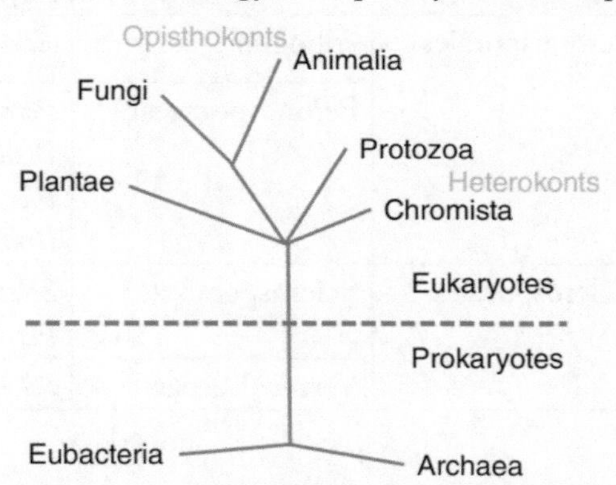

The Tree of Life Opisthokonts is a higher-level phylogenetic category that includes Animalia and Fungi; Chromistan taxa are heterokonts.

Representatives of important plant pathogens from the different classes are described here

Domain: EUKARYA

Kingdom:	1. PROTOZOA	2. CHROMISTA	3. FUNGI

1. Kingdom: PROTOZOA

Phylum: *Plasmodiophoramycota*

Class	Order	Family	Genus
Phytomyxae Slime moulds	*Plasmodiophorales*	*Plasmodiophoraceae*	*Plasmodiophora*

2. Kingdom: CHROMISTA

Phylum: Cercozoa/ Endomyxa

Class	Order	Family	Genus
Phytomyxae Slime moulds	*Plasmodiophorales*	*Plasmodiophoraceae*	*Plasmodiophora*

Phylum: Oomycota

No.	Class	Order	Family	Genus
1.	*Oomycetes*	Pythiales	Pythiaceae	*Pythium* *Phytophthora*
		Peronosporales	Albuginaceae	*Albugo*
			Peronosporaceae	*Peronospora* *Pseudoperonospora* *Plasmopara* *Bremia*
		Sclerosporales	Sclerosporaceae	*Sclerospora* *Peronosclerospora*
			Verrucalvaceae	*Sclerophthora*

3. Kingdom: FUNGI

PHYLUM: 5 (Plant Associated Fungi). More than 140,000 species of true fungi are known, and they are classified in five phyla. In four of the phyla we find plant pathogenic species.

Chytridiomycota Water moulds	*Glomeromycota* Symbionts-Mycorhizas	*Zygomycota* Zygote fungi	*Ascomycota* Sac fungi	*Basidiomycota* Club fungi

PHYLUM: Zygomycota

Class	Order	Family	Genus
Zygomycetes	*Mucorales*	*Mucoraceae*	*Mucor, Rhizopus*

PHYLUM: Ascomycota

Class	Order	Family	Genus
Taphrinomycetes	Taphrinales	Taphrinaceae	*Taphrina*
Dothidiomycetes	Capnodiales	Capnodiaceae	*Capnodium*
		Mycosphaerellaceae	*Mycosphaerella* *Septoria*
	Pleosporales	Venturiaceae	*Venturia*
		Pleosporaceae	*Cochliobolus* *Lewia (Alternaria* *Septoria)*
	c.) Botryosphaeriales	Botrosphaeriaceae	*Botryodiplodia* *Macrophomina*
Eurotiomycetes	Eurotiales	Trichocomaceae	*Eurotium* *Sartorya* *(Aspergillus)* *Emericella* *Talaromyces* *Eupenicillium* *(Penicillium)*
Leotiomycetes	Erysiphales	Erysiphaceae	*Erysiphe* *Uncinula* *Leveillula* *Phyllactinia*

Sordariomycetes	Hypocreales	Hypocreaceae	*Hypocrea (Trichoderma)*
		Nectriaceae	*Nectria Gibberella (Fusarium)*
		Clavicipitaceae	*Claviceps Ustilaginoidea*
		Cordycipitaceae	*Verticillium*
	Microascales	Ceratocystaceae	*Ceratocystis*
	Glomerales	Glomerellaceae	*Glomerella*
	Xylariales	Amphisphaeriaceae	*Pestalosphaeria Pestalotia*
	Incertae sedis	Magnaporthaceae	*Magnaporthe*
Pezizomycetes			

PHYLUM: Basidiomycota

Class	**Order**	**Family**	**Examples**
Agaricomycetes	Agaricales	Agaricaceae	*Agaricus*
		Amanitaceae	*Amanita*
		Tricholamataceae	*Armillaria*
		Pleurotaceae	*Pleurotus*
		Pluteaceae	*Volvariella*
		Lyophyllaceae	*Calocybe*
	Atheliales	Atheliaceae	*Athelia (Sclerotium)*
	Cantharellales	Ceratobasidiaceae	*Thanatophorus (Rhizoctonia)*
	Polyporales	Ganodermataceae	*Ganoderma*
	Corticiales	Corticiaceae	*Corticium*
Pucciniomycetes	Pucciniales (Rust)	Pucciniaceae	*Puccinia*
		Melampsoraceae	*Melampsora*
		Phragmidiaceae	*Phragmidium*
		Incertae sedis	*Hemileia*
Ustilaginiomycetes	Urocystales (Bunt)	Urocystidaceae	*Urocystis*
	Ustilaginales (Smut)	Ustilaginaceae	*Ustilago Sporosorium*

Exobasidiomycetes	Exobasidiales	Exobasidiaceae	*Exobasidium*
	Tilletiales (Smut)	Tilletiaceae	*Tilletia*

Practice Questions

1. ……….. published *Systema Mycologicum* for naming fungi
2. ……… recognized Fungi as separate kingdom
3. Scientific categorization of the organisms in a hierarchical series of groups is referred as……..
4. …………..order consists of rust fungi
5. …………..order consists of smut fungi
6. Describe the taxonomy of fungi
7. Elloborate the different phyla in fungal kingdom
8. Briefly describe rules of nomenclature of fungi.
9. List the taxon(s) mentioning suffixes used in binomial system of nomenclature.
10. Give general characters and classification of fungi

Suggested Readings

Catalogue of Life (http://www.catalogueoflife.org/): Roskov Y., Ower G., Orrell T., Nicolson D., Bailly N., Kirk P.M., Bourgoin T., DeWalt R.E., Decock W., Nieukerken E. van, Zarucchi J., Penev L., eds. (2019). Species 2000 & ITIS Catalogue of Life, 2019 Annual Checklist. Available at www.catalogueoflife. org/annual-checklist/2019. Species 2000: Naturalis, Leiden, the Netherlands. ISSN 2405-884X.

Index Fungorum (http://www.indexfungorum.org/names /names.asp), linked to Species fungorum available at: http://www.speciesfungorum.org/ (accessed 16 April 2020).

International Committee on Systematics of Prokaryotes. Available at: http://www.the-icsp.org/ (accessed 22 Nov, 2022).

Rosselló-Mora, R. and Amann, R. (2001) The species concept for prokaryotes. FEMS Microbiology Reviews 25(1), 39–67, doi:10.1111/j.1574- 6976.2001.tb00571.x (accessed 22 Nov, 2022).

International Committee on Taxonomy of Viruses. Available at: https://talk.ictvonline.org/ (accessed 22 Nov, 2022).

Decraemer, W. and Hunt, D.J. (2013) Structure and Classification. In: Perry, R. and Moens M. (eds) Plant Nematology 2nd ed. CAB International, Wallingford, UK, pp. 3–39.

Chapter - 8

Protozoa

Protozoa (sing. Protozoan) derived from Greek words "*protos*" and "*zoan*" mean 'first animal'. It includes animal, fungi, flagellate and plant like eukaryotes.

The kingdom Protozoa contains the acellular (plasmodial), free-living, slime moulds that are colourful, non-parasitic organisms, move like amoebae that can resemble plant pathogens. Protozoa are at the base of the eukaryotic branch of the 'tree of life' and are believed to be the first eukaryotes to arise. This group consists of a variety of single-celled, primarily colourless, mobile, cell-wall-less organisms. Several taxa of Protozoa were moved to Chromista in 2015.

Flagellate protozoa (plant-infecting trypanosomatids) E.g., *Phytomonas*

- » The promastigote trypanosomatid flagellates are present in latex bearing cells (lactifers) of the laticiferous plant, *Euphorbia pilulifera* was first found out by Lafont during 1909 in island of the Mauritius. He named the organism *Leptomonas davidi*, after his lab technician David, the individual who first observed it.
- » The discovery of these trypanosomatids, initially named *Leptomonas davidi*, was confirmed in *E. pilulifera* in Madras (4,400 kilometres away) later the same year by Donovan.
- » Donovan proposed that a new genus of trypanosomatids, *Phytomonas*, had to be created. They are mostly single celled, eukaryotic parasite, microscopic, motile, wall less microorganism covered by thin flexible membrane; move either by flagella or by pseudopodia or by the movement of the cell itself and reproduce by longitudinal fission. The flagellates have one or more long slender flagella at some or all stages of their life cycle.

» The two main plant families that contain the most Phytomonas hosts are *Euphorbiaceae* and *Asclepiadiacae.*

Genus: *Phytomonas* (trypanosomatids)

» Phytomonas is believed to have arisen from a single monoxenous lineage of insect parasitizing trypanosomatids some 400 million years ago. They are followed by Phytomyxea, which are part of the Rhizaria supergroup and include agriculturally important plant pathogens, vectors of phytoviruses and species that infect marine plants and algae.

» Currently, the genus *Phytomonas* includes more than 200 species that colonize over 20 plant families.

» *Phytomonas* species are obligate parasitic flagellates transferred between plant hosts by insect vectors (*Lincus* sp.) in the Heteroptera suborder as a form of dixenous parasitism.

» The first reports of *Phytomonas* infections of plants initially described the host plants as exhibiting poor growth and wilt. Later, reported that infected plants exhibited a depletion of starch granules from the latex and surrounding parenchyma and a reduction in the latex viscosity.

» The first species, *Phytomonas staheli*, is the aetiologic agent of "*hartrot* (fatal wilt) of coconut palm (*Cocos nucifera*), "*marchitez sorpresiva*" (sudden wilt), and slow wilt of oil palm (*Elaeis guineensis*) both diseases are acute lethal wilts that begin in the leaves and progress to the rotting of the spear and root.

» The second pathogenic species, *Phytomonas leptovasorum*, causes coffee phloem necrosis in both Liberica and Arabica coffee.

» A third species, *Phytomonas françai*, inhabits the latex ducts of cassava (*Manihot esculenta*) and has been linked with yield loss diseases known as "*chochamento de raizes*," or empty root syndrome. This disease is characterised by poor root development and chlorosis of the leaves.

» *Phytomonas* have structures that are characteristic of the family, including the flagellum-associated kinetoplast, subpellicular microtubules, the paraxial rod, and glycosomes. Within a host plant, *Phytomonas* exhibits a fusiform structure twisted 2-5 times along the longitudinal axis. Within the plant, the organisms can be in several flagellated stages: mostly promastigote with some paramastigotes in the phloem and lacticiferous tubes, and amastigote form in the latex.

» They have definite long, oval or spherical body of size ranging between 5×250μm.

» It contains promastigote trypanosomatid flagellates and has a life cycle completed in the hosts, a plant and an insect.

- The genus *Phytomonas* reside in the phloem sieve tubes of nonlactiferous plants like coconut, oilpalm, red ginger, coffee, and are definitely pathogenic.
- They can be grown on specialized culture media.
- The plant-infecting Phytomonas seem to be transmitted by root graft and by insects of the families *Pentatomidae*, *Lygaeidae*, and *Coreidae*.

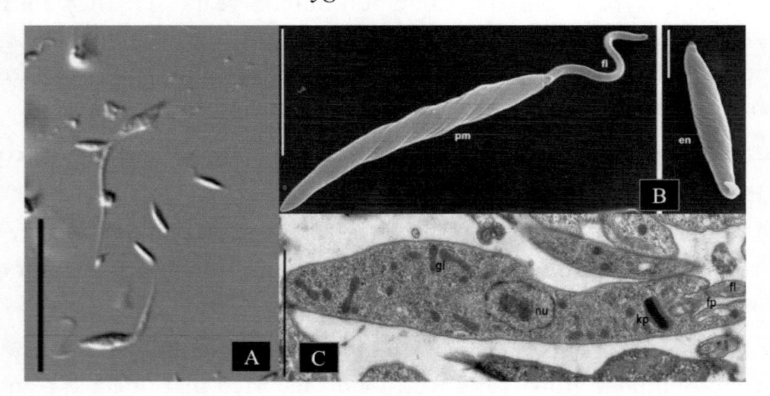

Phytomonas sp. **(A. light microscopy, B. SEM and C. TEM)** cells in culture.

Source: Frolov et al. 2019, PloS One.

Taxonomy

Domain	:	*Eukarya*
Kingdom	:	Excavata (unranked)
Phylum	:	*Euglenozoa*
Class	:	*Kinetoplastea*
Order	:	*Tripanosomatida*
Family	:	*Tripanosomatidae*
Genera	:	Phytomonas

Protozoan diseases

Pathogens	**Diseases**
1. Phloem restricted phytomonas	
Phytomonas leptovasorum	Phloem Necrosis of Coffee
Phytomonas staheli	Hartrot, fatal wilt, bronze leaf wilt, lethal yellowing, and Coronie wilt of Coconut. Sudden wilt (marchitez sorpresiva) of Oil palms.

2. Lactifer restricted phytomonas	
Phytomonas francai	Empty root of cassava

Diseases

1. Phloem necrosis of coffee

- Infected trees show sparse yellowing and drooping of leaves.
- Finally, only the young top leaves remain on bare branches.
- Roots show die-back and the trees die.
- Trees wilt and die within 3-6 weeks during dry season.

2. Hart rot of coconut

- Yellowing and browning of the tips of the older leaves that subsequently spread to the younger leaves.
- Fresh inflorescence become black, unripen nut fall off and roots begin to rot.
- Petioles of older leaves may break and spear becomes necrosis.
- Apical region of the crown rots and emit foul odour.

3. Sudden wilt of oil palm

- Browning of the tips of the older leaves, this subsequently spread to the younger leaves and eventually become ashen grey.
- Reduced plant growth, discoloured bunches, fall off of leaves and rotting of tissues and deterioration of root tips and root system.

4. Empty root of cassava

- Diseased root has poor root system and small slender roots with no starch.
- The above aerial parts of the plants exhibit general chlorosis and decline.
- Diseased plants contain numerous protozoa in the lactifer ducts.

Practice Questions

1. ………. is plant-infecting trypanosomatids
2. ……….. pathogen causes phloem necrosis of coffee

Suggested Readings

Jaskowska, E., Butler, C., Preston, G., & Kelly, S. (2015). Phytomonas: trypanosomatids adapted to plant environments. *PLoS pathogens*, *11*(1), e1004484.

Frolov, A. O., Malysheva, M. N., Ganyukova, A. I., Spodareva, V. V., Yurchenko, V., & Kostygov, A. Y. (2019). Development of Phytomonas lipae sp. n.(Kinetoplastea: Trypanosomatidae) in the true bug Coreus marginatus (Heteroptera: Coreidae) and insights into the evolution of life cycles in the genus Phytomonas. *Plos One*, *14*(4), e0214484.

Chapter - 9

Chromista

Kingdom: *Chromista/ Straminopiles*

The name *Chromista* was introduced by Thomas Cavalier-Smith in 1981. The name Chromista means "colored", and although some chromists, like mildews. This kingdom comprises of eukaryotic, walled microorganisms that produce heterokont, wall less cells in their life cycles. They have closer phylogenetic association with brown algae and diatoms than true fungi belonging to kingdom *Fungi*.

» It possesses uni-or multicellular somatic phase with absorptive type of nutrition.

» The cell walls rich in cellulose and synthesize amino acid by a α, β-diaminopimelic acid lysine biosynthetic acid.

» The asexual spore is biflagellate zoospore with posteriorly oriented whiplash and anteriorly oriented tinsel flagella (heterokonts).

Phylum: ***Cercozoa*** (endoparasitic slime mould)

The phylum Cercozoa consists of unicellular creatures with no shared morphological characteristics. It contains the key Phytomyxea class and Plasmodiophorida order.

» The parasites exist as intracellular plasmodium in the host cell and derive food by absorption.

» The class *Plasmodiophoromycetes* are necrotrophic endoparasites causing abnormal enclargement of host cells.

» They are distinguished by the production of motile cells (zoospores) with two unequal anterior flagella.

» There are 2 types of plasmodia.

Sporangiogenous plasmodium - formed asexually containing many thin-walled zoosporangia and each zoosporangium produce a single or many secondary zoospores or sporangial zoospores.

Cystogenous plasmodium - formed sexually consisting of thick-walled cysts and each cyst gives rise to a single primary zoospore / cyst zoospore. Cysts may be free or united. Cysts act as resting spores.

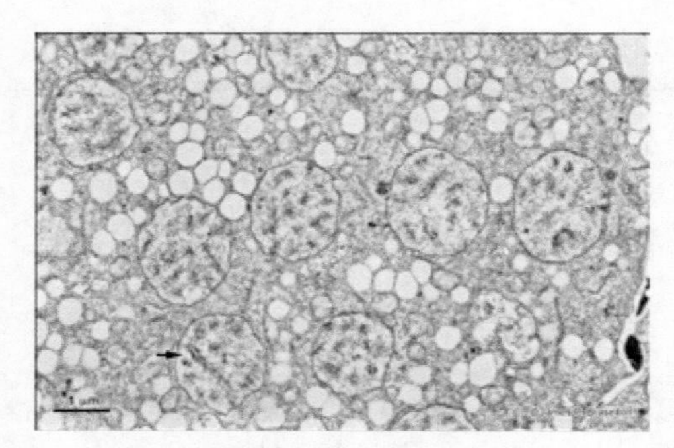

Sporogenic plasmodium

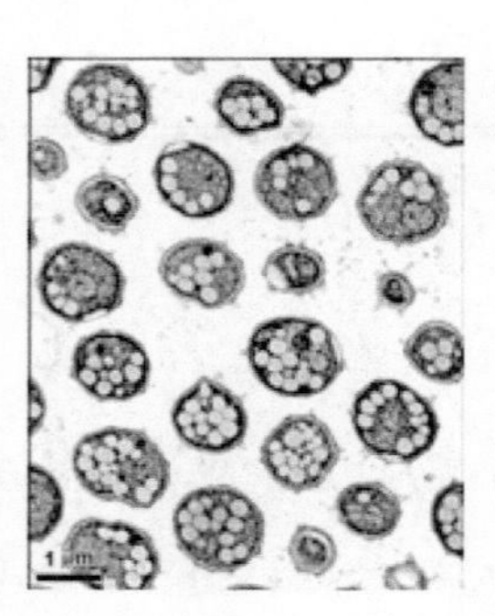

Cystogenous plasmodium

Source: Plasmodiophorid Images

» Zoospores are anteriorly biflagellate, whiplash type, unequal in sizes which are called as Heterokont zoospores. After swimming for some time, the zoospore encysts on the root hair of the host. A cylindrical sharp pointed body, called **stachel**, is formed in a specialized pouch or sheath called **Rohr**.

» Nuclear division is by cruciform division.

» Sexual reproduction is by isogamous planogametic copulation.

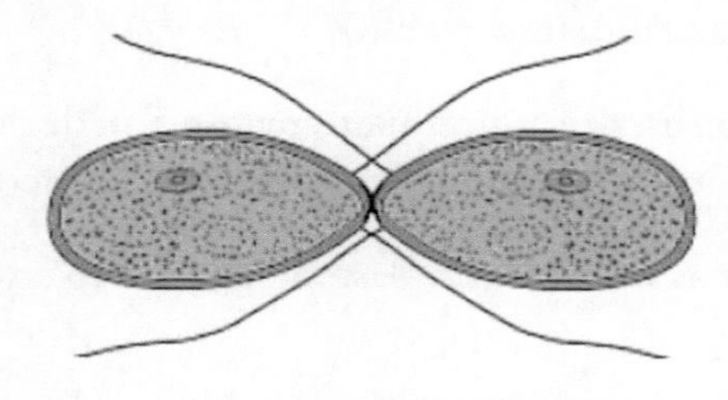

Isogamy

» Members cause abnormal enlargement and multiplication of host cells i.e., hypertrophy and hyperplasia. E.g., *Plasmodiophora brassicae, Spongospora subterranea.*

Genus: *Plasmodiophora brassicae*

Plasmodiophora brassicae is soilborne obligate parasite that causes club root disease of crucifers. The disease was first reported in the United States of America in 1852. In the late 19th century, a severe epidemic of club root destroyed large propotions of the cabbage crop in St. Petersburg, Russia. Woronin, a Russian scientist successfully identified the cause of club root as a "plasmodiophorous organism" in 1875, and gave it the name *Plasmodiophora brassicae.* There are three stages in its life cycle: (i) survival in soil; (ii) infection of root hairs; and (iii) infection of cortical cells.

- They are unicellular organism, produce fruiting body that have a resemblance to those of true fungi and hence called plasmodial slime moulds.
- The slime moulds are heterotrophic, obtain food by absorption.
- It possesses naked, unicellular plasmodial somatic phase.
- They don't produce hyphae and lack cell wall.
- The Plasmodiophoromycota are a group of obligate (i.e. biotrophic) intracellular parasites. The best-known examples attack higher plants.

Pathogen: Obligate endoparasite.

Thallus - naked motile multinucleate mass of protoplasm that feeds in an amoeboid fashion, called **plasmodium** (pl. plasmodia; Gr. *plasma* = moulded object).

Resting spores are spherical with spiny walls.

Zoospores are anteriorly biflagellate, unequal in length and uninucleate, both the flagella are of whiplash type.

i. Primary phase occurs in root hairs

ii. Secondary phase occurs in cortical cells of roots

Primary phase (Root hair)

- After the decay of infected roots, the resting spores are released into the soil.
- The resting spores are tiny, hyaline, spherical and uninucleate. These resting spores germinate and produce zoospores (primary) with two flagella.
- Encysted zoospores attach with root hair and enter.
- These uninucleate, amoeboid zoospores are called myxamoeba.
- Then naked, multinucleate intracellular protoplasm constituting the thallus, plasmodium (primary plasmodium or sporangial plasmodium) forms.
- Cleavage of plasmodium leads to zoosporangium which contains 4-8 uninucleate

anteriorly biflagellate zoospores.

» These zoospores escape in the soil, fuse as gametes and form quadriflagellate binucleate zoospores (secondary zoospores).

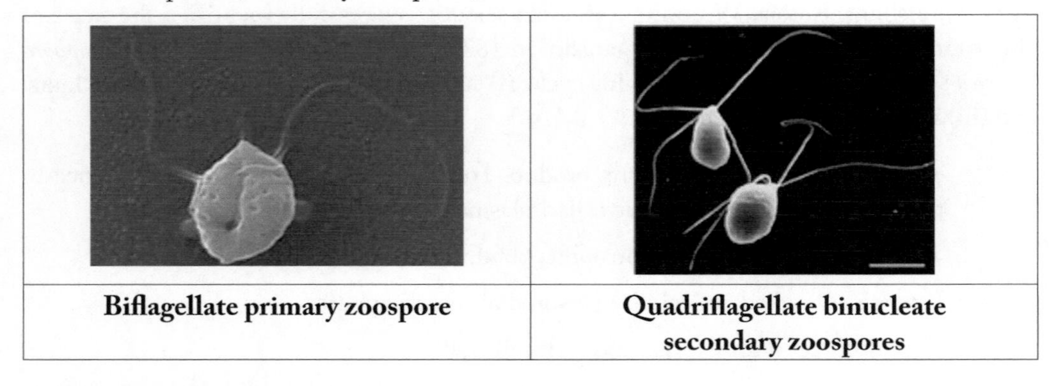

Biflagellate primary zoospore	**Quadriflagellate binucleate secondary zoospores**

Secondary phase (Root cortex)

» Quadriflagellate binucleate zoospores (secondary) reinfect the root and form binucleate plasmodium, which penetrates the root cortex.

» Plasmodium repeatedly divides to form multinucleate secondary plasmodium or sporogenic plasmodium and the host cells hypertrophied.

» The haploid nuclei associated in pairs, karyogamy and meiosis occurs which leads to the formation of resting spores.

» These masses of resting spores released into soil as the root decay.

» The pathogen survives as resting spores lying freely in soil or in crop debris.

» They may remain viable for as long as 10 years.

Cercozoa diseases

Pathogens	**Diseases**
Plasmodiophora brassicae	Clubroot of cabbage and other brassicaceous crops
Spongospora subterranea *Spongospora nasturtii*	Powdery scab of potato, Alfalfa wart Crook root of watercress
Polymyxa betae	Vector for rhizomania (*Beet necrotic yellow vein virus*) of sugar beet
Polymyxa graminis	Vector for *Barley yellow mosaic virus* (BaYMV) and *Soil-borne wheat mosaic virus* (SBWMV)

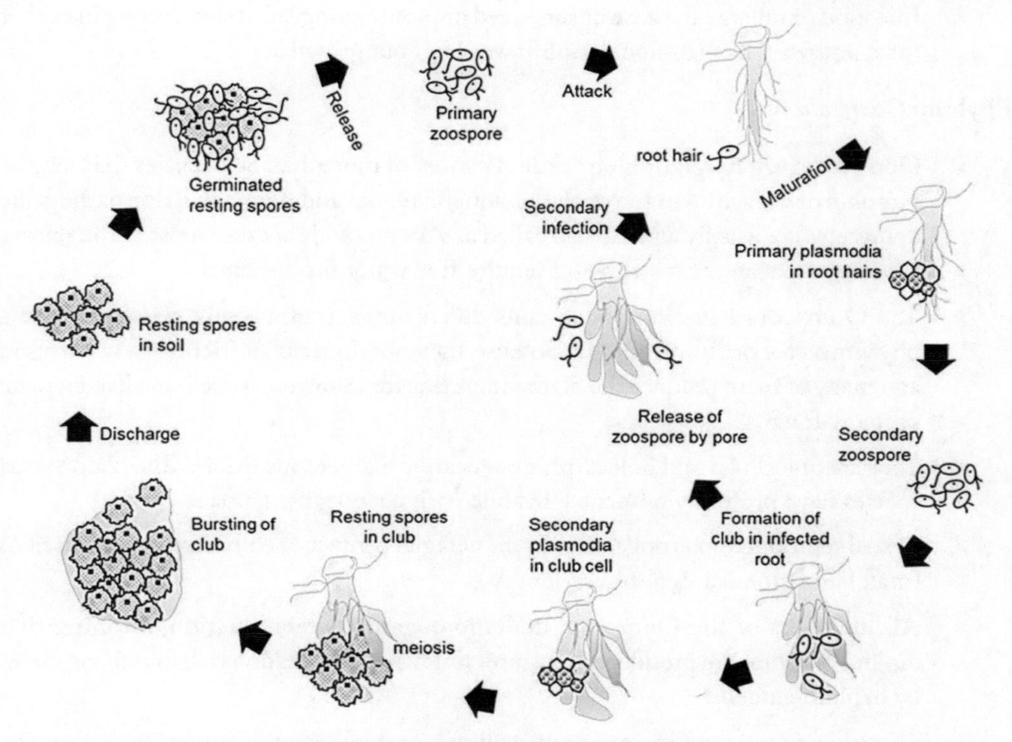

Life cycle of *Plasmodiphora brassicae*

Source: Mehraj et al. 2020, Plants.

Major diseases

1. Clubroot of cabbage

- » Enlargement of roots, club-shaped roots due to hyperplasia (multiplication of cells) and hypertrophy (enlargement of cells), gradual and inconspicuous stunting, yellowing and wilting of plant.
- » It is also named as finger and toe disease. Woronin called hypertrophied cells as **Krankheitsherd**.

2. Powdery scab of potato

- » Small lesions appear in the tubers, progress to raised white pustules and then become brown.
- » The periderm of the tuber then bursts producing dark, erumpent pustules containing a powdery mass.

- Infection of tubers can cause unsuberized or proliferaring lenticels resulting in canker form and wart like (tumour/ cauliflower like) out growth.

Phylum: *Oomycota*

- Oomycota (egg fungi) or Heterokonta consist of more than 800 species that may be saprobic or parasitic on terrestrial or aquatic plants and animals. Plant pathogenic oomycetes are usually aquatic also called as water moulds because most of the species have a spore stage that swim and require free water for dispersal.
- The Oomycota have long been considered fungi, traditionally classified in the phycomycetes or "lower fungi." Because they obtain their nutrients via absorption and many of them produce the filamentous threads known as mycelium characteristic of many fungi.
- They are unicellular and holocarphic or eucarpic filamentous species. The filamentous species have profusely branched, hyaline with coenocytic hyphae.
- Sexual reproduction is oogamous by gametagial contact of club shaped antheridium (male) and globular oogonium (female).
- All members of the Oomycota undergo oogamous reproduction, meaning that diploid oospores are produced as zygotes following fertilization of haploid oospheres by haploid gametes.
- These oospores may be large and solitary or smaller and numerous inside the oogonium. None of the true Fungi produce oospores.
- The cell walls are composed of beta glucans with amino acid hydroxyproline and small amount of cellulose.
- The Oomycota produce motile zoospores with two kinds of flagella, one of which is a short whiplash flagellum oriented posteriorly while the other is called a long tinsel flagellum, which has a fibrous, ciliated structure and is oriented anteriorly.
- Many oomycetes produce one kind of zoospores (monoplanetic) and some produce diplanetic zoospores (both pear shaped and kidney shaped) in *Saprolegnia*.
- The occurrence of two kinds of flagella places these organisms in a group known as heterokonts. Although some true Fungi, namely the Chytridiomycota, produce stages with motile zoospores, their flagella are only of one kind, the posterior whiplash type.
- The vegetative cells of the Oomycota generally consist of coenocytic hyphae (hyphae without septa, i.e., without cross-walls), which contain diploid nuclei, these organisms exist primarily in a **diploid** state.

Major distinctions between the Oomycota in the Chromista and the true Fungi

Character	Oomycota	Fungi
Sexual reproduction	Heterogametangia. Fertilization of oospheres by nuclei from antheridia forming oospores.	Oospores not produced; sexual reproduction results in zygospores, ascospores or basidiospores
Nuclear state of vegetative mycelium	Diploid	Haploid or dikaryotic
Cell wall composition	Beta glucans, cellulose	Chitin. Cellulose rarely present
Type of flagella on zoospores, if produced	Heterokont, of two types, one whiplash, directed posteriorly, the other fibrous, ciliated, directed anteriorly	If flagellum produced, usually of only one type: posterior, whiplash
Mitochondria	With tubular cristae	With flattened cristae

Genus: *Saprolegnia* sp.

- Infects gills of fishes and cause salmon disease.
- The pathogen *Saprolegnia* sp possesses dimorphism, diplanetism and sporangial profiferation.

Genus: *Pythium* sp.

The oomyceteous order *Pythiales* includes aquatic, amphibious and terrestrial fungi and many of them cause serious disease of plants as facultative parasites.

- Mycelium is well developed, hyaline, coenocytic and grows inter or intra cellularly.
- This group is morphologically polymorphic, physiologically unique and ecologically versatile.
- The *Pythium* produces either filamentous or simple or branched (*P. monospermum, P. gracile*), lobed (*P. aphanidermatum, P. graminicolum*) or globose (*P. debaryanum*).

Pythium aphanidermatum

Pythium aphanidermatum is a cosmopolitan pathogen with a wide host range. It is an aggressive species of *Pythium*, causing damping off, root and stem rots, and blights of grasses and fruit. It is of economic concern on most annuals, cucurbits, and grasses.

- It is considered one of the water molds because it survives and grows best in wet soils. Warm temperatures favor the pathogen, making it an issue in most greenhouses.

» *Pythium* is an Oomycete in the order Peronosporales. The hyphae are hyaline and the mycelium has no cross walls.

» To differentiate *P. aphanidermatum* from other *Pythium* species requires examination under a microscope of the sporangia, oogonia and antheridia. Sporangia are the asexual spores and in the case of *P. aphanidermatum*, they are lobate (inflated). The apluerotic oogonium, (oospore doesn't fill the oogonium) and the intercallary (rarely terminal) attachment of the antheridia further distinguishes *P. aphanidermatum* from other *Pythium* species.

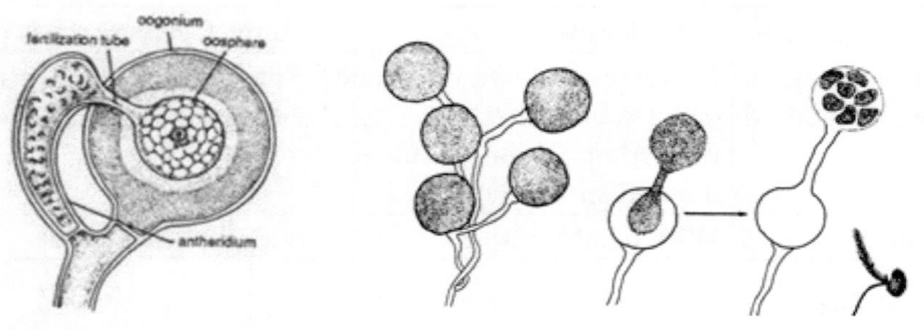

Gametangial contact **Asexual reproduction**

» *Pythium* survives in the soil as oospores, hyphae and sporangia.

» The fungus can survive unfavorable soil moisture and temperatures for several years as oospores.

» Oospores may form a germ tube directly to infect the plant or they may form sporangia.

» The sporangia produce zoospores, which are the motile form of the fungus.

» The zoospores swim around briefly before encysting and forming a germ tube, which can cause infection.

» Sporangia that have developed on plant tissue can germinate in a similar manner as the oospores, by either germinating directly or by forming zoospores.

» The pathogen is dispersed when infected debris is transported to uninfested areas and when the soil moisture is enough to allow the zoospores to swim freely.

» *Pythium aphanidermatum* infects seeds, juvenile tissue, lower stems, fruit rot and roots.

Life cycle

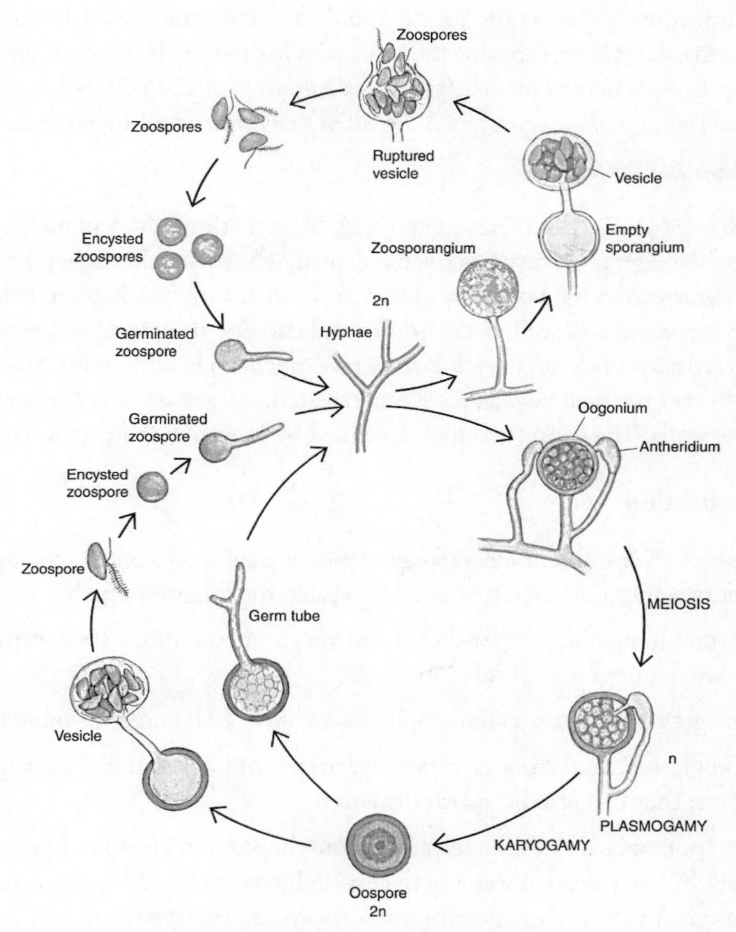

Life cycle of *Pythium* Source: H. Karlsen

Damping-off - *Pythium aphanidermatum*

- *Pythium aphanidermatum* is responsible for pre- and post-emergence damping off.
- Pre-emergence damping off is when the seed is infected prior to germination. This can result in poor or no germination, and is observable as a browning or rotting of the seed.
- Post-emergence damping off takes place after germination and results in a thinning, water-soaked stem near the plant collar, which eventually causes the collapse of the plant. *P. aphanidermatum* can also cause root rot.

- Symptoms of root rot include stunted growth, chlorotic leaves, leaf drop, and wilting.
- The infection begins at the root tip, and can cause the infected region to lose its protective outer layer, exposing the inner root to other pathogens. Other species cause major diseases in crop plants: *Pythium aphanidermatum, Pythium irregulare, Pythium oligandrum, Pythium myriotylum, Pythium graminicola* and *Pythium debaryanum.*

Genus: *Phytophthora infestans*

Phytophthora - (*phytón*), "plant" and (*phthorá*), "destruction"; "the plant-destroyer" is a genus of plant-damaging Oomycetes (water molds), whose member species are capable of causing enormous economic losses on potato and tomato disease known as late blight or potato blight worldwide, as well as environmental damage in natural ecosystems. The cell wall of *Phytophthora* is made up of cellulose. It has long been held that organisms producing zoospores with two, unequal flagella are closely related. *P. infestans* is a coenocytic oomycete with rare cross walls. The genus was first described by Heinrich Anton de Bary in 1875.

Asexual reproduction

- *Phytophthora infestans* produces microscopic, asexual spores called sporangia. Within a day or two after the lesion first becomes visible, the fungus is capable of sporulation.
- Moderate temperatures (10-25°C) and very wet conditions (leaf wetness or 100% RH) are required for sporulation.
- Sporangia are borne on sporangiophores within 8-12 h during favourable conditions.
- Sporangia secede during changing relative humidity and can be captured in air currents; they can also be splash dispersed.
- These sporangia are hyaline (clear), lemon-shaped and 20-40 µm long with a small pedicel. When placed in water or in very high relative humidity, the cytoplasm in the sporangia divide and many swimming zoospores emerge from each sporangium.
- Zoospores can swim for some minutes, after which time they encyst and germinate.
- Sporangia are formed on specialized branches called sporangiophores.
- The branched sporangiophore, with swellings at the points where sporangia were attached is distinctive for *Phytophthora infestans* and useful for identification of this pathogen.
- In the absence of sufficient water or with temperatures above 24°C, no zoospores form. However, sporangia germinate by producing germ tubes that penetrate the host.
- Zoospores (7-12 per sporangium) have two flagella, one anteriorly-directed tinsel type and a posteriorly-directed whiplash type (heterokont).

» Zoospores are usually uninucleate, but binucleate zoospores have been detected. The disease can thus progress rapidly under cool, wet conditions.

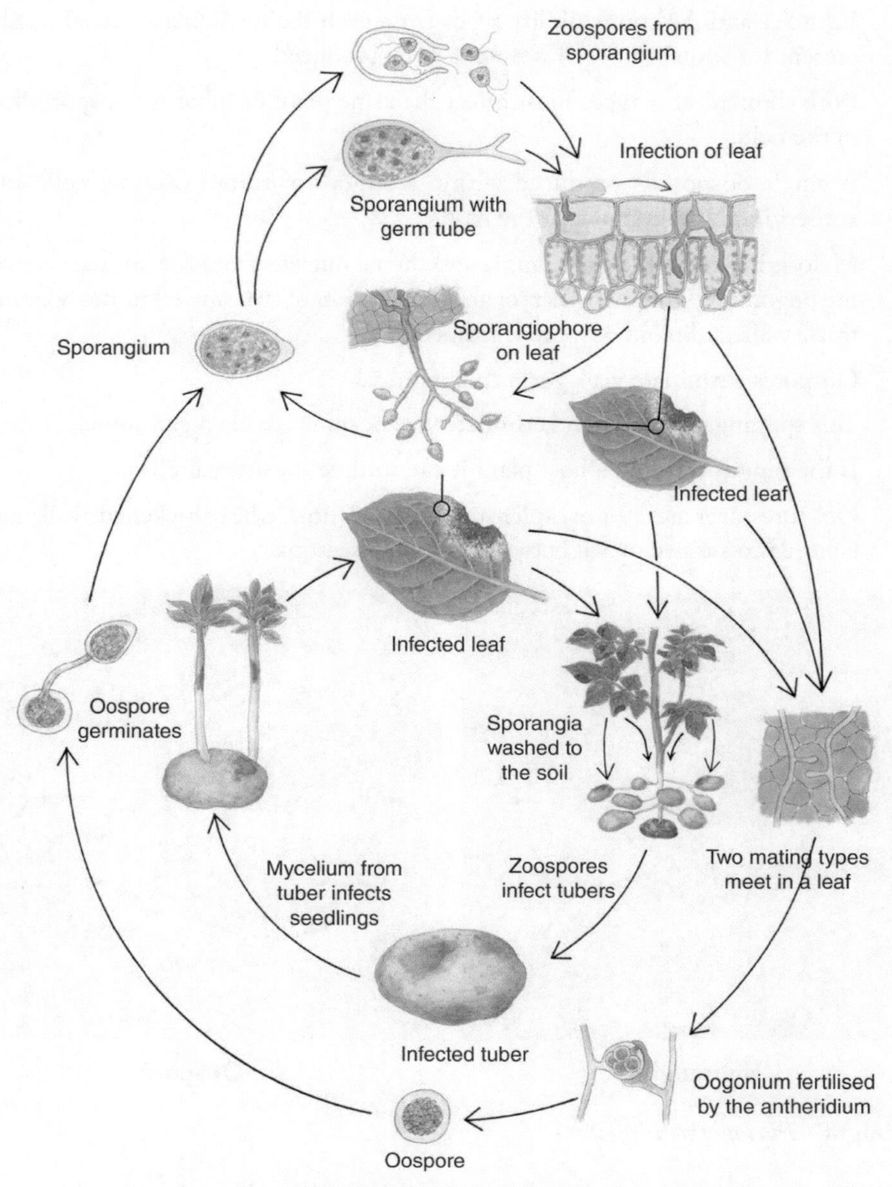

Life cycle of *Phytophthora* sp. Source: H. Karlsen

Sexual reproduction

- Antheridia amphigynous, elongated cylindrical, 17 (max. 22) × 16 µm.
- Both A1 and A2 compatibility types from each thallus (heterothallus) need to be present for sexual spores or oospores to be produced.
- Both compatibility types must infect the same plant or tuber for oospores to form in the field.
- A single oospore is produced within a larger oogonium (mother cell) and the antheridium (male cell) is at the base.
- Meiosis occurs in the gametangia and then a nucleus from the antheridium enters the oogonium. Following karyogamy (the fusion of two nuclei) in the oogonium, a thick-walled, diploid oospore is formed.
- Oospores germinate via a germ sporangium.
- This sporangium can then germinate via zoospores or via a germtube.
- If the fungus contacts a host plant it can initiate the asexual phase.
- Oospores average 30 µm, aplerotic, wall 3-4 µm, it has thickened walls and are believed to survive in soil between cropping seasons.

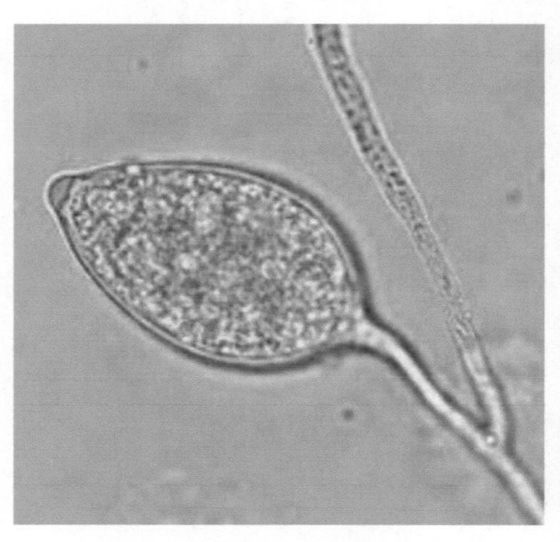

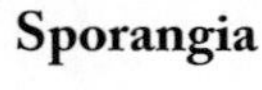

Sporangia

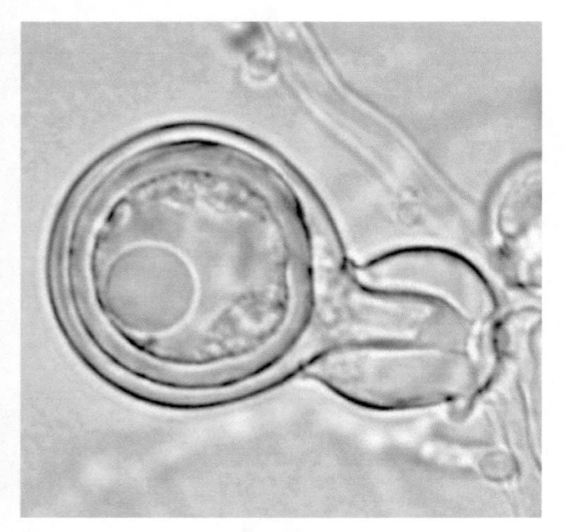

Oospore

Late blight - *Phytophthora infestans*

Symptoms

- Late blight of potato is identified by small water-soaked blackish/brown lesions on leaves or have chlorotic borders but expanded rapidly and the entire leaf becomes become necrotic. In humid conditions, *P. infestans* produces sporangia and sporangiophores on the surface of infected tissue and the resulting white sporulation can be seen at the margins of lesions on abaxial (lower) surfaces of leaves.

» Potato tubers can become infected in the field when sporangia are washed from lesions on the foliage and enter into the soil. Infections generally begin in tuber cracks, eyes or lenticels. Infected tuber tissues are copper brown, reddish or purplish in color. Sporulation may occur on the surface of infected tubers in storage or on discarded cull piles. Infected tubers are often invaded by soft rot bacteria which rapidly convert adjoining healthy potatoes into a smelly, rotten mass that must be discarded.

Differences between *Pythium* and *Phytophthora*

Pythium	***Phytophthora***
Facultative parasite	Facultative saprophyte
Hyphal wall contain greater amount of protein	Hyphal wall contain little amount of protein
Mycelium is both inter and intracellular. When intracellular, no haustoria are produced. During intercellular, haustoria are produced.	Mycelium is both inter and intracellular. When intracellular, no haustoria are produced. Only intercellular. Haustoria are produced.
Production of sporangia on somatic hyphae. Sporangiophores are indistinct from hyphae.	Sporangiophores can be distinguished by sympodial (zig-zag) branching and nodal swellings.
Sporangia are globose or elongated or lobed. They are produced intercalary or terminally. No papillum.	Sporangia are lemon or pear or oval shaped, produced terminally. Papillum is present.
Sporangia germinate by forming vesicle. Differentiation of zoospores takes place in the vesicle.	No vesicle is seen. Zoospores differentiate in the sporangium.
Antheridium is of paragynous type.	Amphigynous type.
Homothallic	Heterothallic
Asexual reproduction is by Zoospores in sporangia.	Zoospores in sporangia and chlamydospores.
Appresorium and haustoria are absent	Appresorium and haustoria are present.
Oogonial wall is colourless, smooth and spiny.	Oogonial wall is brown, rough and warty.
Oospore is pleurotic, hyaline and smooth.	Oospore is apleurotic, brown and warty.
Direct germination of oospore is common.	Indirect germination of oospore.

Order: *Peronosporales*

- Many species are destructive obligate parasite causing very serious diseases in some important crop plants.
- The diseases caused by these members include white rusts, downy mildews, damping off, leaf blights, and seedling blights.
- Members are mostly terrestrial.
- Mycelium is coenocytic, and produce inter and intra-cellular hyphae.
- If inter-cellular, it produces well-branched lobed haustoria.
- Sporangia are produced on well-developed, distinct sporangiophores and sporangia are deciduous (fall off at maturity).
- Sporangiophores may be indeterminate / indefinite type (sporangiophores continue to grow indefinitely producing sporangia at the tip as they grow.ie., sporangia of different ages are seen on sporangiophores) or determinate/definite type (sporangia are not produced until sporangiophores complete their development and maturity and all the sporangia are produced at one time.ie., single crop of sporangia are produced.)
- Zoospores are monomorphic (producing morphologically one type of zoospores i.e., reniform zoospores) and monoplanetic (only one swarming period.).
- Zoospores are reniform ie., kidney-shaped and biflagellate. Some species exhibit highly specialized parasitism i.e., obligate parasites.
- Oogonium produces a single oosphere /egg surrounded by conspicuous periplasm except in Family: Pythiaceae, in which periplasm is inconspicuous. Periplasm serves as a source of nutrients to the oosphere.

Family: *Albuginaceae*

- All are obligate parasites causing diseases of vascular plants.
- There are several species of Albugo. Of these, *Albugo candida* attacks all kinds of cruciferous plants, such as radish, cress, mustard, cabbage and cauliflower.
- Some of the other species commonly found are *Albugo ipomoeae-panduranae* on sweet potato and morning glory, *Albugo portulaceae* on portulaca, *Albugo occidentalis* on spinach and *Albugo bliti* on various members of the Amaranthaceae.
- In *Albugo candida* the mycelium is intercellular and coenocytic. The parasite sends short, knob-like haustoria into the cells. The parasite thus shows heteratrophic mode of nutrition and absorbs food from the host tissue.

Genus: *Albugo*

Albugo (derived from a Latin word means white), the only genus of family Albuginaceae is represented by more than 25 species. It is an obligate parasite reported which attacks mostly crucifers like turnip, mustard, radish, cabbage, cauliflower etc.

- » Thallus is eucarpic and mycelial. Hyphae are intercellular, coenocytic, aseptate and profusely branched.
- » Cell wall is composed of fungal cellulose.
- » Some mycelium is intracellular in the form of knob-like haustoria for the absorption of food material from the host cells.
- » Each sporangiophore gives rise to several sporangia which are produced in succession, one below the other, so that a chain of sporangia is formed with the oldest at the tip and youngest at the base (basipetal manner).
- » Sporangia are globose. Successive sporangia are connected by isthmus or disjunctor cell or separation disc.
- » Periplasm is conspicuous.

Asexual reproduction

- » The asexual reproduction takes place by means of biflagellate zoospores formed inside the sporangia. In the very beginning the hyphae accumulate just beneath the epidermis of the infected leaf. From these hyphae, certain thick-walled, clavate aerial sporangiophores come out.
- » The terminal end of the sporangiophore becomes constricted and sporangium contains 5-8 nuclei and cytoplasm. Successively the sporangia develop by constriction method, in basigenous chains. In between each two sporangia a gelatinous pad develops acting as a separator of two sporangia from each other.
- » The sporangium is smooth, double- walled and rounded. When the sporangia are formed in abundance on innumerable sporangiophores, the pressure is caused; the host epidermis ruptures and hundreds of sporangia are seen on the surface of the host in the form of white creamy powder forming pustules. The sporangia are transferred from one place to another by various agencies such as wind, insects, water, etc.
- » On the maturation of the sporangium the protoplast is cleaved into uninucleate protoplasts. Each protoplast metamorphoses into a naked, biflagellate, uninucleate, reniform and vacuolate zoospore. The sporangium bursts anteriorly and the zoospores liberate in the film of water.
- » On the maturation of the sporangium the protoplast is cleaved into uninucleate protoplasts. Each protoplast metamorphoses into a naked, biflagellate, uninucleate, reniform and vacuolate zoospore. The sporangium bursts anteriorly and the zoospores liberate in the film of water.

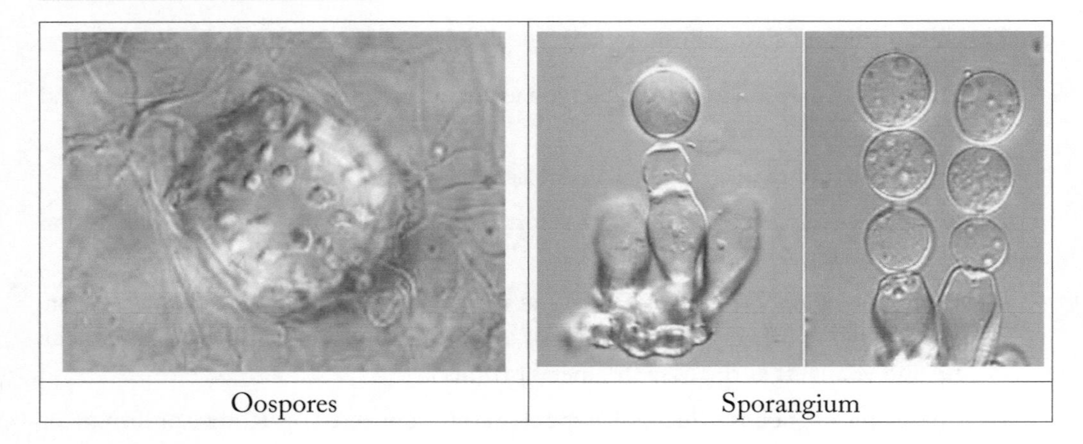

Oospores	Sporangium

Sexual reproduction

- The sexual reproduction is **oogamous**. The sex organs develop on the hyphal ends in the intercellular spaces of the deeper tissues of petioles and stems. The female sex organs are **oogonia** and the male sex organs are **antheridia**. The oogonium is rounded and the antheridium club-shaped. The developing oogonia and antheridia are separated from rest of the mycelium by septa.
- The cytoplasm, vacuoles and nuclei are uniformly distributed in the young oogonium. On the maturation of the oogonium the protoplasm of the oogonium differentiates into two regions. The outer region is called the periplasm containing thin cytoplasm, many nuclei and many vacuoles.
- The central protoplasm with denser consistency surrounded by periplasm is called the oosphere or the egg. The dense cytoplasm within the oosphere contains one female nucleus in it and called the ooplasm. In the beginning of the development of the oogonium there are many nuclei, which degenerate soon leaving one functional female nucleus.
- The antheridium develops on the terminal end of another hypha lying very close to the oogonium. The hyphal end swells, becomes club-shaped and separates from rest of the mycelium by a septum. This swollen multinucleate club-shaped portion is called the antheridium.
- The antheridium attaches itself to the oogonial wall and at the point of contact a fertilization tube develops from the antheridium. The fertilization tube penetrates the oogonial wall and reaches the oosphere through the periplasm. One functional male nucleus transfer through the tube, reaches the egg, fuses with the female nucleus and the rest of the nuclei of the antheridium degenerate.
- The oospore is thick-walled and three-layered. The outermost thick layer of the

wall is warty in *Albugo candida*. The oospore contains a large diploid (2n) nucleus. The reduction division (meiosis) is not yet seen in *Albugo candida* but it has been observed in the other species of *Albugo*.

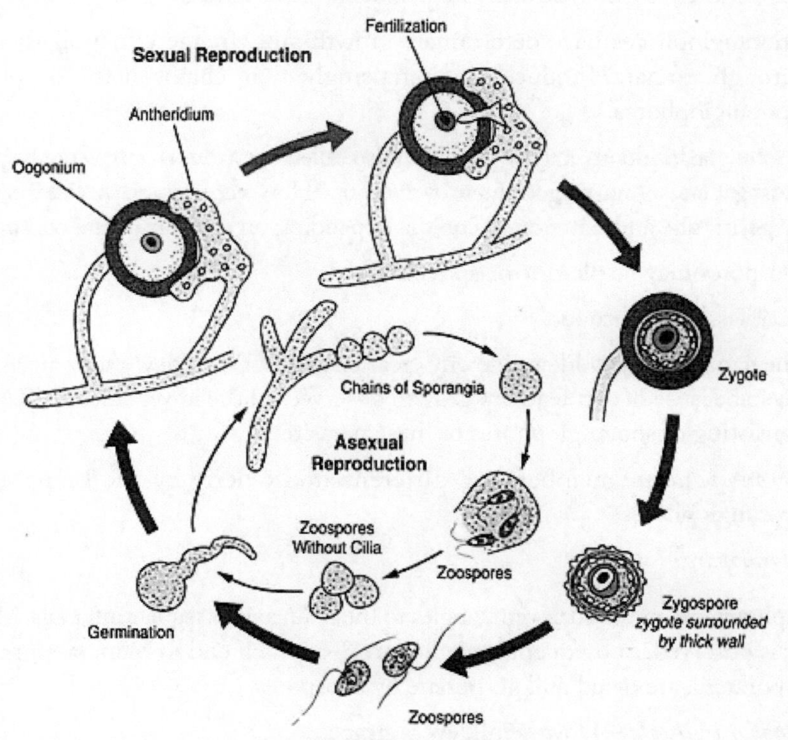

Life cycle of *Albugo* sp.

White blister - *Albugo candida, Albugo bliti*

- » The disease caused by Albugo is commonly known as white rust because it appears in the form of shiny, white, smooth irregular patches (pustules) or blisters on the leaves, stems and other aerial parts of the plant.
- » The pustules are initially formed on the lower surface of the leaf but in several cases they may be present on both the surfaces.
- » Increase in the size of the cells (hypertrophy) and organs takes place.
- » It results in the formation of large galls on the various parts of the host.
- » Severe infection causes proliferation of the lateral buds, discoloration of flowers, malformation of floral parts, and sterile gynoecium.

Family: *Peronosporaceae* (Downy mildew parasite/ hairy parasite)

Members of this family caused downy mildew disease in dicots. Generally, this is called as hairy parasite, as the sporangiosphores look like hairs on the lower surface.

» Mycelium is coenocytic and intercellular with haustoria

» Sporangiophores have determinate growth and emerge either singly or in tufts through stomata. Produce sporangia singly or in chains at the tip of branched Sporangiophores.

» Sporangia deciduous, may be papillate (also called operculum) or may not be papillate. In most genera, sporangia germinate by zoospore. However, in some species they germinate by germ tube and function as conidia depending on environmental conditions.

» Oospores may be plerotic or aplerotic.

» Periplasm conspicuous.

» The name downy mildews (downy= feathery or soft + mildew= superficial growth) is given because of soft feathery growth observed on the lower side of affected foliage consisting of sporangiophores of these parasite.

» *Peronosporaceae* members are differentiated chiefly by the branching of the sporangiophores

Genus: *Plasmopara*

Sporangiophores are branched at right angles to the main axis at regular intervals. Monopodial branching is observed. Subsequent branches are 3-6 which end in blunt sterigmata of 3 in number. Sporangia are ovoid and germinate by zoospores.

E.g., *Plasmopara viticola* – Downy mildew of grapes

Downy mildew of grapes

» Foliar symptoms appear as yellow circular spots with an oily appearance (oilspots). Young oilspots on young leaves are surrounded by a brownish-yellow halo. A white downy fungal growth (sporangia) will appear on the underside of the leaves and other infected plant parts.

» Infected shoot tips become thick, curl, and eventually turn brown and die. Young berries are highly susceptible, appearing grayish when infected. Berries become mummified when mature. Eventually, infected berries will drop.

Genus: *Peronospora*

Sporangiophores are dichotomously branched 2-7 times at acute angles and tips of branches are curved and pointed bearing sporangia on sterigmata. Sporangia are hyaline, ovoid, non-papillate and always germinate by germ tube. i.e., sporangia behave like conidia.

E.g., *Peronospora destructor* – Downy mildew of onion

Downy mildew of onion

» The affected leaves turn yellow and die off from the tip downwards.

» In moist conditions a white and later purplish mould develops on affected parts of the leaf.

» Bulbs can also be infected and often sprout prematurely or shrivel in store.

Genus: *Pseudoperonospora*

Sporangiophores are branched at acute angles with curved, blunt tips, bearing sporangia on sterigmata. Sterigmata are unequal (1 big and 1 small). Sporangia are greyish, ovoid, papillate and germinate by zoospores.

E.g., *Pseudoperonospora cubensis* – Downy mildew of cucurbits

Downy mildew of cucurbits

» Downy mildew symptoms first appear as small yellow spots or water-soaked lesions on the topside of older leaves.

» The centre of the lesion eventually turns tan or brown and dies.

» The yellow spots sometimes take on a "greasy" appearance and do not have a distinct border.

Genus: *Bremia*

Sporangiophores are dichotomously branched; tips of branches are expanded to cup shaped apophysis with four sterigmata bearing sporangia. Sporangia are ovoid, papillate and germinate by zoospores.

E.g., *Bremia lactucae* – Downy mildew of lettuce.

Downy mildew of lettuce

» Downy mildew causes light green to yellow angular spots as the spots are vein-limited on the upper surfaces of leaves. The white fluffy growth of the pathogen develops on the lower sides of these spots. With time these lesions turn brown and dry up.

» Older leaves are attacked first. Severely infected leaves may die.

» On rare occasions the pathogen can become systemic, causing dark discoloration of stem tissue.

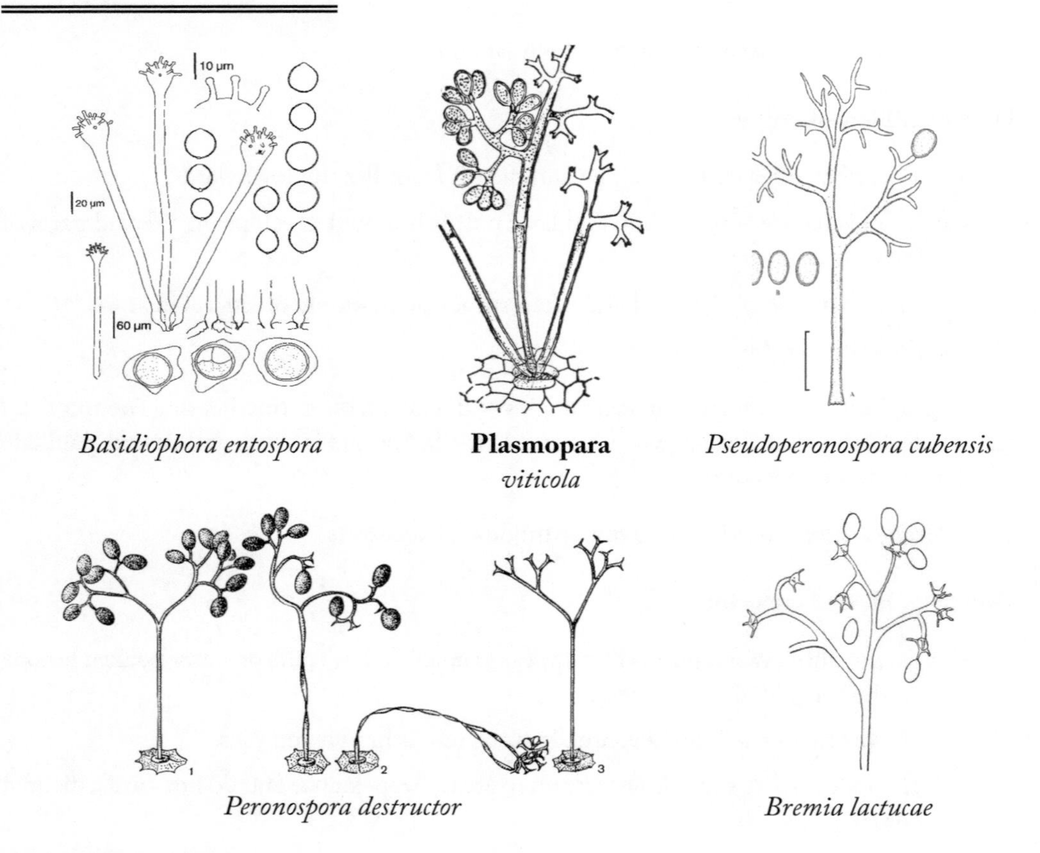

Basidiophora entospora **Plasmopara** *viticola* *Pseudoperonospora cubensis*

Peronospora destructor *Bremia lactucae*

Life cycle of *Plasmopara viticola*

P. viticola is heterothallic with two mating types. An antheridium fertilizes an oogonium to form the sexual oospore in fallen leaves infected in the previous season.

- » Oospores are spherical, covered by two inner oospore membranes and an outer wrinkled oospore wall.
- » They germinate in spring when temperatures reach 10°C and vineyard soils are wet.
- » The germ tube terminates in a macrosporangium which releases an average of 8-20 and up to 60 zoospores.
- » Zoospores require surface wetness to infect the host and infection takes place only through the stomata.
- » Zoospores swim on the tissue surface, encyst near stomata, and each spore forms a single germ tube which penetrates the stomata.
- » In the substomatal cavity the germ tube swells, forming a substomatal vesicle from

which a single hypha arises growing intercellularly.

» In as little as 3.5 hours the first haustorium forms where the pathogen contacts the host cells. Later additional haustoria form parasitizing the mesophyll cells. The incubation time, the period between infection and the first appearance of symptoms, depends on temperature and ranges from 4 to 21 days, with an average of 7-10 days.

» The pathogen sporulates through stomata during warm, humid nights.

» The sporangiophores are branching monopodially in the upper third at right angles to the main axis, and with a base tapering to a conical point.

» The sporangia are ovoid, colorless, each producing 1-6 zoospores.

» For sporulation, *P. viticola* requires at least 95-98% RH, temperatures between 10 and 30°C and at least 4 hours of darkness.

» Individual lesions resporulate a number of times under favorable conditions, and can retain the potential to sporulate for several months.

» Secondary cycles of infection occur repeatedly throughout the growing season if weather conditions are favorable.

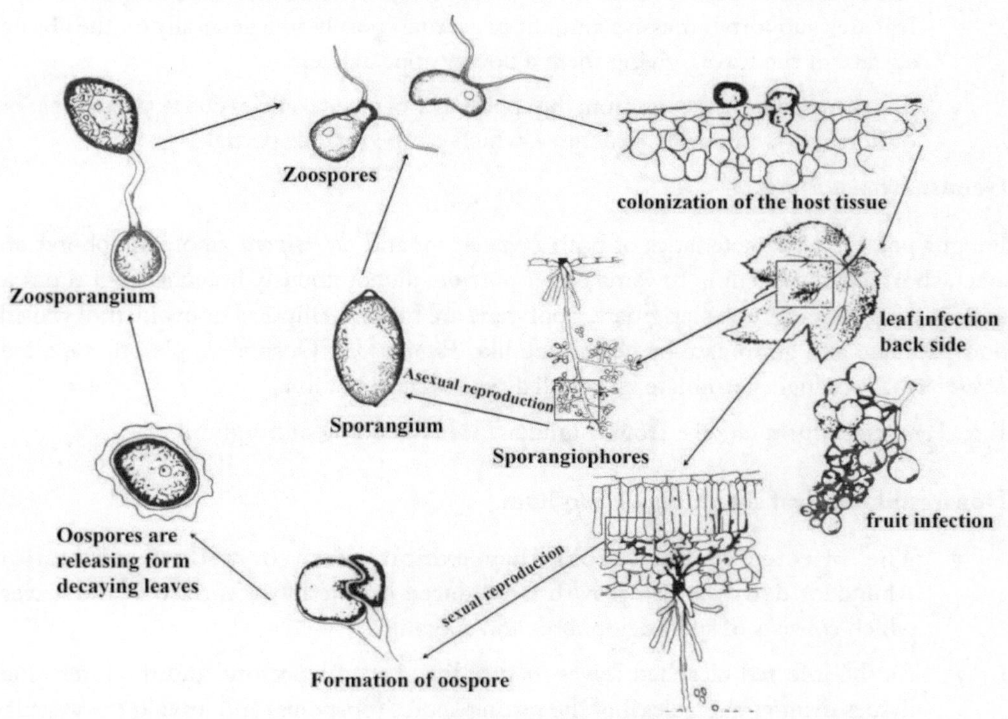

Life cycle of *Plasmopara viticola*

Source: Kassemeyer et al. (2015)

Family: *Sclerosporaceae* (Downy mildew parasite/ hairy parasite)

Members are obligate parasites causing downy mildew and root rots in monocots.

- » Mycelium is intercellular and enters the host cell through peg-like haustoria.
- » Indeterminate sporangiophores emerge singly or in clusters through stomata.

Genus: *Sclerospora*

Sporangiophores are stout, having upright branches, bearing sporangia on sterigmata. Sporangia are hyaline, ovoid, smooth walled, papillate and germinate by zoospores. Oospore is plerotic, smooth and golden brown.

Sporangia germinate either directly or indirectly.

E.g., *Sclerospora graminicola* – Downy mildew/ Green ear of pearl millet.

Downy mildew/ Green ear of pearl millet

- » Leaf symptoms begin as chlorosis at the base of the leaf lamina. The infected chlorotic leaf area supports a massive amount of asexual sporulation, generally on the abaxial surface of the leaves, giving them a downy appearance.
- » The name 'green ear' stems from the appearance of green panicles due to transformation of floral parts into leafy structures, which can be total or partial.

Genus: *Peronosclerospora*

Fungus possesses characteristics of both *Peronospora* and *Sclerospora*. Sporangiophores are erect, short, stout, widening towards upper portion, dichotomously branched 2-5 times at apex bearing sporangia on sterigmata. Sporangia are hyaline, elliptical or ovoid, thin walled, non-papillate and germinate by germ tube like *Peronospora*. Oospore is plerotic type like *Sclerospora*. Sporangia germinate by conidial type of germination.

E.g., *Peronosclerospora sorghi* - Downy mildew/ leaf shredding of Sorghum.

Downy mildew/ leaf shredding of Sorghum

- » The infected leaves unfold they exhibit green or yellow colouration Abundant downy white growth is produced on the lower surface of the leaves, which consists of sporangiophores and sporangia.
- » As the infected bleached leaves mature they become necrotic and the interveinal tissues disintegrate, releasing the resting spores (oospores) and leaving the vascular bundles loosely connected to give the typical shredded leaf symptom.

Genus: *Sclerophthora*

Sporangia are produced sympodially in groups (like *Phytophthora*) of between two and six in a basipetal succession on sporangiophores, which arise from hyphae congregated in the substomatal cavities; sporangial production occurs superstomatally. Oospore is hyaline and thick walled with an oil globule. E.g., *Sclerophthora rayssiae* var. *zeae* - Downy mildew of maize; *Sclerophthora macrospora* - Crazy top

Downy mildew of maize

- The most characteristic symptom is the development of chlorotic streaks on the leaves. Plants exhibit a stunted and bushy appearance due to shortening of the internodes. White downy growth is seen on the lower surface of leaf.
- Downy growth also occurs on bracts of green unopened male flowers in the tassel. Small to large leaves are noticed in the tassel.
- Proliferation of auxillary buds on the stalk of tassel and the cobs is common **(Crazy top).**

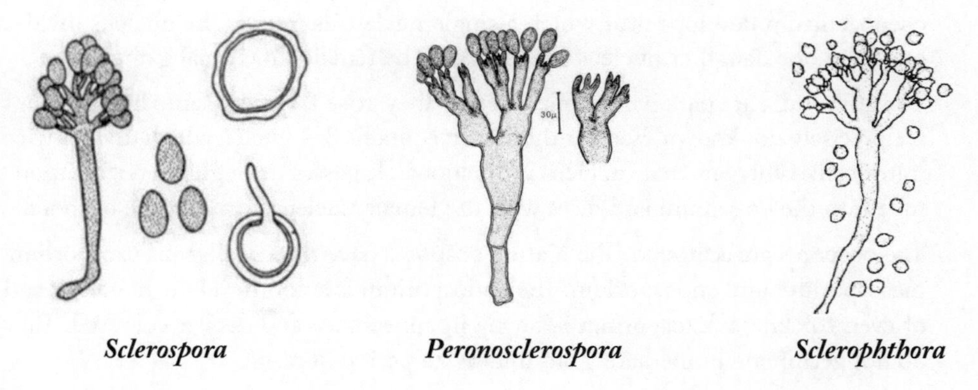

Sclerospora *Peronosclerospora* *Sclerophthora*

Life cycle of ***Sclerospora graminicola***

Asexual reproduction

- Sporangiophores are developed from the mycelium and emerge either singly or in groups through the stomata, in the early hours of the morning.
- The sporangiophores are produced on normal leaves, never in the inflorescence. They are hyaline, broad, unseptate and unbranched in the lower part, but a few thick branches are produced di-or trichotomously at the upper end. The tips of the final branches, or the sterigmata are slightly swollen forming sporangia which are cut off by a septum.

- The sporangia are hyaline, broadly elliptical (sometimes broadly cylindrical) and bear a small papilla at the free end with a thin smooth wall.
- The mature sporangia, which are multinucleate, fall off soon and germinate in the surrounding film of water by producing reniform and biflagellate zoospores.
- The zoospores are liberated between the temperature range of 18-23°C. After liberation they swim for a while, come to rest, and then germinate forming germ-tubes which show chemotactic response to roots of bajra.

Sexual reproduction

- Sexual reproduction is oogamous. The sex organs (antheridia and oogonia) develop within the host tissues, mostly in leaves and malformed spikelets.
- The oogonia are produced in large numbers in the leaf tissues. They are usually terminal, occasionally intercalary. The protoplasm with about 60 nuclei enters the oogonium from the mycelium. After the oogonium has expanded it is cut off by a septum. The nuclei pass through two mitotic divisions, before there is differentiation into ooplasm and periplasm. Then, the nuclei pass through a meiotic division. A coenocentrum develops near which a single nucleus is found. The nucleus divides once and one daughter nucleus later serve as the female functional gamete.
- The antheridia are paragynous but whether they arise from the same hypha or not are precisely not known. Each antheridium contains 3-4 nuclei, which divide twice mitotically. Only one male nucleus is functional. It passes through the fertilization-tube into the oogonium and fuses with the female nucleus, forming the oospore.
- The oospores are scattered. The mature oospores have three walls viz., exosporium, mesosporium and endosporium. The endosporium is smooth, yellow in colour and of even thickness. Exosporium is tawny in appearance and deeper coloured. They do not germinate immediately but undergo a period of rest.
- They retain their viability for at least five years after they fall to the ground with the debris of the plant.
- Germination takes place when favourable conditions appear, i.e. temperature between 20°-25°C, abundant air supply or oxygen, and low soil moisture contents.
- During germination of oospore, the diploid nucleus of the oospore divides into several nuclei involving meiosis.
- This followed by the bursting of its walls and the formation of a germ tube which is hyaline, non-septate and capable of infecting young seedlings.
- The green-ear disease is primarily soil borne. When the conditions are favourable, the oospores which fall to the ground infect young seedlings on germination.

Life cycle of *Peronosclerospora sorghi*

Pernosclerospora sorghi has a polycyclic disease cycle. It is capable of causing secondary infections of susceptible hosts throughout the growing season. Its resting structures, the structures that allow the pathogen to overwinter, are the oospores. They are often disseminated by wind.

» The oospores have very thick walls, which makes them capable of surviving in the soil for years under many different weather conditions.

» It is also possible for oospores and mycelium to overwinter in the seeds of maize.

» The oospores and mycelium that are present in the seed, it is possible for these infected seeds to become a source of inoculum, infecting the maize plant as it grows.

» The oospores are the main source of the primary inoculum of this disease. They are present in the soil when the host seedlings are germinating.

» The oospores then infect the roots of the seedlings. This type of infection is a systemic infection of the plant.

» The pathogen grows throughout the plant, infecting the leaves as they grow, leading to chlorosis. The chlorotic leaves develop white streaks. These white streaks are the location of oospore production. This only occurs in plants that were systemically infected as a seedling. When the oospores become mature, the white streaks on the leaves turn brown and become necrotic. These necrotic areas become shredded over time, which is how the mature oospores are disseminated.

» The spores are carried by the wind, and they become the source of inoculum in subsequent generations. As the pathogen continues to develop in the host plant, there may also be production of conidia on the leaf surface.

» It is the conidia and the conidiophores that cause the white, downy growth on the undersides of the leaves.

» Conidia develop quickly and are released from their conidiophores within five hours of maturation. After they are released, they are wind disseminated. If the conidia land on mature host plants, they infect and cause local lesions on the leaves.

» These lesions do not systemically infect the plants, and there is little overall damage from them. If the conidia land on a plant that is only a few weeks old or less, the plant is susceptible to systemic infection by the conidia.

» The conidia can be the main cause of the infection if they are being produced on a host plant that is in the same area as other susceptible host plants that are just emerging from the ground.

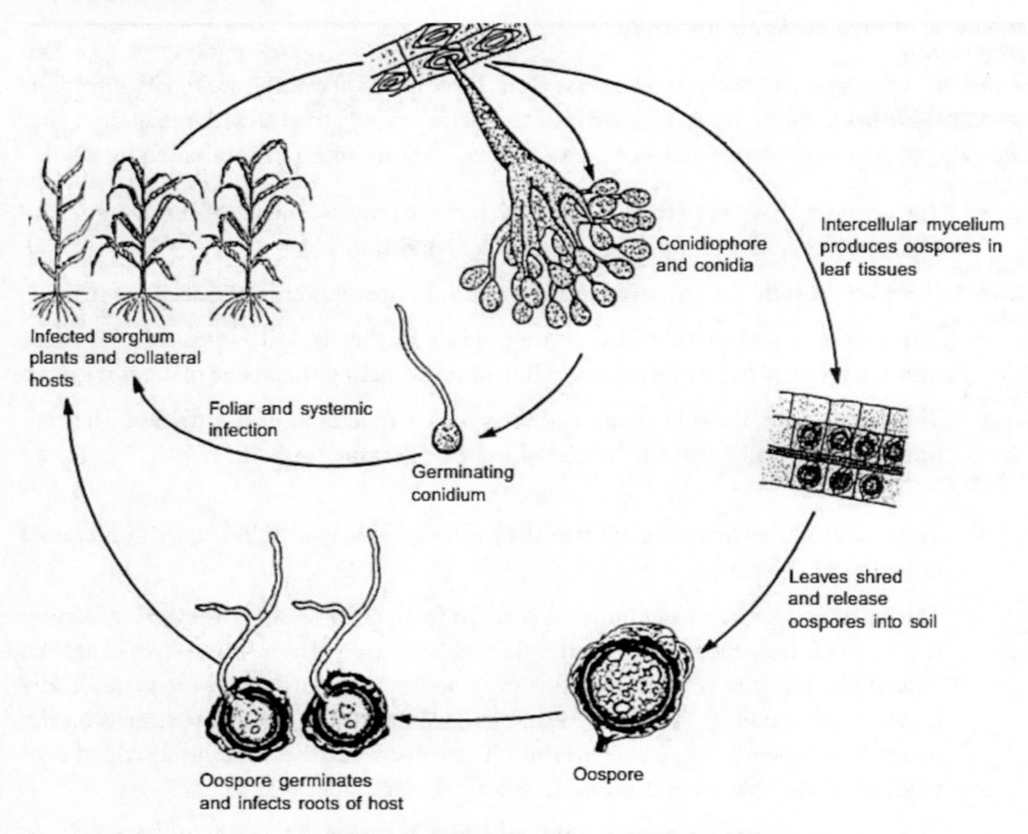

Life cycle of *Peronosclerospora sorghi*

Source: Bock and Jeger

Practice Questions

1. White rust in crucifers is caused by..........
2. Hypertrophied inflorescence of mustard represents on set of which stage of *Albugo candida*.........
3. The name Chromista means......
4. ....number of zoospore produced by *Plasmodiophora brassicae*
5. .... is vector for *Barley yellow mosaic virus* (BaYMV) and *Soil-borne wheat mosaic virus* (SBWMV)
6. Excessive enlargement of cell due to the infection of pathogen is.........
7. White blister of amaranthus is caused by..........

8. Explain the disease cycle of club root of cabbage
9. Define plasmodium and elaborate its types
10. Difference between damping-off and root rot
11. Explain the differences between *Pythium* and *Phytophthora*
12. Explain the differences between the Oomycota in the Chromista and the true Fungi
13. Explain the types of sporangiophores produced by downy mildew parasites

Suggested Readings

Kageyama, K., & Asano, T. (2009). Life cycle of Plasmodiophora brassicae. *Journal of Plant Growth Regulation*, *28*(3), 203-211.

Dixon, G. R. (2004). The biology of Plasmodiophora brassicae Wor.-A review of recent advances. In *IV International Symposium on Brassicas and XIV Crucifer Genetics Workshop 706* (pp. 271-282).

Dixon, G. R. (2009). The occurrence and economic impact of Plasmodiophora brassicae and clubroot disease. *Journal of Plant Growth Regulation*, *28*(3), 194-202.

Diederichsen, E., Frauen, M., Linders, E. G., Hatakeyama, K., & Hirai, M. (2009). Status and perspectives of clubroot resistance breeding in crucifer crops. *Journal of Plant Growth Regulation*, *28*(3), 265-281.

Constantinescu, O., & Fatehi, J. (2002). Peronospora-like fungi (Chromista, Peronosporales) parasitic on Brassicaceae and related hosts. *Nova Hedwigia*, *74*(3-4), 291-338.

Alexopoulos, C. J., Mims, C. W., & Blackwell, M. (1996). *Introductory mycology* (No. Ed. 4). John Wiley and Sons.

Dyer, P. S., & Kueck, U. (2017). Sex and the imperfect fungi. *Microbiology spectrum*, *5*(3), 5-3.

Voglmayr, H. (2008). Progress and challenges in systematics of downy mildews and white blister rusts: new insights from genes and morphology. *The Downy Mildews-Genetics, Molecular Biology and Control*, 3-18.

Berlin, J. D., & Bowen, C. C. (1964). The host-parasite interface of Albugo Candida on Raphanus sativus. *American Journal of Botany*, *51*(4), 445-452.

Thakur, R. P., & Mathur, K. (2002). Downy mildews of India. *Crop Protection*, *21*(4), 333-345.

Singh, S. D. (1995). Downy mildew of pearl millet. *Plant Disease*, *79*(6), 545-550.

Johnson, D. A., Engelhard, B., & Gent, D. H. (2009). Downy mildew. *Compendium of hop diseases and pests. The American Phytopathological Society, St Paul, MN*, 18-22.

Michelmore, R. W., Ilott, T., Hulbert, S. H., & Farrara, B. (1988). The downy mildews. In *Advances in plant pathology* (Vol. 6, pp. 53-79). Academic Press.

Chapter - 10

Fungi: *Chitridiomycota*

Kingdom: Fungi

In the kingdom Eumycota, true fungi are a varied and ubiquitous group of microorganisms. Fungal microorganisms are eukaryotic and heterotrophic. The vast majority of fungi are small, although some, such as mushrooms and bracket fungi, have huge fruiting bodies, f ew are anaerobic. Either they are unicellular (yeasts) or they grow as threadlike structures (hyphae) to produce mycelium. The life cycle of the majority of fungus includes both sexual (teleomorph stage) and asexual (anamorph stage) reproduction. Spores are the primary method of reproduction for fungi.

» The majority of organisms that cause plant diseases are fungi. However, the vast majority of fungi are saprophytic and play an essential role in decomposition.

» In numerous fungi, the sexual stage might be uncommon or lacking. The "Catalogue of Life 2019" (CoL) classifies fungi into five phyla: Ascomycota, Basidiomycota, Chytridiomycota, Glomeromycota, and Zygomycota.

» Plant pathogens are found in four of these phyla, Chytridiomycota, Zygomycota, Ascomycota and Basidiomycota.

Phyum: *Chitridiomycota* (Chitrids)

The name Chitrids derived from the greek word chitridion, meaning "little pot" describing the structure containing unreleased spores. They occur commonly in aquatic habitats, although many inhabit the soil. Genera like *Olpidium*, *Physoderma* and *Synchytrium* occur parasitically on many plants of economic importance to man.

» It is saprophytic and absorptive nutrition. Some genera lack mycelium, and are unicellular and holocarpic.

» Some genera thallus is coenocytic with oval multinucleate cell or elongated hyphae or a well-developed non-mycelial type (possess rhizoid), it may be holocarpic or eucarpic and monocentric or polycentric.

» Eucarpic thalli with one reproductive body are called monocentric, whereas those containing many reproductive bodies are called polycentric. Rhizoids are integral part of eucarpic thallus.

» The thallus may be epibiotic (attached externally but sending rhizoids inside the host cell) or endobiotic (growing inside the host cell) or interbiotic (attached to many host).

» The cell wall mainly consists of chitin and keratin, but in some genera of hypochitrids the cellulose is said to be present.

» The vegetative plant body is coenocytic. However, a septum develops at the base of each reproductive body.

» A distinctive characteristic of this order can be observed in the ultrastructure of the zoospore. Ribosome is loosely aggregated around the nucle that is not enclosed in a nuclear cap.

» This phylum is unusual among the fungi; their gametes are the only fungal cells known to have a flagellum. The present of posteriorly single whiplash flagellate on diploid zoospores is the distinguishing feature of the class. Zoospores released from a zoosporangium will swim for a period of time until a suitable substrate. The zoospore attaches itself by producing rhizoids that will grow into the substrate and anchor it.

» Sexual reproduction in Chytridiomycetes varies from simple isogamy to oogamy through anisogamy. Plasmogamy is by planogametic copulation. The sexual spores (resting spores) are thick-walled and germinate to produce a sporangium after a dormant period.

» The members have alteration of generations (E.g., *Allomyces*). Haploid thallus forms male and female gametagia from which flagellated gametes released. Gametes and female gametangia attract the opposite sex by producing sex hormones.

» The germinated zygote produces a diploid zoospores resulting in a diploid thalli and thick walled sporangia which after meiosis release haploid zoospores which form haploid thalli.

» In genera like *Olpidium* and *Synchrtrium* conjugation of two isogamous planogametes takes place between two anisogamous planogametes.

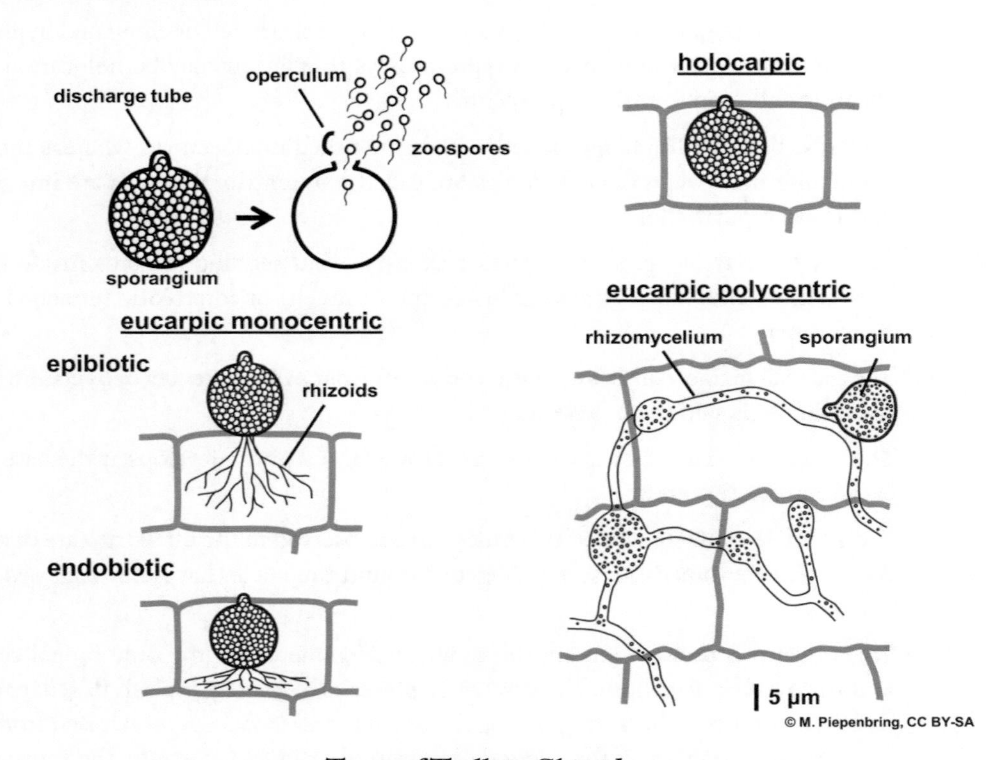

Types of Thalli in Chitrids

Source: Piepenbring

Order: *Chitridiales*

- » Most of them are water or soil inhabiting. Some are endobiotic inside the cell (*Synchytrium*), while others are epibiotic on host surface (*Rhizophydium*).
- » Many forms are endobiotic, i.e. live completely within the cells of the host (*Synchytrium*) whereas some are also epibiotic, i.e. having their reproductive bodies on the host surface (*Rhizophydium*).
- » Zoospores are released through papilla (*Rhizophydium*) or through pore (*Synchytrium*) in the sporangial wall.
- » Nutritional requirements of the majority of Chytridiales include mineral salts and carbohydrates in the form of sugar, starch or cellulose.

Diseases

Pathogens	Diseases
Physoderma zea-maydis	Brown spot of maize
Physoderma alfalfa	Crown wart of lucerne
Urophlyctis leproides	Crown wart of sugarbeet
Synchytrium endobioticum	Potato wart
Olpidium brassicae	Tobacco root burn, Root rot of cabbage and strawberry
Olpidium virulentus	Vector for *Tobacco necrosis virus, Tobacco stunt virus, Mirafiori lettuce big vein virus* and *Lettuce big-vein associated virus*
Olpidium cucurbitacearum	Vector for *Cucumber necrosis virus*

Genus: *Synchytrium endobioticum*

It is an obligate parasite in the epidermal cells of plant tissues. Early infection shows rapid cell division in the host tissues. The cells become enlarged due to hypertrophy and assume the shape of galls and forms warty out growth. Resting spores (winter sporangia) are mostly spherical, thick-walled, about 30-80 μm diameter. They tend to be integral components of aggregates or crumbs of soil.

Life cycle

» *S. endobioticum* is an obligate, holocarpic, endobiotic parasite.

» It is long-cycled chytrid which does not produce hyphae but a thallus comprised of sporangia.

» Two forms of sporangia exist, so-called summer and winter sporangia (resting spores), which contain 200-300 motile zoospores. The summer or swarm stage results from host infection by haploid zoospores in which a sorus of one to nine sporangia form, and the winter or resting stage results from infection by conjugated (diploid) biflagellated zoospores.

» Both sporangial types germinate to release pear-shaped (1.5-2.2 μm diam.) zoospores.

» Motility is by means of posterior flagella.

» The resting (meio-)sporangia are golden brown, ridged and spheroidal (ca 35-80 μm diam).

» If infection conditions are suitable, i.e. soil temperature and water, the rapidly reproducing summer sporangia release their zoospores thus setting up repeated infection cycles.

» At the same time, (meio-)sporangia (resting spores) are formed and, while conditions no longer favour the summer stage, the resting spores will overwinter in the infection zones of the potato.

» The resting spores induce hypertrophy of the infected tissue resulting in the so-called warts. This tissue will rot down in the soil during the winter months to release the resting spores into the soil.

» Resting spores can remain viable for decades. The ultimate lifespan has yet to be determined.

» The commonest means of spreading resting spores are by wart or soil distribution. Other limited means of distribution are by wind (over dried infested soil) or through animal droppings.

» The chitinous/ melaninized wall of the resting spore is extremely chemo-resistant to common soil agents. Its resistance and longevity impact directly on control measures.

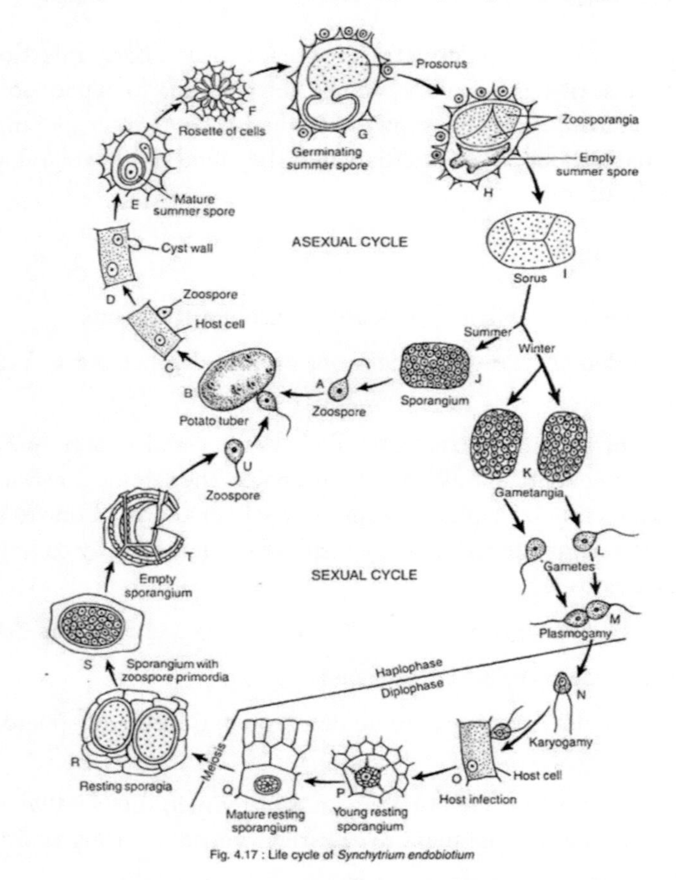

Life cycle of *Synchytrium endobioticum*

Symptoms

- » Galls vary in shape but are mostly spherical, with corrugated surfaces, and range from pea-size to fist-size <1 cm to >8 cm diam.).
- » Below ground galls are white to brown, turning black as they decay.
- » These galls appear at stem bases, stolon tips and tuber eyes.
- » They may not be evident until harvest-time. At harvest, galls may desiccate or decay. Tubers may be disfigured or completely replaced by galls.
- » Tuber galls may develop after harvest, in storage.

Practice Questions

1. Chitrids derived from the greek word ....
2. Sexual reproduction in Chytridiomycetes varies from simple .... to .... through anisogamy
3. .... chitrid fungi have alteration of generations
4. ....is typical endobiotic fungi
5. Potato wart disease is caused by.....
6. Describe the symptoms and pathogenesis of wart disease in potato

Suggested Readings

Barr, D. J. S. (2001). Chytridiomycota. In *Systematics and evolution* (pp. 93-112). Springer, Berlin, Heidelberg.

Longcore, J. E., & Simmons, D. R. (2012). Chytridiomycota. *eLS*.

Karling, J. S. (2013). *Synchytrium*. Elsevier.

Lange, L., & Olson, L. W. (1978). The zoospore of Synchytrium endobioticum. *Canadian Journal of Botany*, *56*(10), 1229-1239.

van de Vossenberg, B. T., Prodhomme, C., Vossen, J. H., & van der Lee, T. A. (2022). Synchytrium endobioticum, the potato wart disease pathogen. *Molecular Plant Pathology*, *23*(4), 461-474.

Chapter - 11

Fungi: *Zygomycota*

Phylum: ***Zygomycota***

Zygomycota are commonly thought of as bread molds, but there are many species of fungi within this classification that form symbiotic relationships with plants or infect animal hosts. Two other common names for Zygomycota are pin molds and sugar molds. These fungi are mainly terrestrial, and their spores are dispersed by air. They are saprophytes or weak plant parasites that can cause soft rot or moulds.

- The term "pin mold" refers to the appearance of certain species, while "sugar molds" refers to the sugar-rich fruit that is often affected by zygomycota.
- The most familiar is the mold that affects strawberries and other fruits.
- They share many characteristics with flagellated fungi, and therefore were once thought to be related to acquatic fungi.
- However, differences in cell-wall structure and a lack of flagellated spores or gamets indicate that there is no relation.
- The zygomycota are usually fast growing fungi characterized by primitive coenocytic (mostly aseptate) hyphae.
- Asexual spores include chlamydoconidia, conidia and sporangiospores contained in sporangia borne on simple or branched sporangiophores.
- Sexual reproduction is isogamous (fusion of two specialized similar hyphae called gametangia) producing a thick-walled sexual resting spore called a zygospore.
- The production of a thick-walled resting spore (ie, zygospore) within a commonly ornamented zygosporangium, formed after

- Most isolates are heterothallic i.e. zygospores are absent, therefore identification is based primarily on sporangial morphology. This includes the arrangement and number of sporangiospores, shape, color, presence or absence of columellae and apophyses, as well as the arrangement of the sporangiophores and the presence or absence of rhizoids.

Genus: *Mucor* sp. (Pin mould)

Mucorales are large, spherical, non-apophysate sporangia with pronounced columellae and conspicuous collarette at the base of the columella following sporangiospore dispersal.

- Colonies are very fast growing, cottony to fluffy, white to yellow, becoming dark-grey, with the development of sporangia.
- Sporangiophores are erect, simple or branched, forming large (60-300 μm in diameter), terminal, globose to spherical, multispored sporangia, without apophyses and with well-developed subtending columellae.
- A conspicuous collarette (remnants of the sporangial wall) is usually visible at the base of the columella after sporangiospore dispersal.
- Sporangiospores are hyaline, grey or brownish, globose to ellipsoidal, and smooth-walled or finely ornamented.
- Chlamydospores and zygospores may also be present.

Genus: *Rhizopus* sp. (Black bread mould)

- The genus *Rhizopus* is obligate mutualistic biotrophs, characterised by the presence of stolons and pigmented rhizoids, the formation of sporangiophores, singly or in groups from nodes directly above the rhizoids, and apophysate, columellate, multispored, generally globose sporangia, rarely produce chlamydospores. It is heterothallic in nature.
- Mycelium is coenocytic, colored, branched mycelium.
- After spore release the apophyses and columella often collapse to form an umbrella-like structure.
- Sporangiospores are globose to ovoid, mitosporangium, one-celled, hyaline to brown and striate in many species.
- Colonies are fast growing and cover an agar surface with a dense cottony growth that is at first white becoming grey or yellowish brown with sporulation.

Life cycle of *Rhizopus stolonifer*

Asexual reproduction

- » In asexual reproduction, special non-motile hyphae called sporangiophores produce sporangia that form in an upright fashion.
- » The sporangia at the tips of the upright hyphae develop as bulbous black portions.
- » Branching structures, called rhizoids, anchor the fungus into the substrate, releasing digestive enzymes and absorbing nutrients for the fungus.
- » When conditions are good the sporangia, containing numerous haploid spores of which are produced through mitosis, release the spores into the surrounding atmosphere.
- » These spores then may land on a moist surface and the life cycle repeats.
- » During unfavourable condition, thick-walled, nutrition-rich, intercalary mycelium segments arise by septation of mycelium, termed as chlamydospores. With the onset of favourable condition, the chlamydospore germinates and gives rise to a new mycelium.

Sexual reproduction

- » In sexual reproduction, resistant spherical spores are formed, called zygospores.
- » Zygospores are thick-walled, which make them highly resistant to environmental hardships.
- » The word zygospore comes from the Greek word zygos, which means joining.
- » The zygospores are the only diploid phase of *Rhizopus stolonifer* reproduction.
- » They are composed of two suspensor cells, which are the former gametangia or hyphae.
- » There is a suspensor cell on either side of a large, rough, dark brown spore. The suspensor cells are present to provide support.
- » The zygospore forms from two special haploid hyphae of opposite mating types that touch due to hormones and being in close proximity of each other.
- » The two cytoplasms intermingle, also known as plasmogamy. As this occurs the nuclei of both parents enter the conjunction, causing the resting spore to develop. Karyogamy is the term used to describe the fusion of the two nuclei.
- » After the zygospore has fully formed, meiosis occurs and haploid spores are formed and dispersed.

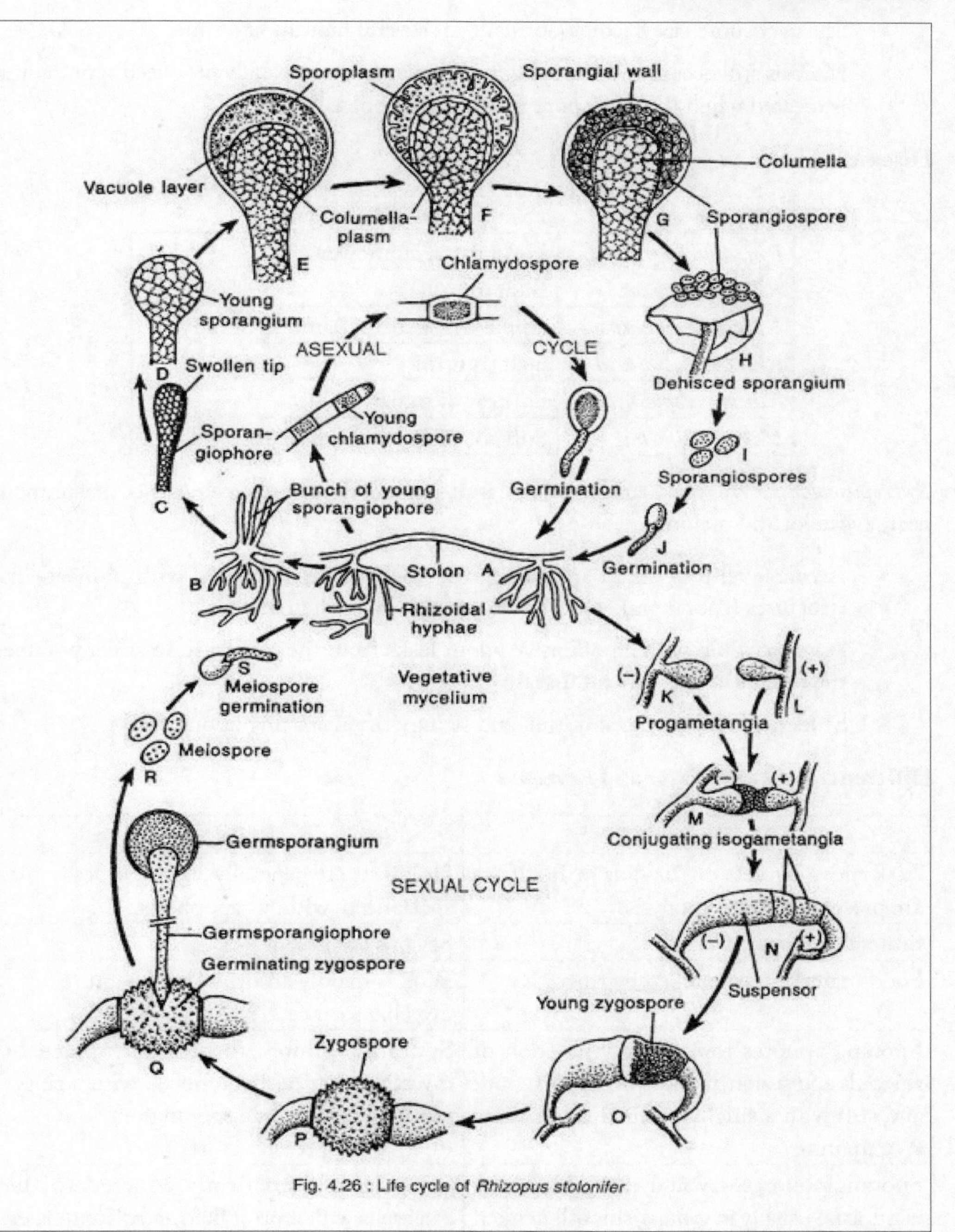

Fig. 4.26 : Life cycle of *Rhizopus stolonifer*

Life cycle of *Rhizopus*

» The zygospore can become dormant for several months at a time.

» Meiosis still occurs and a sporangium similar to the asexually produced sporangium is created when the zygospore finally cracks open.

Diseases

Rhizopus solonifer	Fruit rot and bread mould
Rhizopus nigricans	Head rot of sunflower
Rhizopus oryzae	Bulb rot
Rhizopus arrhizus	Apple whisker rot, Almond of hull rot
Rhizopus atrocarpi	Jack fruit rot
Mucor racemosus	Soft rot of tomato
Mucor piriformis	Soft rot of apple and pear

Rhizopus solonifer is a weak parasite on ripe fruit - peach, fig, strawberries, citrus, persimmon, pear, avocado, and melons.

» A coarse cottony mold appearing over the surface is covered with pinhead-like structures (sporangia), white when young, black when mature.

» A watery fluid with an offensive odour leaks from the soft fruit. In sweet potatoes the fungus causes soft rot disease.

» Storage roots become soft, wet, and stringy, often starting at one end.

Difference between *Mucor* and *Rhizopus*

Mucor	***Rhizopus***
Absorptive hyphae or rhizoids or holdfasts are present, definite apopysis.	Holdfasts are generally absent, or less specialized, without apophysis.
Stolons are present	Stolons are absent
Food materials is absorbed by rhizoids	Food is mainly absorbed by the entire mycelial surface.
Sporangiophores formed only junction of rhizoids and stolon, unbranched, short, stout and stiff with a single terminal columellate sporangium	Sporangiophores formed any place of mycelium, variously branched with a single terminal columellate sporangium
Sporangiophores dry and easily blown by wind, arise usually in groups, smooth marked by longitudinal striations.	Sporangiophores firmly adhered to the columella with drop of fluid and disseminated by insects and ants.

Cause contamination in laboratory culture	Don't cause contamination in laboratory culture

Practice Questions

1. Columella is a specialized structure found in the sporangium of ....
2. ....are commonly thought of as bread molds
3. ....is called as pin mould
4. ....causes jack fruit rot disease
5. The sexual spore of rhizopus is called as ....
6. ....is anchoring organ in rhizopus fungi
7. Difference between rhizopus and mucor
8. Difference between stolen and rhizoids

Suggested Readings

Benny, G. L., Humber, R. A., & Morton, J. B. (2001). Zygomycota: zygomycetes. In Systematics and evolution (pp. 113-146). Springer, Berlin, Heidelberg.

Zheng, R. Y., Chen, G. Q., Huang, H., & Liu, X. Y. (2007). A monograph of Rhizopus. SYDOWIA-HORN-, 59(2), 273.

Bautista-Baños, S., Bosquez-Molina, E., & Barrera-Necha, L. L. (2014). Rhizopus stolonifer (soft rot). In Postharvest decay (pp. 1-44). Academic Press.

Nieuwenhuis, M., & Van den Ende, H. (1975). Sex specificity of hormone synthesis in Mucor mucedo. Archives of Microbiology, 102(1), 167-169.

Chapter - 12

Fungi: *Ascomycota*

Phylum: ***Ascomycota***

» It is commonly called as "sac fungi, since the sexual spores (ascospores) are produced in sac-like body called ascus. Sometimes refered as cup fungi, earth tongues, cramp balls, dung buttons, truffles or moulds.

» Largest phylum in the kingdom Fungi (approximately 70% of all described fungi). Ascomycota has been shown to be the largest phylum of fungi, as compared to the other phyla (*Chytridiomycota, Zygomycota,* and *Basidiomycota*), with well over 87,000 species identified and named. This phylum is also morphologically diverse with species that range from single celled organism to multicellular cup fungi.

» This phylum is a monophyletic group; it contains all descendants of one common ancestor. The body of an ascomycete consists of single cells as in yeasts or of hyphae as in filamentous ascomycetes. Some species may also be dimorphic, having both forms. Most ascomycetes have both a sexual and an asexual stage in their life cycles. The life cycle is dominated by haploid cells (mycelium). Flagellate stage is completely absent.

» All are not pathogenic. Includes mushroom fungi like morels (sponge mushroom), summer truffles, french truffles, white truffles, yeast, fungi used for industrial purpose, genetic experiments (*Neurospora*), etc.

» Vegetative stage may be single celled, mycelial or dimorphic. Mycelium is well developed, profusely branched and septate with simple pore. Septal pore is plugged by membrane bound electron dense bodies, called woronin bodies. Cell wall is made up of chitin and glucans. Anastomosis is common. Somatic hyphae often organized into somatic tissues like sclerotia, stroma and mycelial strands (rhizomorph).

» Two distinct reproductive phases *viz.*, teleomorph (sexual stage) and anamorph (asexual stage) are observed in most cases.

» Asexual reproduction takes place by exogenously produced non motile thallospores called conidia and are non-motile (absence of motile spores) through fission, budding, fragmentation and by conidiogenesis. The conidiophore bearing is formed in the compound conidiophore or fruiting bodies (conidiomata) like acervuli, pycnidia, sporodochia and synnemata.

» Sexual ascospores are produced endogenously inside the ascus and the ascospores are usually 8 in an ascus. The plasmogamy occurs either by gametangial contact or gametangial copulation or spermatization or somotogamy. Gametangial contact is the common method of sexual reproduction. But the karyogamy doesn't occur immediately followed by plasmogamy.

» Ascomycetes are characterized by septate hyphae with simple pores. Asexual reproduction by conidia. Sexual reproduction by ascospores, typically eight, in an ascus. Asci are often housed in a fruiting body or ascocarp E.g., Cleistothecia, apothecium, naked asci or perithecia.

» Ascomycetes are 'spore shooters'. They are fungi which produce microscopic spores inside special, elongated cells or sacs, known as 'asci', which give the group its name.

» Prior to sexual reproduction, compatible haploid mating-type hyphae (+ and -) fuse to form a dikaryotic hypha. In contrast to the basidiomycetes, ascomycetes have a more limited dikaryotic stage. The dikaryotic stage eventually gives rise to an ascocarp and sexual ascospores.

» Ascus may be globose, oval, elongated, club or cylindrical shaped with or without stalk (sessile) and are unitunicate or bitunicate.

a. Prototunicate ascus is the primitive type. The wall is very thin, dissolves at maturity and spores are released. The ascospores are released in a mucilaginous substance. E.g., *Eurotium.*

b. Unitunicate ascus contains one wall more or less uniformly all around and the ascospores are released one by one through the pore at the tip. E.g., *Ophiostoma ulmi, Gibberella fujikorai* and *Nectria galangina.*

c. Bitunicate ascus has two distinct wall layers. Outer wall breaks at the tip and the inner wall come out. Ascospores are released forcibly by a puffing action through the pore at the tip of the inner wall. Some ascus has a lid / operculum at one end or slit made through two walls and ascospores are released individually. E.g., *Magnaporthe grisea, Capnodium ramosum* and *Botryodiplodia theobroame.*

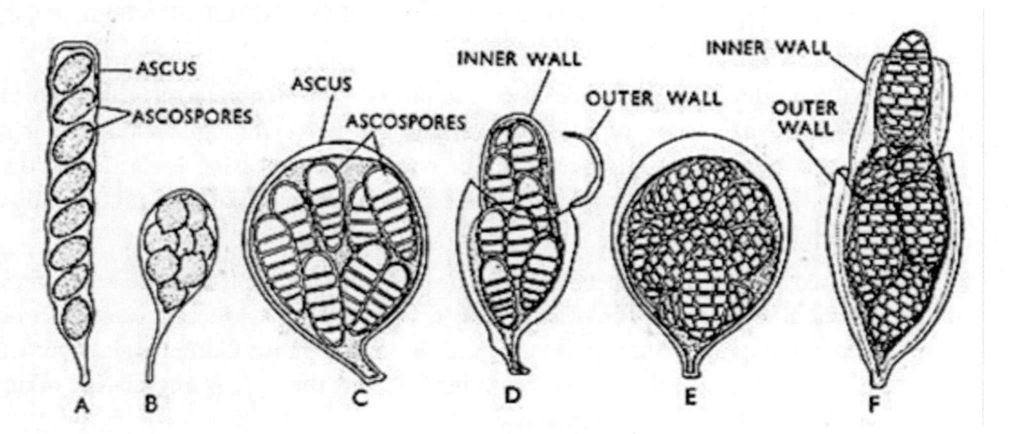

A-B: Unitunicate
C-F: Bitunicate
C & E: Intact bitunicate
D: Upon maturity expanding innerwall and splitting the outer wall
F: Release of ascospores through a pore at the tip of expanded inner wall

Ascospores are uni-or multinucleate; uni-or multicellular with cross wall either transversely or longitudinally or in both directions (*Pleospora multiseptata*)

Sexual reproduction

- The sexual part of the life cycle, in which two compatible haploid hyphae become intertwined and form an **ascogonium** and an **antheridium are** multinucleate elongated structures. In this case, the ascogonium acts as a "female" and accepts nuclei from the antheridium after plasmogamy has occurred.
- Male nuclei pass from antheridium to ascogonium through a pore developed at the point of contact or through a trichogyne (structure to receive the male nuclei) in ascogonium but no fertilization tube is formed.
- The resultant dikaryon is then capable of forming a cup-shaped ascocarp. Asci begin to form on the surface of the ascocarp at the tips of the dikaryotic mycelium, and karyogamy occurs to form the highly transient diploid nucleus.
- The diploid nucleus immediately undergoes meiosis, yielding four, genetically distinct, haploid nuclei. After an additional round of mitosis, the ascus now contains eight haploid nuclei.
- These eight nuclei will eventually develop into eight ascospores (haploid resting sexual spore), which are released from the ascus on the surface of the ascocarp.

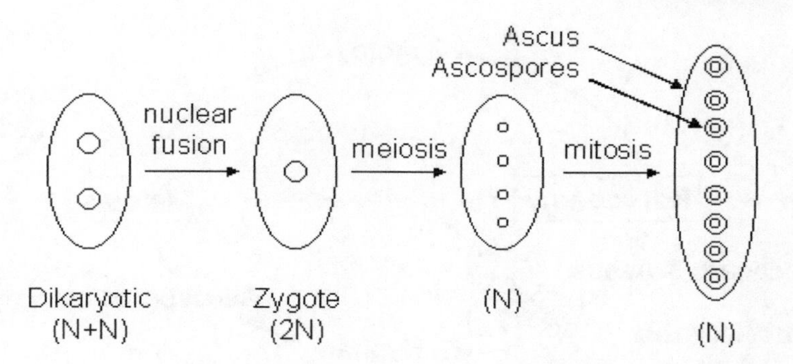

» In the final step in the sexual cycle, haploid mycelia arise from the aforementioned ascospores as the sexual cycle begins again.

» The binucleate ascogenous hyphae elongate bends over to form crozier / crosier (hook).

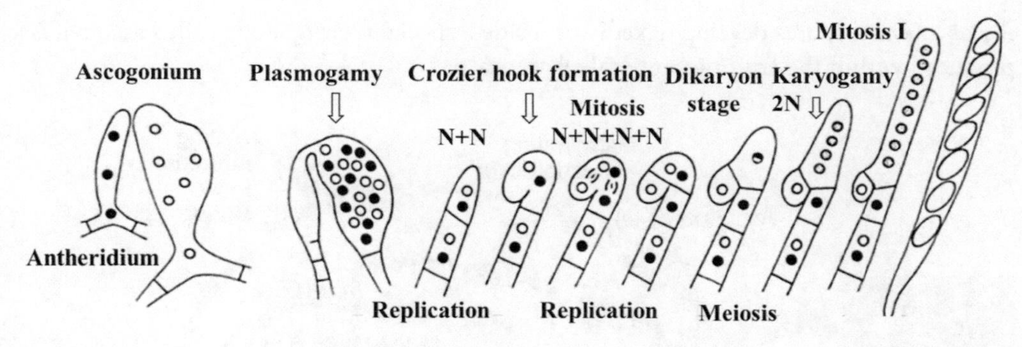

Crozier formation (Ascospore formation)

Asexual reproduction

» Asexual reproduction is the dominant form of propagation in the Ascomycota, and is responsible for the rapid spread of these fungi into new areas.

» Here, a compatible haploid partner is not present and the haploid mycelium is capable of producing asexual spores (conidia) by segmentation of its hyphae. Furthermore, Ascomycota also reproduce asexually through budding. These segments/ budded spores will compartmentalize into conidia, and wind or water dispersal will follow.

» The most important and general is production of conidia, but chlamydospores are also frequently produced.

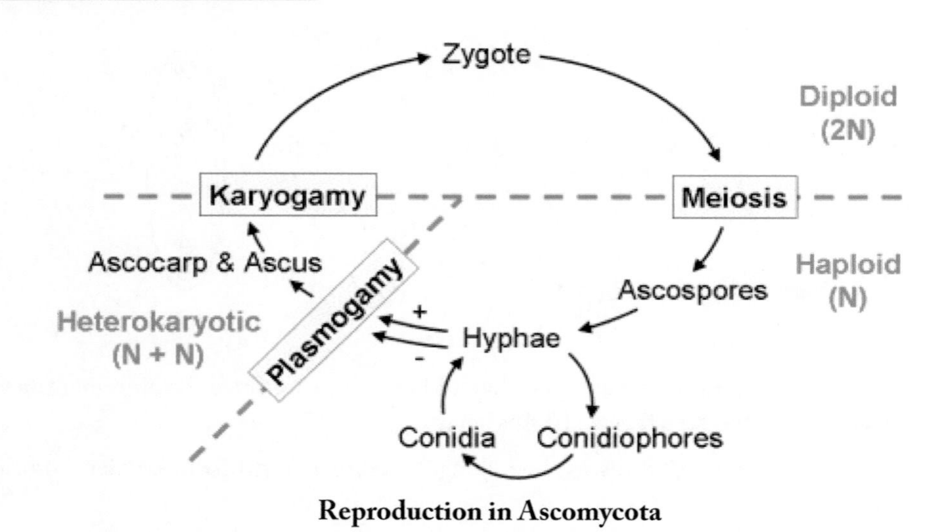

Reproduction in Ascomycota

Ascus and ascospores develop nakedly or inside a special fruiting body called ascoma. Asci produced within the fruiting body called ascocarp.

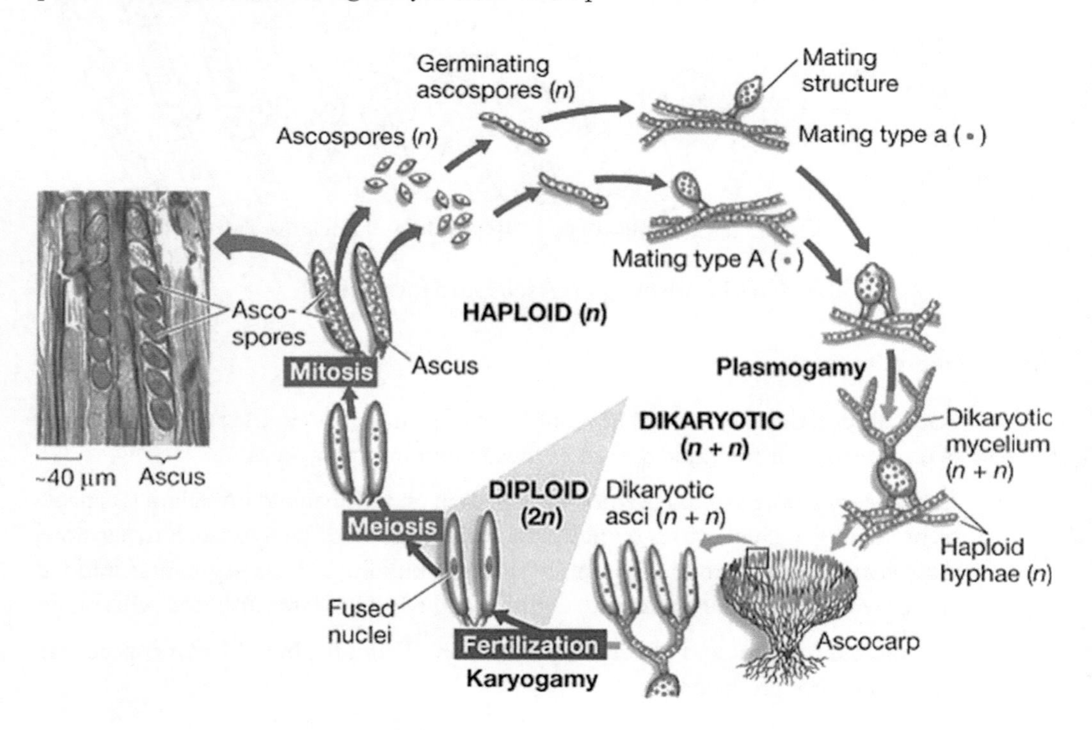

The life cycle of ascomycota

Several types of ascocarp/ ascoma

1. Closed ascocarp

» Cleistothecia (*sing.* Cleistothecium) is a closed non-ostiolate ascocarp with a peridium contains well organized membrane like layer of cells, having no predefined opening; with diversified hyphal appendages for anchoring and dessimination. Asci not regularly arranged, asci discharged ascospores violently by irregular splitting or chasm of ascus and peridium wall - *Aspergillus* and *Penicillium.*

» Chasmothecia (*sing.* Chasmothecium) is a closed non-ostiolate ascocarp with a peridium contains well organized membrane like layer of cells, also lack a preformed opening, Asci not regularly arranged, but ascospores are released by a chasm in the ascomatal wall – *Erysiphales.*

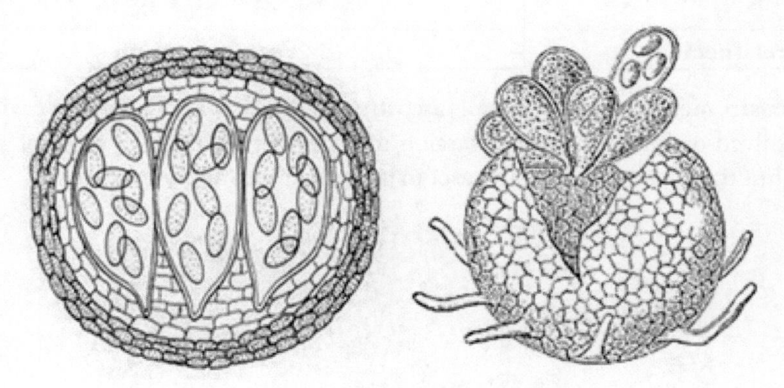

2. Flask shaped with a ostiole

» **Perithecia** (*sing.* Perithecium) subglobose or flask-like ostiolate ascocarp with a preformed opening (ostiole) through which ascospores are discharged. The asci are either arranged in a hymenium or in a basal bush. Most fungi producing perithecia also have unitunicate asci and are classified in Sordariomycetes. Sometimes limited to ascohymenial types formed from the development of an ascogonium – *Claviceps, Glomerella, Magnaporthe, Cryphonectria* and *Gibberella.*

» **Pseudothecia** (*sing.* pseudothecium) look similar to perithecia, but they differ in development. Asci form in locules (openings) inside vegetative fungal tissue called ascostroma; formation of bitunicate asci, and many members of this group produce darkly pigmented, multiseptate ascospores or conidia. E.g., *Venturia, Mycosphaerella, Alternaria*, *Cladosporium*, *Phoma*, and *Stemphylium.*

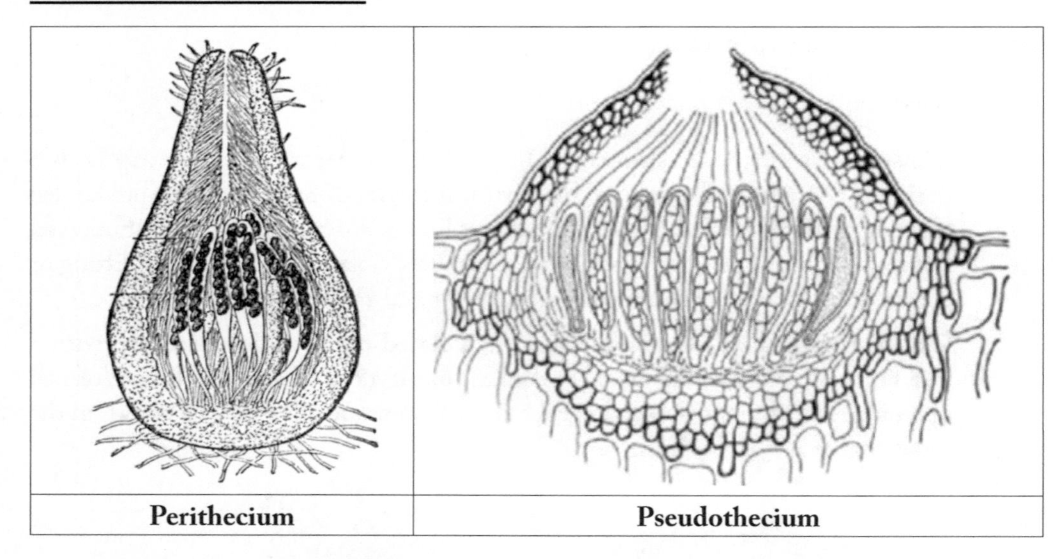

Perithecium	Pseudothecium

» **Ascostroma.** Ascostromata (*sing*, ascostroma) are ascocarp in which a cavity is either dissolved or formed via compression in a stroma, after which the asci are formed within the cavity (locule). The asci in an ascostroma are bitunicate.

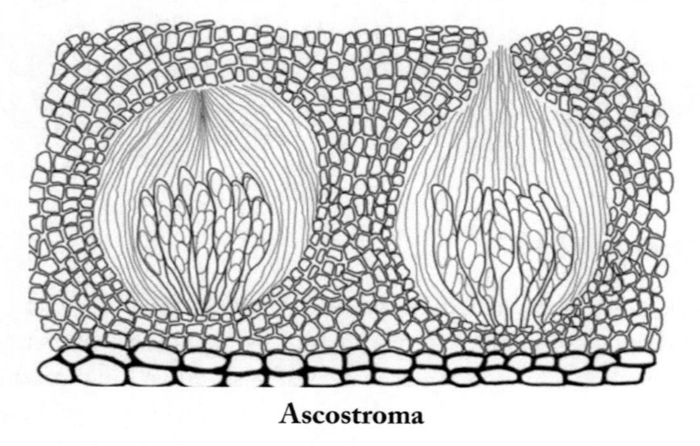

Ascostroma

3. Saucer shaped /open type

» **Apothecia** (*sing.* apothecium) are an open disc or cup-like or saucer-like ascocarp that may be sessile or stalked, with asci developing on the exposed hymenium – *Sclerotinia* and *Monilinia* spp.

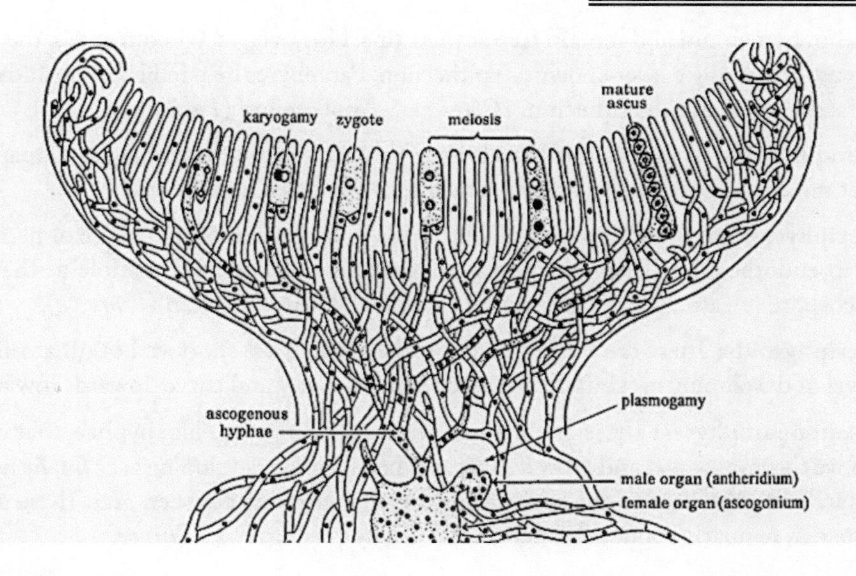

4. Wall less ascostroma

- Pseudothecium an ascostromatic ascocarp having asci in numerous unwalled locules. Sterile elongated hairs present in between the asci inside the ascocarp are known as paraphyses.
- In lower ascomycota, asci are produced nakedly called as **Naked asci** (E.g., *Taphrina*)

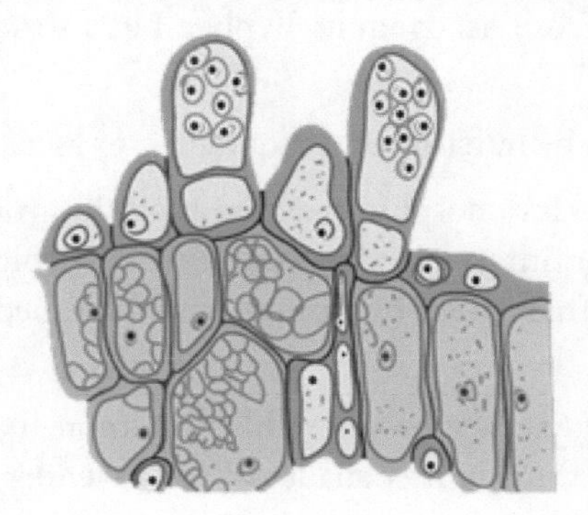

Naked Asci

Sterile Thread like Structures in Ascocarp

Ascocarps contain sterile thread like structures of various types.

1. **Paraphyses**: These are elongated, cylindrical, club shaped or sometimes branched threads arising from bottom of ascocarp. They may be septate or aseptate. They grow among

asci in hymenium and remain free at their tips. However, in Discomycetes, the tips fuse together forming a layer known as epithecium. Paraphyses help in liberation and dispersal of ascospores. E.g., Perithecium (*Claviceps*), Apothecium (*Peziza*).

2. **Paraphysoids**: These are inter ascal tissue that stretch and resemble pseudoparaphyses, but remotely septate, very narrow, anastomose and tips remain free.
3. **Periphyses**: These are short, hair like threads lining in side of an ostiole of perithecium or pseudothecium. Their function is to direct the asci towards ostiole at the time of ascospore release. E.g., Perithecium (*Claviceps*), Pseudothecium (*Venturia*).
4. **Periphysoids:** These are the lateral periphyses which are short and originate above the level of developing asci but do not reach base of cavity and curve upwards towards apex.
5. **Pseudoparaphyses:** These are distinct, vertical, paraphyses -like hyphae, that originate above the level of asci and grow downwards between the developing asci, finally becoming attached to the base of the cavity, thus forming curtains between asci. These are often broader, regularly septate, branched and anastomosing. E.g., *Elsinoe.*

Important genera

Subphylum: *Taphrinomycotina*

1. Mycelium is pseudomycelium or dikaryotic mycelium.
2. Ascocarp is absent i.e., asci are naked.
3. Asci are not formed from ascogenous hyphae but formed directly from zygote or ascogenous cells.
4. Asci release ascospores by bursting or deliquescing of ascus.
5. *Taphrina* and *Protomyces* is dimorphic in nature, typically grow as yeasts during one phase of their life cycles, then infect plant tissues in which typical hyphae are formed, and ultimately they form a naked layer of asci on the deformed, often brightly pigmented surfaces of their hosts.
6. Species of *Protomyces* generally have a reddish, salmon-like color in culture whereas, *Taphrina* species often show lighter shades of red, lavender or yellow.

Genus: *Taphrina*

» *Taphrina* is a fungal genus within the Ascomycota that causes leaf curl diseases and witch's brooms of certain flowering plants.

Diseases: *Taphrina deformans* - peach leaf curl

Taphrina maculans – leaf blotch of turmeric.

Symptoms: Leaf puckering and distortion, acquiring a characteristic downward and inward curl. The distorted, reddened foliage that it causes is easily seen in spring. When severe, the disease can reduce fruit production substantially.

Life style: Obligate parasite

Mycelium: Sub cuticular, mycelium is septate containing typical thick walled binucleate cells called ascogenous cells. Hyphae may be intercellular, sub cuticular, or may grow with in walls of epidermis.

Asexual reproduction: The ascospores are released into the air, carried to new tissues, and bud (divide) to form bud-conidia. It is by round or ovoid haploid blastospores produced by budding of ascospores or blastospores.

Sexual reproduction: It is by gametangial contact. No ascocarp is produced. Unitunicate asci are produced directly from the mycelium (Naked asci). Sexual spores are ascospore and produced eight numbers per ascus. Cells of the fungus break through the cuticle of distorted leaves and produce elongated, sac-like structures called asci that produce sexual spores called ascospores, which give the leaf a grayish white, powdery or velvetlike appearance.

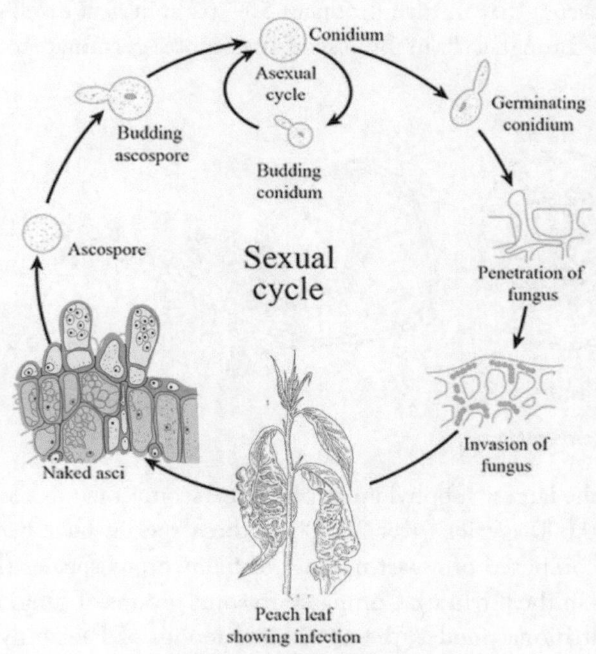

Life cycle of *Taphrina*

Genus: *Protomyces*

- *Taphrina* is a fungal genus within the Ascomycota that causes stem gall disease.

Diseases: *Protomyces macrosporus* – Stem gall of coriander

Symptoms: *Protomyces* species are plant parasitic and cause galls on stems, leaves and fruits.

Life style: Obligate parasite

Mycelium: Septate, hyaline, irregular, branched with intercellular and multinucleate mycelium.

Resting structure: Chlamydospores, which are globose to ellipsoid, thick-walled, three-layered membrane and smooth with brownish colour.

Asexual reproduction: it is seen that cell division is by budding on a narrow base near the poles of the cell. The cells are ovoid to elongate, and elongate cells may be asymmetrical. Pseudohyphae and true hyphae are not formed in culture.

Sexual reproduction: Heavy walled spherical structures, variously termed ascogenous cells, synasci, and "chlamydospores," are formed in infected tissue. The ascogenous cells germinate to form a vesicle (ascus) that in turn produces 50–200 spherical or ellipsoidal ascospores. Ascospore release is through a slit in the vesicle. Ascospores germinate to form asaprophytic, yeast-like budding stage.

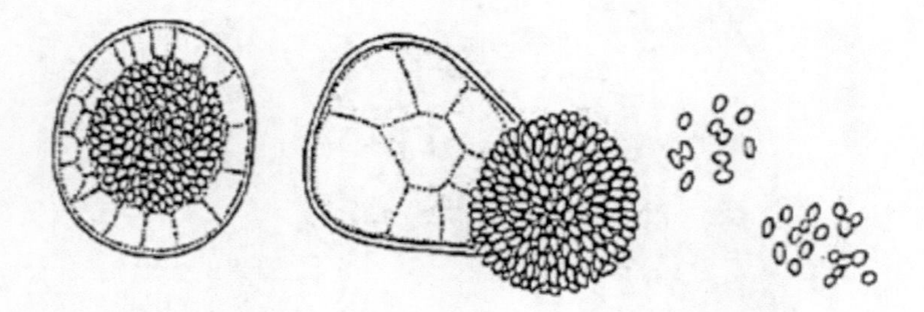

Subphylum: *Pezizomycotina*

Pezizomycotina is the largest subphylum of phylum Ascomycota (also known as sac fungi). With well over 100,000 species (over 30,000 of these species have been well described), this subphylum is composed of a vast number of filamentous species (highest number of filamentous species in the phylum). Compared to some groups of fungi that largely rely on specific, limited nutritional modes, the nutritional modes of Pezizomycotina range from pathogenic and parasitic modes to symbiotic to saprophytic modes making the subphylum one of the most successful group of fungi.

» They are cup fungi (they produce a fruiting body in the asci)

» A majority of the species in the subphylum produce hyphal threads

» Mostly reproduce through fission

» Septa with Woronin bodies

» Septa of the fungi have wide central pores

» The resting structures are sclerotia, chlyamydospore and rhizomorph.

» The life cycles are highly pleomorphic, with single species producing 0-1 sexual stage and 0-many asexual stage.

» Most of the species have both sexual and asexual reproduction.

» Asexual reproduction: by the formation conidia, oidia and thallospores. The conidia (mitospores) are produced from specialized conidiophores singly or in chains. The conidiophores are produced on fruiting structures like acervuli, sporodochia, pycnidia, synnemata.

» Sexual reproduction: sexual ascospore is produced inside the ascomata/ sporocarp. Plasmogamy occurs by gametangial contact or spermatization.

A sporocarp refers to the structure that releases spores. This structure has been identified in some of the species.

The following types of sporocarps have been identified in *Pezizomycotina*:

» **Apothecia** - The apothecium is found in many lichens and species of the subphylum and consists of a cup or disc shaped body on the concave surface. The asci (that contains the spore) are exposed to the environment. E.g., *Leoteomycetes, Pezizomycetes*

» **Perithecia** - Compared to apothecia, the perithecia are partly closed. However, the ostiole is exposed through a small opening. E.g., *Sordariomycetes*

» **Cleitothecia** - Compared to apothecia and perithecia, the Cleistothecia is closed completely while the ascospores are arranged randomly across the sporocarp. E.g., *Eurotiomycetes*

» **Ascostromata** - The ascospores of the Ascostromata are contained in a large cavity that resembles other sporocarps. E.g., *Dothidiomycetes*

The asci of *Pezizomycotina* are classified as:

» Inoperculate

» Operculate

» Prototunicate

- Unitunicate
- Bitunicate

Genus: *Capnodium*

- Capnodium is a genus of sooty molds in the family Capnodiaceae.
- It was circumscribed in 1849 by French mycologist Camille Montagne.

Disease: *Capnodium mangiferae* - Sooty mould of mango

Capnodium citri - Sooty mould of citrus

Capnodium thea - Sooty mould of Tea

Capnodium braziliensis - Sooty mould of coffee

Symptoms

- Black encrustation is formed on leaves, flowers, stem and fruit.
- Mycelium is superficial and lives on the sugary secretion of the plant hoppers.
- Photosynthetic ability is reduced which results in reduced fruit set and fruit fall

Life style: Obligate parasite

Mycelium: Dark septate, superficial hyphae with mucilaginous coating

Asexual reproduction: The fungus produces 4 types of conidia such as torula, trichothecium, coniothecium and brachysporium as well as micro and macro-conidia inside the pycnidia. Pycnidia are usually developed in the superficial pseudothecia.

Sexual reproduction: Ascocarp is pseudothecium, pseudothecia are ovoid to vertical, stalked and setose with sticky mucilaginous covering. Lack of pseudoparaphyses. Asci are ovoid and arranged in fasciciles. Ascospores are hyaline to dark, one to multisept ate ascospores.

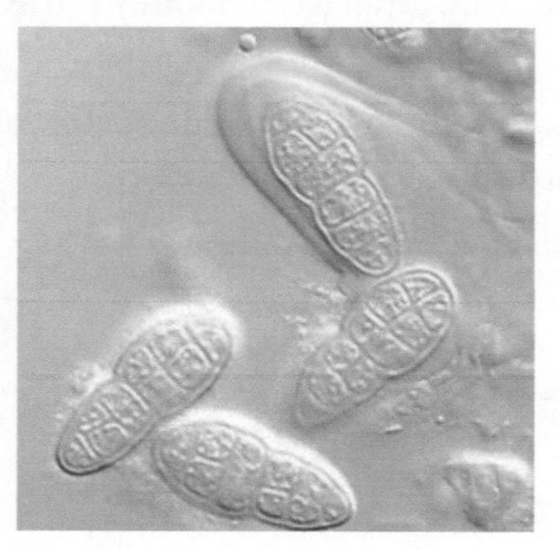
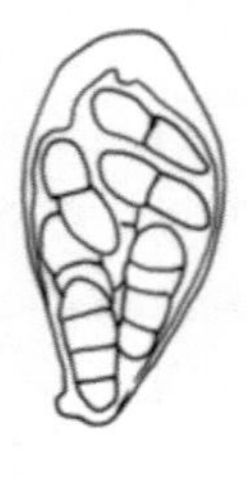

Genus: ***Elsinoe***

Disease: *Elsinoe fawcettii* – Citrus scab

Elsinoe mangiferae – Mango scab

Elsinoe ampelina - Grapevine anthracnose

Symptoms

» Minute water-soaked spots which subsequently evolve into amphigenous, creamy-yellowish or variously bright-coloured pustules on leaf and fruits surface.

Life style: -

Mycelium: hyaline, scanty, septate and short-branched.

Asexual reproduction: Acervuli intra-epidermal or sub-epidermal and scattered. Conidiogenous cells are the hyaline or pale-brown phialidic conidiophores, which have 2-4 septa. Conidia are hyaline, unicellular, ellipsoid, biguttulate.

Sexual reproduction: Ascomata globose and dark. Asci are up to 20 per locule, subglobose or ovoid, bitunicate, eight-spored. Ascospores hyaline, ellipsoidal or oblong, with two to four cells, usually constricted at the central septum.

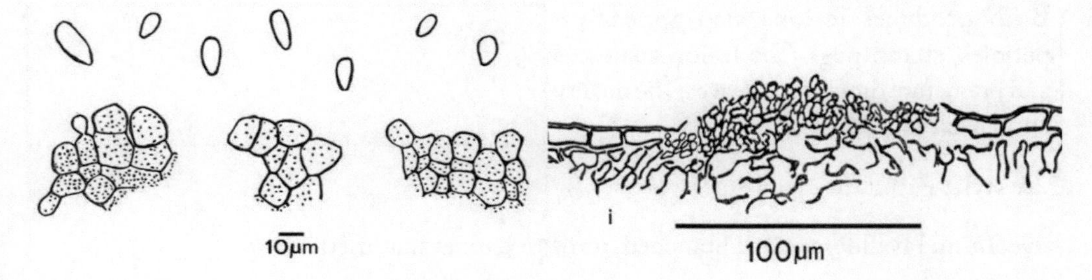

Genus: ***Mycosphaerella***

» *Mycosphaerella* is a genus of ascomycota. With more than 10,000 species, it is the largest genus of plant pathogen fungi.
» Its anamorphic stage is *Cercospora*.
» Most species of this genus cause plant diseases, and form leaf spots.
» It was circumscribed in 1849 by French mycologist Camille Montagne.

Disease: *Mycosphaerella rabiei* (Asexually: *Aschochyta rabiei*) – Chickpea blight

Mycosphaerella arachidis (Asexually: *Cercospora arachidicola*) – Early tikka of peanut

Mycosphaerella berkeleyii (Asexually: *Cercospora personatum*) – Late tikka of peanut

Mycosphaerella musicola (Asexually: *Cercospora musae*) – Yellow sigatoka of banana

Mycosphaerella fijiensis (Asexually: *Pseudocercospora fijiensis*) – Black sigatoka of banana

Mycosphaerella areola (Asexually: *Ramularia areola*) – Greymildew of cotton

Symptoms: Ciricular or irregular necrotic spot on the surface of the leaves.

Early leaf spot - *Cercospora arachidicola = Mycosphaerella arachidis*
Late leaf spot - *Cercospora personata = Mycosphaerella berkeleyii*

Difference between early and late leaf spot

Early leaf spot	Late leaf spot
Infection starts about a month after sowing.	Infection starts around 5 – 7 week after sowing
Circular or irregular reddish brown lesion with yellow halo	Small circular dark brown lesion without yellow halo
On lower surface lesion light brown colour. Conidia whip like pale yellow 3-12 septum	Lower surface lesion carbon black colour Conidia obclavate 4-12 septa
Both produces lesions also appear on petioles, stems, pegs. The lesion coalesces and premature dropping of leaves. The quality and yield reduced	

Life style: Facultative parasite

Mycelium: Hyaline, septate, branched, turns to dark at maturity.

Asexual reproduction: Conidia are elongated, narrow and multiseptate. The condiospores are pale to a light olive-brown, and are smooth, long, and have three or more septa.

Sexual reproduction: Sexual spores are ascospores which are hyaline, septate produced in erumpent, ostiolate dark brown pseudothecia. Plasmogamy occurs by spermatization.

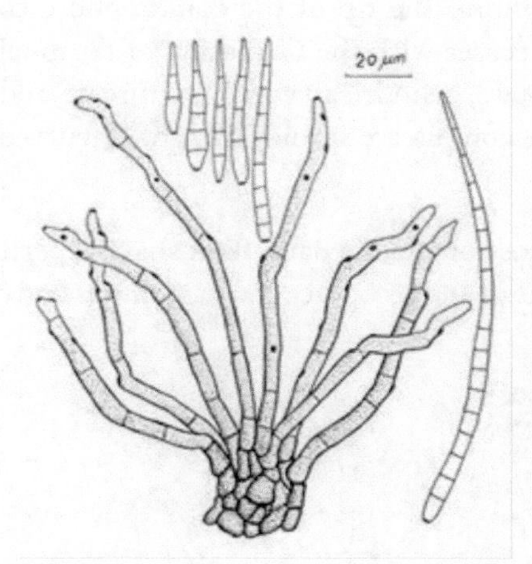

Genus: *Cochliobolus* (Anamorphs: *Helminthosporium, Bipolaris, Dreshlera*)

The gramincolous genera *Bipolaris* and *Dreshlera* are considered as offshoot of the lignicolous genera *Helminthosporium.*

Disease: *Cochliobolus miyabeanus* (Asexually: *Bipolaris oryzae*) – Brown spot of rice

Cochliobolus heterostropus (Asexually: *Bipolaris maydis*) – Southern corn leaf blight

Symptoms

» Brown, sesame like spots are formed on leaves.

» Initially minute dark brown to black dots are formed later become oval or rectangular.

» Spots coalesce to form drying of leaves.

» Grains are discolored and shriveled.

Life style: Facultative saprophyte

Mycelium: Well branched, septate, both inter and intra cellular.

Asexual reproduction: Conidiophore is dark brown, short, erect, and branched at base. Conidia are golden brown, cylindrical or worm like, 6-15 septate.

Helminthosporium: conidia are subhyaline to brown, obclavate, thick walle, 9-11 transverse septate conidia singly from pore at the tip on the lateral sides in the upper part of the determinate conidiophores. The conidia are dark brown to black protruding basal scar.

Dreshlera* and *Bipolaris: Conidiophore is a typical sympodula with indeterminate growth and form conidia in whorls (resembling bunches of banana) above and below the septa of simple branched conidiophore. The tip of the conidiophore continues to grow after the formation of conidia and ceases with the formation of terminal conidia. In ***Dreshlera*** the conidia are dark, multiseptate, cylindrical, which germinate and form germtube from any orall cells. In ***Bipolaris*** the conidia are fusoid, straight or curved, bi-walled and germinate only by the tow end cells.

Sexual reproduction: Ascostromata are dark, flask shaped, pertithecia like with an ostiole. Asci are cylindrical to fusiform and ascopores are filamentous and coiled, hyaline, 6-15 septate.

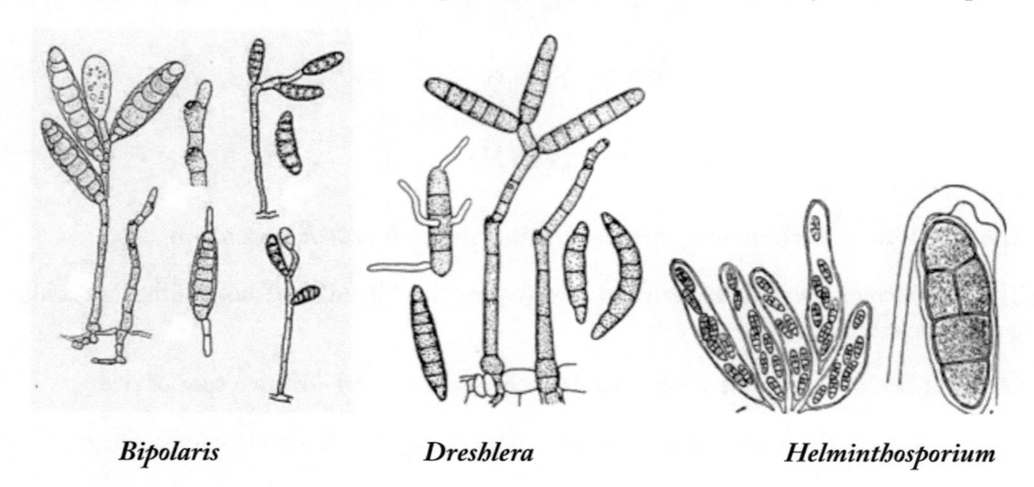

Bipolaris ***Dreshlera*** ***Helminthosporium***

Genus: ***Alternaria***

- *Alternaria* is a genus of ascomycete fungi.
- *Alternaria* species are known as major plant pathogens. They are also common allergens in humans, growing indoors and causing hay fever or hypersensitivity reactions that sometimes lead to asthma.
- At least 20% of agricultural produce spoilage is caused by *Alternaria* species; most severe losses may reach up to 80% of yield.
- It is polyphagus with worldwide distribution and occurs most frequently as a saprobe on dead and decaying organic materials on or in seeds and responsible for causing rapid blighting in economically important crop plants.
- Most species including *A. solani*, no *sexual* stages have been reported.

Disease: *Alternaria solani* – Early blight of potato and tomato

Alternaria padwickii – Stuburn disease of rice

Alternaria brassicae - Leaf spot of crucifers

Symptoms: Brown colour circular spots with concentric rings surrounded by yellow halo. Lesions are coalese and cause drying of leaves. Spindle shaped sunken lesion on the stem results in girdling. Circular or irregular sunken lesions with concentric rings present on fruits.

Life style: Necrotroph

Mycelium: Mycelium is branched, septate, dark brown.

Asexual reproduction: Conidiophores are simple or branched bearing conidia at the apex, straight or curved, 1-3 septate, dark coloured. Muriform conidia (dictyospores), brown coloured with vertical and horizontal septa and a long beak, obclavate with a beak, 3-8 tranversely septate and 1-2 longitudinally or obliquely septate, conidia are produced acropetally in chains (catenulate) through the pores formed at the apex of the beak of conidia.

Sexual reproduction: Ascocarp is dark. Ascospores are brownish, sharply fusoid with both vertical and horizontal septa.

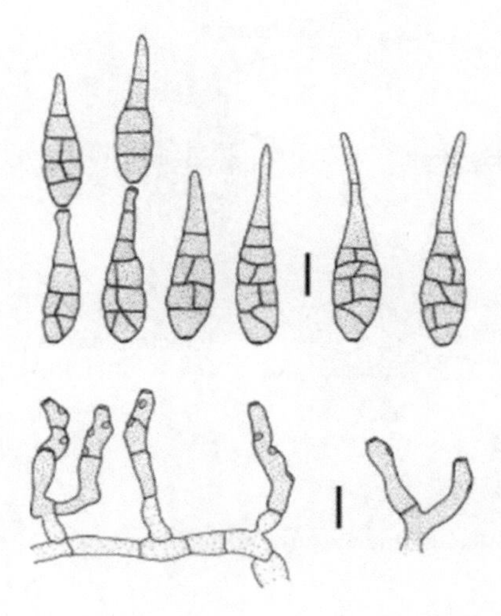

Genus: ***Venturia***

Disease: *Venturia inaequalis* (Asexually: *Spilocaea pomi*) – Scab of apple

Venturia pirina – Scab of pear

Symptoms

Apple scab occurs wherever apples are grown and may be a very serious disease on susceptible varities. The disease can also infect crabapple and rowan. Scab diseases similar to apple scab occur on pear, firethorn, and hawthorn. The scab-like leaf spots and fruit spots, from which the name was developed, may cause defoliation and reduction in fruit quantity and quality.

Life style: hemi-biotrophic

Mycelium: Mycelium is well branched, septate, subcuticular uninucleate mycelium, initially hyaline and become grey later.

Asexual reproduction: Mycelium form stroma to bear conidiophore and conidia. Conidiophores are brown, rarely septate, cut off at its tip successively to form a conidium and elongate after each conidium is formed. Conidia are smoky brown, ovate with truncate base and subacute apex, unicellular but become bicelled later.

Sexual reproduction: It is a heterothallic fungus, overwinters as pseudothecia (sexual fruiting bodies) following a phase of saprobic growth in fallen leaf tissues. Asci are spatulate with a short stalk. Ascospores are boat shaped, bi-celled, yellowish, upper cell is short and wider than lower cell.

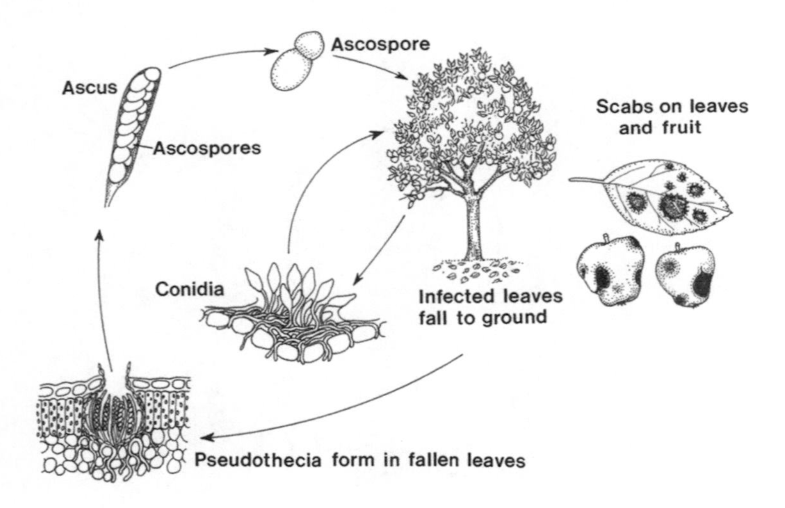

Genus: ***Botryosphaeria***

- Botryosphaeria is a genus of pathogenic fungi in the family Botryosphaeriaceae.
- There are 193 species, many of which are important disease-causing agents of various import ant agricultural crops.
- Its anamorph is *Lasidiplodia theobromae.*
- It causes root rot and collar rot diseases in many crops.

Disease: *Lasidiplodia theobromae* – Lethal blight of coconut

Life style: Facultative saprophyte

Mycelium: Hyaline, septate, branched, turns to dark at maturity.

Asexual reproduction: Conidia are oval in shape, Single celled, hyaline at young and turned, bicelled brown colour at maturity.

Sexual reproduction: Sexual spores are ascospores which are hyaline, septate produced in erumpent, ostiolate dark brown perithecia.

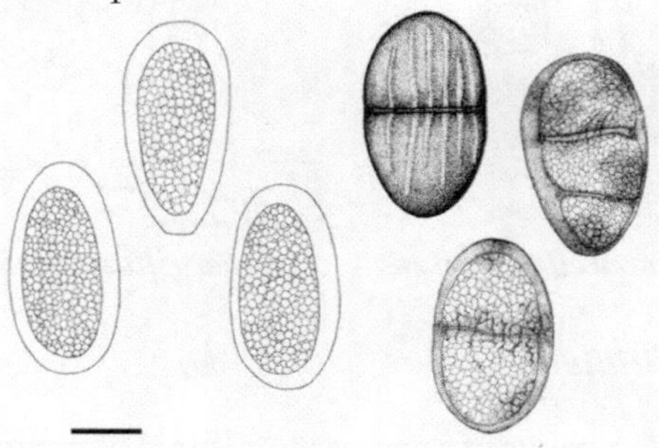

Genus: *Eurotium* spp. (*Aspergillus*)

Eurotium is a genus consisting of a few hundred mold species found in various climates worldwide. *Aspergillus* was first catalogued in 1729 by the Italian priest and biologist Pier Antonio Micheli. It causes black mould disease (E.g., *Aspergillus niger*) in plant products. It is a potential producer of mycotoxin called aflotoxin. (E.g., *A. flavus, A. parasiticus* and *A. nomius*)

Disease: *Eurotium niger* (*Aspergillus niger*) – Collar rot of peanut

Symptoms: Water-soaked lesions on the collar region of young seedlings with formation of black moldy growth which are the spores of pathogen. Severe infection leads to complete rotting of collar region and death of seedlings.

Life style: Facultative parasite

Mycelium: Hyaline, septate and branched mycelium

Asexual reproduction: Conidiphore arises from foot cell, long, branched, aseptate with a vesicle at tip. Vesicle contains metula and phialides. Dark brown coloured single celled conidia (phialospores) are produced in basipetal chain.

Sexual reproduction: Male and female gametangia are produced on same hyphae. Cleistothecia contain asci. Asci is globose and not arranged in a layer or hymenium (scattred asci). Each ascus contains 8 pulley wheel ascospores which are released within the cleistothecia.

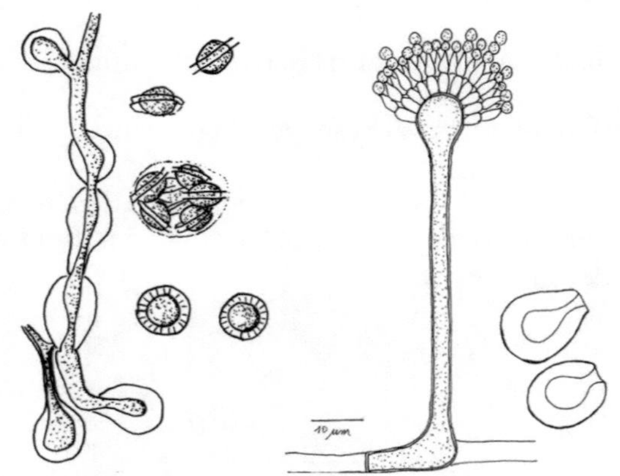

Emericella nidulans *Aspergillus nidulans*

Genus: *Talaromyces* (*Penicillium*)

***Talaromyces* (*Penicillium*)** is a genus of ascomycetous fungi of major importance in the natural environment as well as food and drug production. Some members of the genus produce **penicillin**, a molecule that is used as an antibiotic, which kills or stops the growth of certain kinds of bacteria inside the body. Other species are used in cheese making. It causes **blue mould** and **green mould** disease in post-harvest products. Circular or irregular water-soaked lesions formed on the rind near the stalk end which spreads further and cover a larger area with white mycelia growth. Later, due to sporulation the infected fruits are covered with green/blue moldy growth. Severe infection leads to rotting of fruits.

Disease: *Talaromyces digitatum* (*Penicillium digitatum*) – Greenmould of fruits

Symptoms: Diseased fruit show soft rot with velvety green mouldy growth.

Life style: Facultative parasite

Mycelium: Hyaline, septate and branched

Asexual reproduction: Conidiphore are long, slender, septate and branched arises once or twice or two-third of its total growth with chains of spores held in a brush - like dry cluster; each chain arises from a bottle-like phialide. Conidia are oval, green/blue and arranged in chain as basipetal succession.

Sexual reproduction: Male and female gametangia are produced on different hyphae. Antheridium is coiled, septate and tubular structure. Cleistothecia with loosely interwoven hyphae contain asci. Asci are globose and not arranged in a layer or hymenium (scattred asci). Each ascus contains ascospores which are released within the cleistothecia.

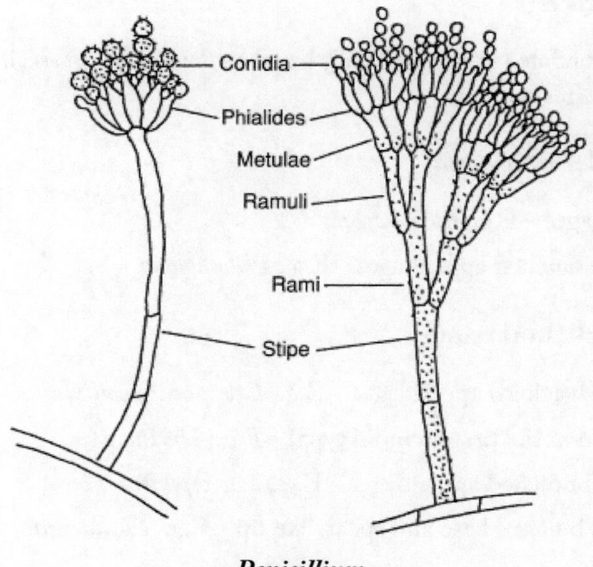

Penicillium

Powdery Mildew

Symptoms: White powdery growth on upper leaf surfaces, on stems, floral parts and fruits leading to premature defoliation. The powdery mildew fungi sexually produce closed ascocarp called cleistothecium. Asexually produces chains or single conidia over conidiophores, which are anchored by haustoria.

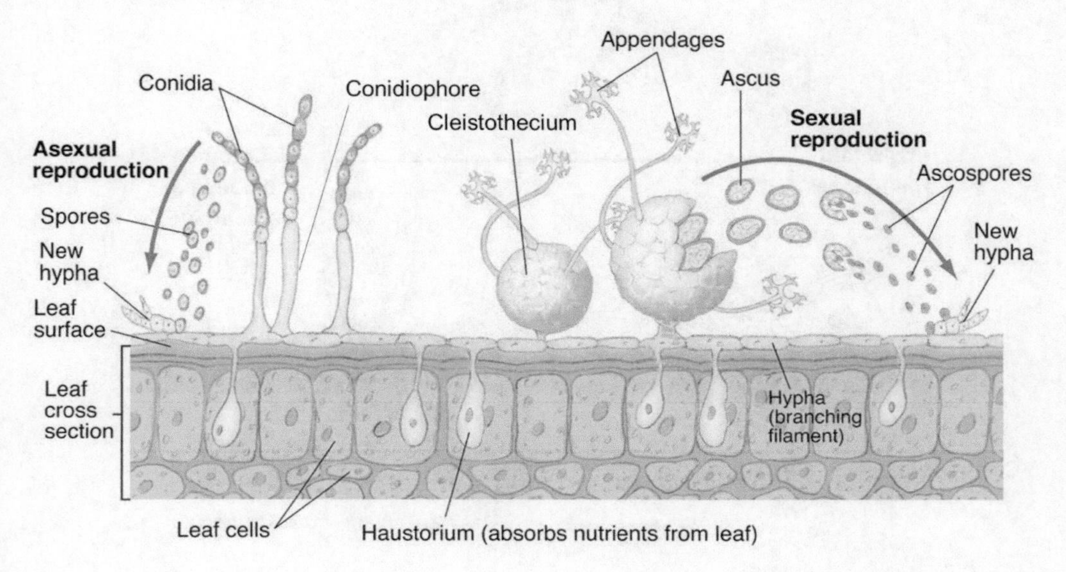

Teleomorphic characters

The genera are differentiated teleomorphically based on the number of asci in the cleistothecium and type of appendages on it.

I. One ascus in a cleistothecium

i. Myceloid appendages - E.g., *Sphaerotheca*

ii. Dichotomously branched appendages - E.g., *Podosphaera*

II. Many asci in a cleistothecium

i. Mycelium-like (Myceloid) appendages - E.g., *Erysiphe, Leveillula*

ii. Appendage coiled at the tip (circinoid type) - E.g., *Uncinula*

iii. Dichotomously branched appendages - E.g., *Microsphaera*

iv. Appendage with bulbous base and spear-like tip - E.g., *Phyllactinia*

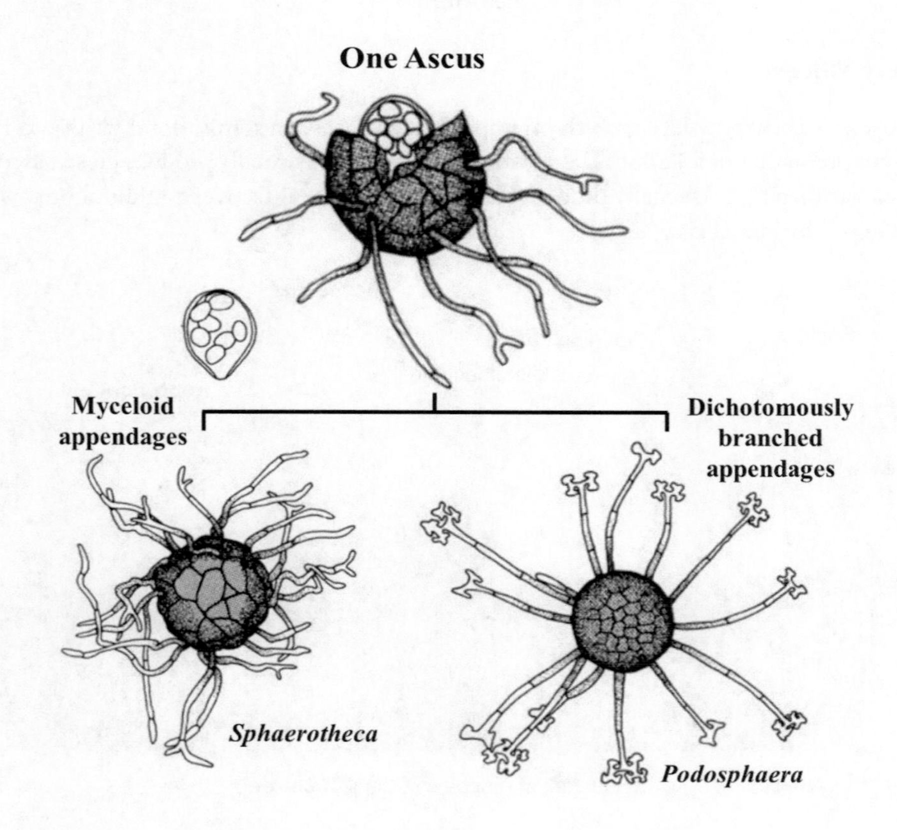

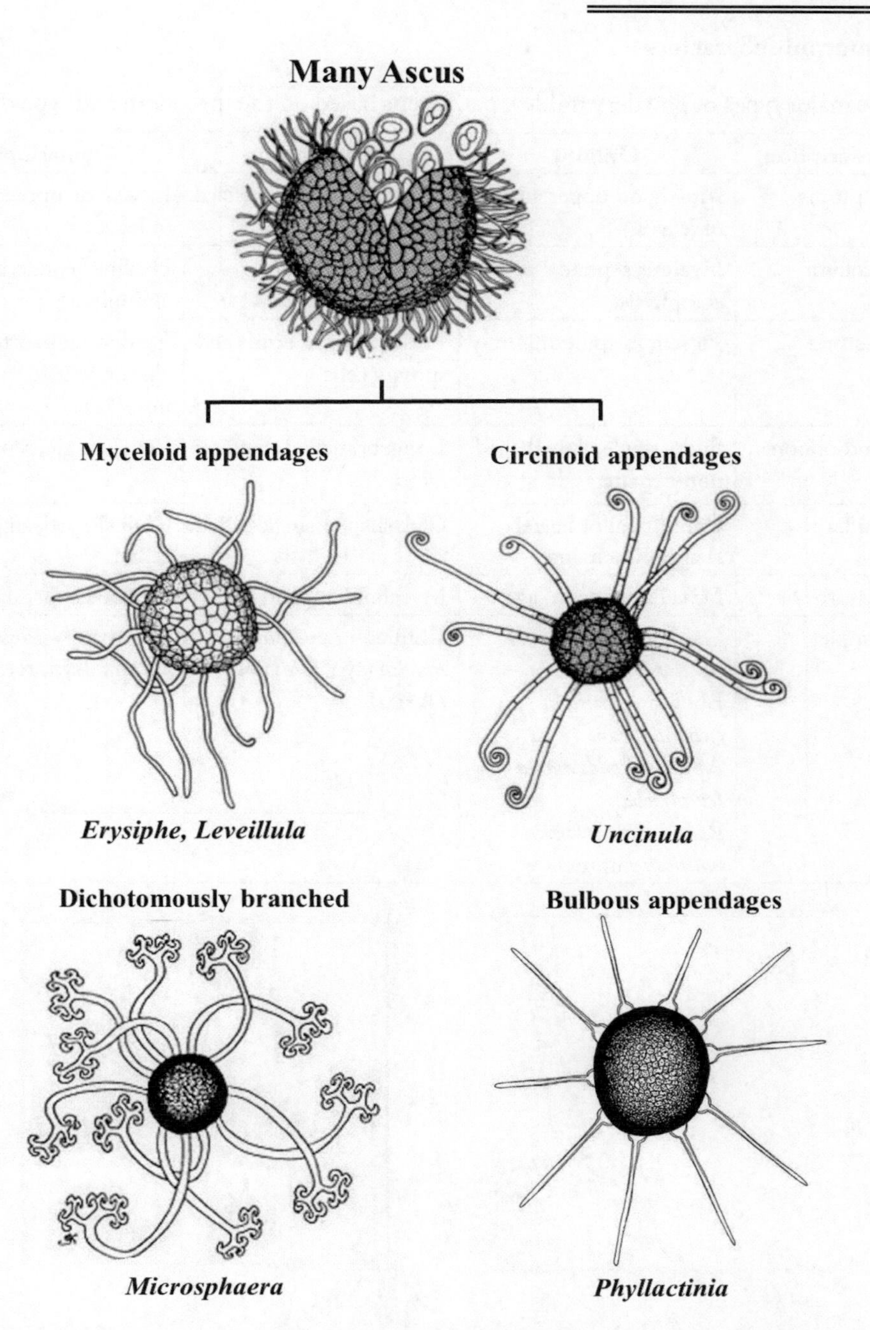

The fungi are classified by the types of appendages on and the number of asci (one or several) within each mature fruiting body (cleistothecium)

Anamorphic characters

Three major types of powdery mildew pathogens based on the mycelium and type of conidia

Description	Oidium	Oidiopsis	Ovulariopsis
Symptoms	Mostly on upper surface of leaves	Mostly on lower surface of leaves	Lower or upper surface of leaves
Mycelium	Hyaline, septate, ectophytic	Hyaline, septate, endophytic	Hyaline, septate, ecto-and endophytic
Haustoria	Present in epidermis only	Present in epidermis and spongy cells	Epidermal haustoria absent, haustoria in inner cells
Conidiophores	Short, single, club-shaped, non-septate	Long, branched, septate	Long, single, septate
Conidia	Cylindrical or barrel shaped, in chains	Club shaped, single celled	Club shaped, single celled
Cleistothecia	Mycelioid appendage	Mycelioid appendage	Bulbous appendage
Examples	Grapevine - *Uncinula necator* Bhendi - *Erysiphe cichoracearum* Apple – *Podosphaera leucotricha* Rose - *Sphaerotheca pannosa* var. *rosae*	Chillies - *Leveillula taurica* (syn. *Oidiopsis taurica)*	Muberry - *Phyllactinia guttata* (syn. *P. corylea*)

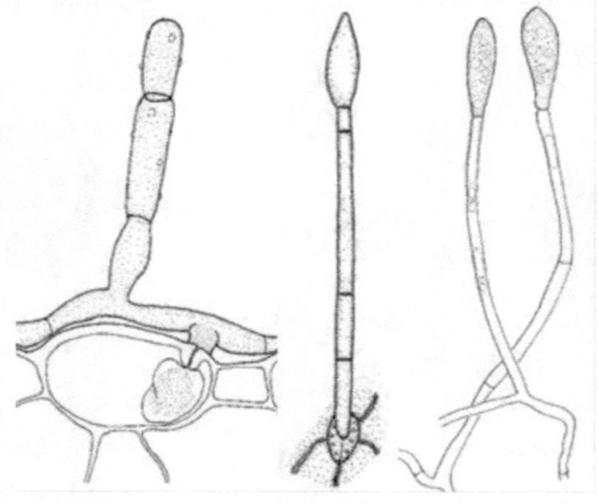

Oidium ***Oidiopsis*** ***Ovulariopsis***

In addition, based on conidiophore types, the family was further classified as *Pseudoidium* type, *Euoidium* type of *Neoerysiphe*, *Euoidium* type of *Golovinomyces*, *Euoidium* type of *Arthrocladiella*, *Euoidium* type of *Podosphaera*, *Cystotheca*, *Euoidium* type of *Sawadaea* with micro-conidia and microconidiophores, *Oidium* type of *Blumeria*, *Oidiopsis* type, *Ovulariopsis* type and *Streptopodium* type.

Genus: *Erysiphe*

Disease: *Erysiphe polygoni* - Powdery mildew of pulses

Erysiphe cichoracearum - Powdery mildew of cucurbits

Symptoms: Appearance of white powdery growth mostly on upper leaf surfaces. Affected leaves become chlorotic and subsequently dry and defoliation occurs. Photosynthetic area is reduced which results in poor yield.

Life style: Obligate parasite

Mycelium: Mycelium is ectophytic, hyaline, septate and branched, haustoria are bulbous and sac-like.

Asexual reproduction: Conidiophore is simple, erect, and hyaline and bear chain of conidia. Conidia are hyaline, thin walled, single celled and ovate or barrel or cylindrical in shape.

Sexual reproduction: Cleistothecia are black, round with myceloid appendages and each cleistothecium bears 2 to 8 asci. Asci are ovate and contain 3 to 8 ascospores. Ascospores are hyaline, elliptical and single celled

Genus: ***Leveillula***

Disease: *Leveillula taurica* - Powdery mildew of redgram, chilli, castor and clusterbean

Symptoms: Whitish fungal growth on the under surface of the leaf and the corresponding upper surface show yellow discolouration or chlorotic patches, later the disease spreads to entire leaf surface cause yellowing and defoliation of leaves. The disease progresses from the older to younger leaves and shedding of foliage is prominent symptom.

Life style: Obligate parasite

Mycelium: Mycelium is endophytic, hyaline, septate and branched.

Asexual reproduction: Conidiophores emerge through stomata, single or in groups, simple or branched, septate and bear single conidium. Conidia are hyaline, single celled and clavate.

Sexual reproduction: Cleistothecia are with myceloid appendages, many asci per cleistothecium. Each ascus contains two curved ascospores.

Genus: ***Phyllactinia***

Disease: *Phyllactinia guttata* - Powdery mildew of mulberry

Symptoms: White fungal growth on lower surface of leaves and corresponding upper surface shows yellow discolouration. Gradually turn to yellowish brown or black, quickly cover entire leaf surface followed by drying and defoliation. Unsuitable for feeding larvae.

Life style: Obligate parasite

Mycelium: Mycelium is hyaline, septate and branched.

Asexual reproduction: Conidiophores are erect, septate, long, hyaline and simple. Conidia are hyaline, single celled, clavate or flame shaped and borne singly on conidiophore.

Asexual reproduction: Cleistothecia are flat, spherical, black and with bulbous base and pointed tip appendages. Asci are clavate, 10-30 asci per cleistothecium. Ascospores are two per ascus and oval to elliptical. More prevalent during cooler months in plains.

Genus: ***Podosphaera***

Disease: *Podosphaera leucotricha* (Powdery mildew disease of apple and pear)

Podosphaera pannosa (Rose powdery mildew)

Symptoms: *Podosphaera leucotricha* causes a range of symptoms. On stems, symptoms include wilting and discoloration. Wilting and leaf curling occur on leaves. Symptoms of the inflorescence include discoloration (non-graminous plants), dwarfing, stunting, and twisting

Life style: Obligate parasite

Mycelium: Ectophytic, septate and branched, produced haustoria

Asexual reproduction: conidiophores – unbranched, septate, hyaline, conidia hyaline and single celled, hyaline and borne in chain

Sexual reproduction: Chasmothecium contains a single ascus with dichotomously branched appendages (*Podosphaera leucotricha).* Chasmothecium contains a single ascus with hyphae like appendages (*Podosphaera pannosa)* (formerly *Sphaerotheca pannosa*).

Genus: ***Claviceps***

Claviceps species infect only the ovary of the cereals (i.e. infection process is high organ specificity (i.e) pathogen infects only the ovary not any other parts of the plants) and entire ovary is converted into sclerotia in place of normal seed development from the ovary. The pathogen infects the florets, grows through the poll en tube to the base of the ovary where it ramifies the entire ovary tissue and depend on the living host plant tissue (biotroph).

Disease: *Claviceps fusiformis* (Asexually: *Sphacelia fusiformis*) – Ergot of pearl millet

Claviceps purpurea – Ergot of rye; *Claviceps sorghi* – Ergot of sorghum

Symptoms: Individual spikelets in the ear are affected. Creamy sticky liquid oozed out from the affected spikelets will turn blackened due to saprophytic fungal growth. Long, straight or curved, and brownish hard sclerotia are produced from the affected spikelets. There is no much yield reduction by this fungus, but the infected seed is convert into ergot (sclerotia) which contains poisonous alkaloids viz., ergotoxine, ergometrine and ergatamine and LSD (Lysergic acid diethylamide).

Life style: Biotroph, infection is not systemic even though the pathogen is biotrophs.

Mycelium: Hyaline, septate and branched.

Survival structures – Sclerotia (Ergot), the affected spikelets are converted in to dark brown coloured, slender, slightly curved sclerotia (regular).

Asexual reproduction: Asexual fruiting body is sporodochia like structure. Conidiophores are short and hyaline. Conidia are single celled, thin walled, hyaline and elliptical to oblong in shape. Sclerotium is brown coloured, cylindrical, slightly curved

Sexual reproduction: Sclerotium (carpogenous) germinate and produce sexual fruiting body called perithecium. Each perithecium contains several asci which contain filiform ascospores; pathogen infects mainly cereals (poaceae family).

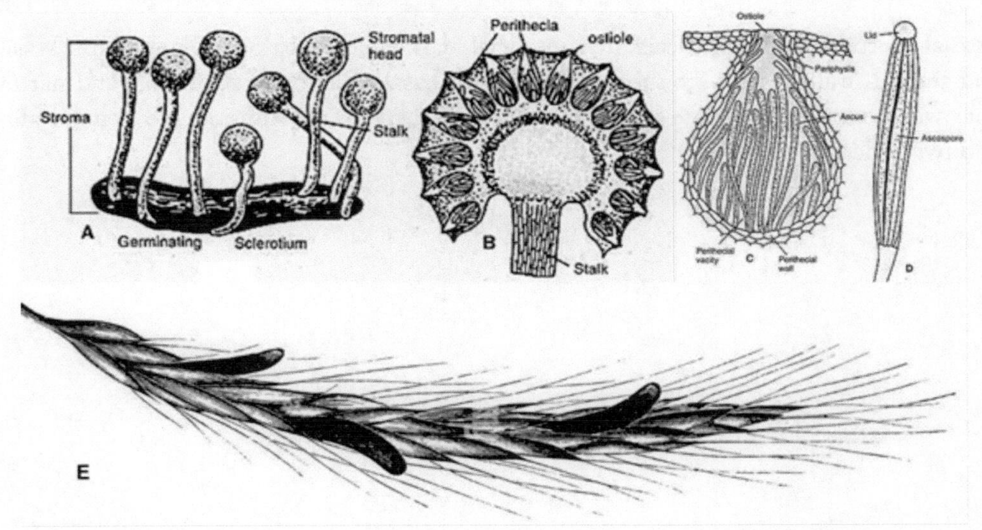

Claviceps: A. Sclerotium forming stroma, B. Stroma consists of perithecia, C. Perithecium, D. Ascus and ascospores

Genus: *Fusarium*

It is vascular colonizing pathogen. They enter the root outer epidermis move through the cortex and finally enters and colonize the xylem vessels causing blockage of water movement leading to wilting and vascular discoloration of the plants.

Disease: *Fusarium oxysporum* f. sp *cubense* – Panama wilt of banana

Fusarium oxysporum f. sp *ciceri* – Wilt of chickpea

Fusarium oxysporum f. sp *lycopersici* – Wilt of tomato

Symptoms: Yellowing and drooping of outer leaf blade around the pseudo-stem results in yellow skirt symptom. Longitudinal splitting of pseudo stem at the base purple streaks in the vascular bundles of infected corm radiating from periphery to centre.

Life style: Necrotroph.

Mycelium: Mycelium is hyaline, septate and branched.

Asexual reproduction: It is polymorphic fungus. The fungus produces microconidia, macroconidia and chlamydospores. Microconidia are hyaline, oval, single celled or one septate produced on sporodochia. Macroconidia are sickle-shaped, 3-5 celled and hyaline. Chlamydospores are thick walled, round resting spores, produced terminally or intercalarily on older mycelium.

Sexual reproduction: Perithecia are superficial, dark violet, subglobose, to globose, sessile and smooth walled. Asci are subcylindrical with basal stalk, broader middle, and narrow apex with central apical spore and eight-spored. Ascospores are ellipsiodel to ovate, hyaline and two celled.

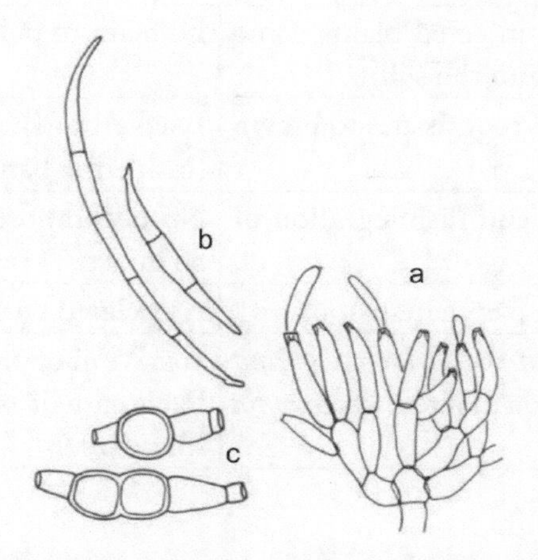

***Fusarium*: A. Sprodochia producing microconidia, B. Macroconidia, C. Chlamydospores**

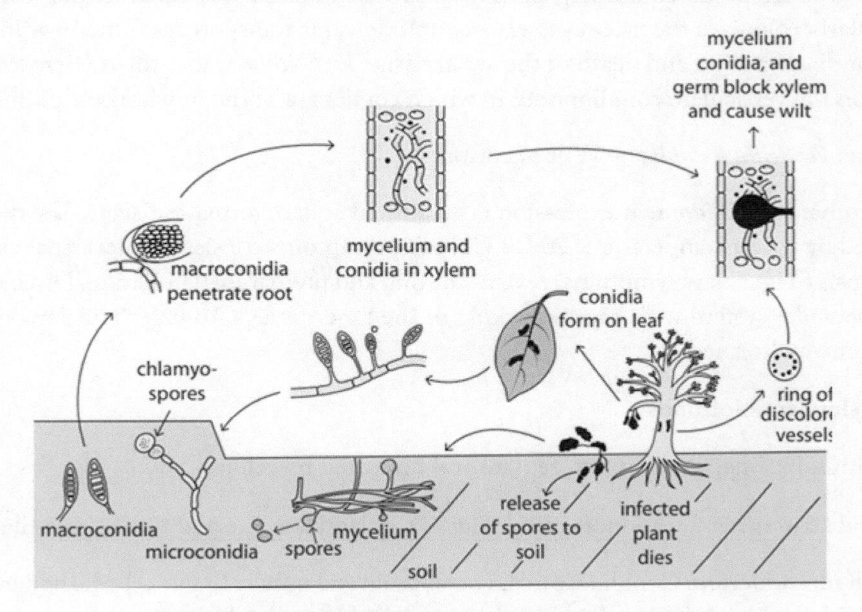

Life cycle of vascular wilt pathogen (*Fusarium*)

Difference between root rot and wilt

No.	Root rot	Wilt
1.	Occur in patches	Occur on isolated individual plants
2.	Sudden death of plant	Gradual wilting of plant

3.	On pulling, the infected plants come off very easily from the soil	Difficult to pull out from the soil
4.	Discoloration of root tissues to brown / black	Black / brown / pink streaks on the stem (inside the bark tissues)
5.	Decomposition and disintegration of bark tissues	No decomposition and the tissues will be intact.
6.	Associated with parenchymatous	Associated with vascular bundles
7.	E.g., Pulses root rot (*Macrophomina phaseolina*); Cotton root rot (*Rhizoctonia solani*)	E.g., Wilt of pigeonpea (*Fusarium udum*) Panama wilt of banana (*F. oxysporum* f. sp. *cubense*)

Genus: *Verticillum*

It is a vascular tissue colonizing pathogen as that of *Fusarium* as described earlier- it particularly colonizes the xylem vessels, disrupting water transport resulting in wilting and vascular discoloration and death of the aerial tissue. *Verticillum* is identified microscopically based on the verticillate conidiophore in which condia are borne in whorls of philides.

Disease: *Verticillum dahliae* – Wilt of cotton

Symptoms: *Verticillium* wilt expression is greatest at square formation stage. The disease is favoured by cooler temperatures 20-22°C. Foliar symptoms consist of interveinal chlorosis or necrosis (Tiger claw symptoms) severe stunting and premature defoliation. Discoloration of the vascular system with brown flecking of the inner tissues. Younger boll s may abscise or become malformed.

Life style: hemibiotrophs

Mycelium: hyaline, filament ous, septate and branched mycelium

Survival structures: It produces microsclerotia, chlamydospore and resting mycelia

Asexual reproduction: Conidia are ovoid or ellipsoid and usually single-celled. They are borne on phialides, which are specialized hyphae produced in a whorl around each conidiophore. Each phialide carries a mass of conidia. *Verticillium* is named for this verticillate (=whorled) arrangement of the phialides on the conidiophore. The fungus forms microsclerotia in dying tissue, which are masses of melanized hyphae.

Sexual reproduction – Not identified

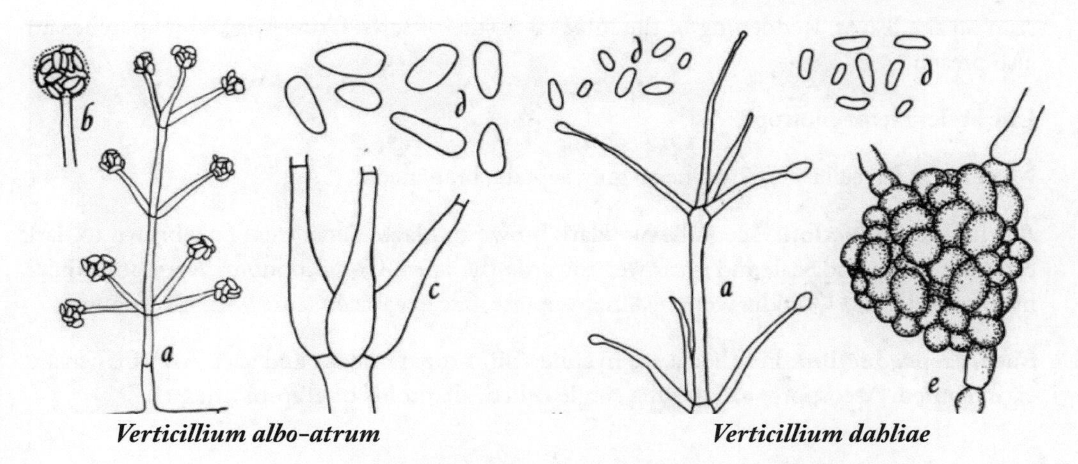

Verticillium albo-atrum ***Verticillium dahliae***

(a) verticillate conidiophores; (b) conidial head; (c) close-up of conidiophore branching; (d) conidia; (e) mature microsclerotium (Source: L. Gray).

Genus: *Ustilaginoidea virens*

Ustilaginoidea virens, perfect sexual stage *Villosiclava virens*, is a plant pathogen which causes the disease "false smut" of rice which reduces both grain yield and grain quality.

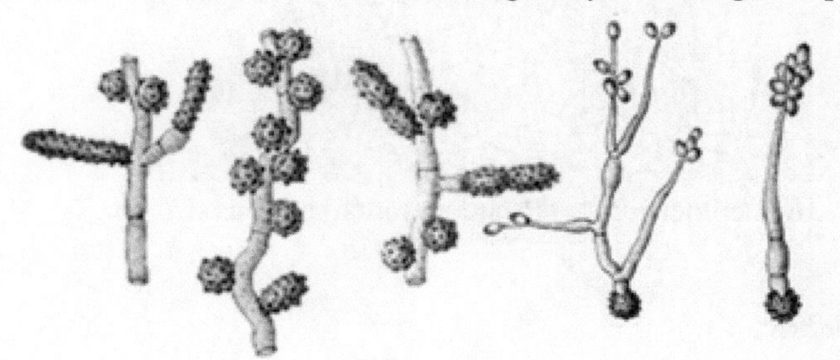

Genus: *Colletotrichum* (Asexual) / *Glomerella* (Sexual)

- The species are usually foliar pathogens and hemibiotroph.
- They attack more than 3000 species of monocots (maize, sorghum) and dicots plants (pulses and fruits as post-harvest disease).
- The disease caused by *Colletotrichum* is called anthracnose disease.
- The typical symptom includes sunken, necrotic and ulcer like lesions on leaves, stem, and fruits/pods.

Disease: *Colletotrichum falcatum* (*Glomerella cingulata*) - Red rot of sugarcane

Symptoms: Yellowing and drying of 2 and 3 young leaves in the whorl. Reddish lesions are seen on the leaves. Reddening of the internal tissues of setts. Cross-wise white patches are also present.

Life style: Hemi-biotroph

Mycelium: Mycelia were hyaline to grey, septate, branched.

Asexual reproduction: Acervuli were dark brown to black. Setae were pale brown to dark brown, base cylindrical, and narrower towards the apex. Conidiophores were unicellular, hyaline, phialidic. Conidia were hyaline, aseptate, falcate, apices acute with oil globules.

Sexual reproduction: Perithecia are hyaline, filiform paraphyses and asci. Asci are clavate, eight spored. Ascospores are hyaline single celled, allantoids or ellipsoid shaped.

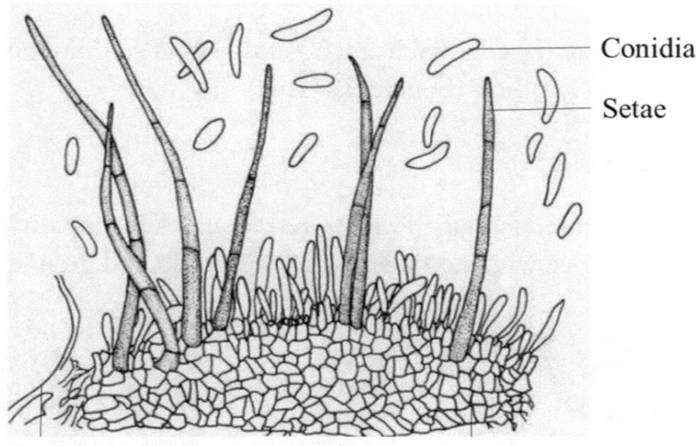

Genus: *Pestalotiopsis*

- *Pestalotiopsis* is an appendage-bearing conidial anamorphic form in the family Amphisphaeriaceae.
- *Pestalotiopsis* are common in tropical and temperate and may cause plant disease.
- *Pestalotiopsis* is a relatively important plant pathogenic genus known mostly from the tropics, where it causing leaf blights in many plant species.
- Species may also cause rots of fruit and other post harvest diseases.

Symptoms: Minute yellow spots surrounded by a greyish margin appear on the leaflets. The centre of the spots turns greyish white with dark brown margins with a yellow halo. Many spots coalesce into irregular grey necrotic patches giving a blighted or burnt appearance.

Life style: Necrotroph

Mycelium: Filamentous, colored, branched mycelium.

Asexual reproduction: Asexual fruiting body is acervulus. Asexual spore, Conidia is spindle shaped with five cells (having 4 septa). The central three cells are dark and terminal cells are hyaline. The upper hyaline cell bears two or three appendages called setae.

Sexual reproduction: The sexual states or t eleomorphs of *Pestalotiopsis* species have been identified as *Pestalosphaeria.* It produces perithecium with ascus and ascospores

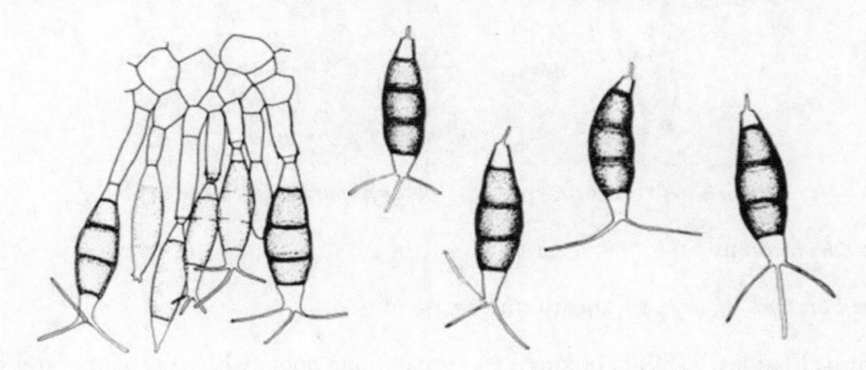

Genus: *Magnaporthe* (*Pyricularia*)

Members of the *Magnaporthe grisea* complex can also infect other agriculturally cereals including wheat, rye, barley, and pearl millet causing diseases called blast disease or blight disease. Rice blast causes economically significant crop losses annually. Each year it is estimated to destroy enough rice to feed more than 60 million people. The fungus is known to occur in 85 countries worldwide.

Disease: *Magnaporthe oryzae* (*Pyricularia oryzae*) also known as rice blast fungus, rice rotten neck, rice seedling blight, blast of rice, oval leaf spot of graminea, pitting disease, ryegrass blast, and Johnson spot.

Life style: hemibiotroph

Mycelium: Mycelium is highly septate, intercellular at first and later on intracellular. Young hyphae are colourless later become olivaceous.

Asexual reproduction: Conidiophores are unbranched and bear three celled conidia at the tips. Usually, 2-3 conidia are borne directly on the conidiophore and there is no special asexual fruiting body. Conidia are teardrop or water drop shaped. The conidium has a slightly projection called hilum at its base where it is attached to the conidiophores.

Sexual reproduction: Fruiting body is perithecia that contain asci and ascospores.

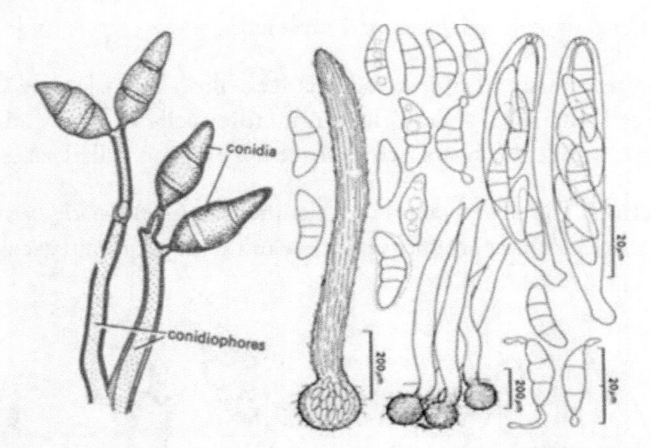

Pyricularia **(Anamorph),** ***Magnaporthe*** **(Teleomorph)**

Genus: *Sarocladium*

Disease: *Sarocladium oryzae* - sheath rot disease of rice.

Symptoms: Flag leaf exhibits oblong or irregular, long spots with grey center and brown margin. Complete choking of ear head or partial emergence of the panicle. Unemerged panicles rot. The glumes are also discoloured.

Life style: Necrotroph

Mycelium: Hyaline, sparsely branched and septate.

Asexual reproduction: Conidiophores are branched once or twice with 3–4 phialides in a whorl. The conidium is single celled (aseptate), hyaline, cylindrical in shape and located on tip of phialides.

Genus: *Macrophomina*

Macrophomina phaseolina is a soil-borne plant pathogen fungus that causes damping off, seedling blight, collar rot, stem rot, charcoal rot, basal stem rot, and root rot on many plant species.

Disease: *Macrophomina phaseolina* (Pycnidial stage); *Rhizoctonia bataticola* (Sclerotial stage) - Charcoal root rot disease of pulses.

Symptoms: Infected seedlings show a reddish-brown discoloration at the soil line extending up the stem that may turn dark brown to black. Foliage of infected seedlings can appear off-color or begin to dry out and turn brown. A twin-stemmed plant may develop if the fungus kills the growing point.

Life style: Saprophyte

Mycelium: Greyish white, sparsely branched, septate and inter and intra cellular.

Asexual reproduction: Dark, brown, globose pycnidia with an ostiole are produced on the surface of the stem. Inner wall of the pycidium is lined with pycnidiophore bearing pycnidiospore. Pycnidiophores are hyaline, short and rod shaped. Pycnidiospores are hyaline, thin, one celled and oval shaped.

Survival structures: The sclerotia are black with mycelial appendages.

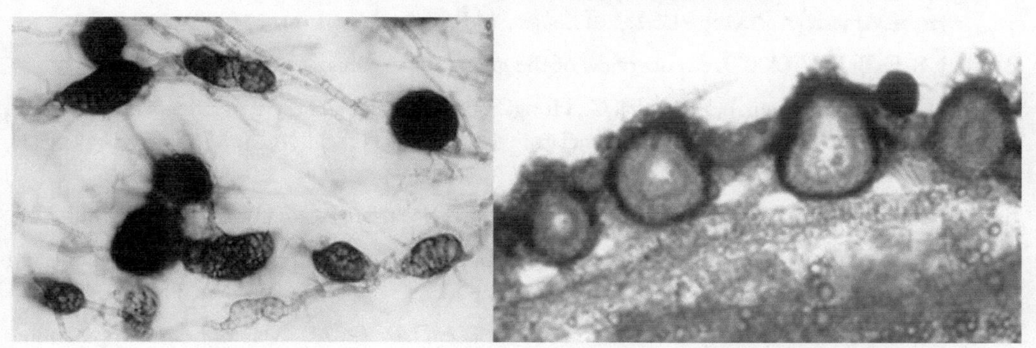

Sclerotia **Pycnidia**

Practice Questions

1. ……… is a largest phylum in the kingdom fungi
2. ……… is a closed non-ostiolate ascocarp
3. ………. flask-like ostiolate ascocarp with a preformed opening (ostiole) through which ascospores are discharged
4. Peach leaf curl is caused by…………
5. ……….is a genus causes sooty molds in crop plants
6. Explain the sexual and asexual fruiting bodies of ascomycetean fungi
7. Explain the differences between early and late leaf spot of groundnut
8. Explain the differences between *Aspergillus* and *Penicillium*
9. Illustrate the systemic infection cycle of Fusarium in crop plants
10. Explain the differences between wilt and root rot

Suggested Readings

Tudzynski, P., & Scheffer, J. A. N. (2004). Claviceps purpurea: molecular aspects of a unique pathogenic lifestyle. *Molecular Plant Pathology*, *5*(5), 377-388.

Miedaner, T., & Geiger, H. H. (2015). Biology, genetics, and management of ergot (Claviceps spp.) in rye, sorghum, and pearl millet. *Toxins*, *7*(3), 659-678.

Braun, U., Cook, R. T. A., Inman, A. J., & Shin, H. D. (2001). The taxonomy of the powdery mildew fungi. *The Powdery Mildews: A Comprehensive Treatis*, 13-55.

Bélanger, R. R., Bushnell, W. R., Dik, A. J., & Carver, T. L. (2002). *The powdery mildews: a comprehensive treatise*. American Phytopathological Society (APS Press).

Bennett, S. E. B. J. W. (2007). An overview of the genus Aspergillus. *The Aspergilli*, 23-34.

Visagie, C. M., Houbraken, J., Frisvad, J. C., Hong, S. B., Klaassen, C. H. W., Perrone, G., & Samson, R. A. (2005). Identification and nomenclature of the genus Penicillium. *Studies in mycology*, *53*(1), 53-62.

Brown, D. W., & Proctor, R. H. (2013). *Fusarium* (Vol. 5). Caister Academic Press: Haverhill, UK.

Cannon, P. F., Damm, U., Johnston, P. R., & Weir, B. S. (2007). Colletotrichum–current status and future directions. *Studies in mycology*, *59*(1), 129-145.

Marquez, N., Giachero, M. L., Declerck, S., & Ducasse, D. A. (2021). Macrophomina phaseolina: General characteristics of pathogenicity and methods of control. *Frontiers in Plant Science*, *12*, 634397.

Chapter - 13

Fungi: *Basidiomycota*

The Basidiomycota contains about 30,000 described species, which is 37% of the described species of true Fungi. The most conspicuous and familiar Basidiomycota are those that produce mushrooms, which are sexual reproductive structures. Basidiomycota are found in virtually all terrestrial ecosystems, as well as freshwater and marine habitats.

Symbiotic lifestyles (intimate associations with other living organisms) are well developed in the Basidiomycota. Symbiotic Basidiomycota include important plant pathogens, such as "rusts" (Uredinales) and "smuts" (Ustilaginales), which attack wheat and other crops. Not all symbiotic Basidiomycota cause obvious harm to their partners, however. For example, some Basidiomycota, as well as a handful of Ascomycota, form ectomycorrhizae, which are associations with the roots of vascular plants. Ectomycorrhizal Basidiomycota help their plant partners obtain mineral nutrients from the soil, and in return they receive sugars that the plants produce through photosynthesis. Other symbiotic Basidiomycota form associations with insects, including leaf-cutter ants, termites, scale insects, woodwasps, and bark beetles.

- » Second largest phylum in kingdom Fungi
- » Lower basidiomycetes cause rust and smut diseases and higher basidiomycetes destroy timbers as wood rotters.
- » The distinct features include extensive dikaryophase, clamp connections and dolipore septum.

General characteristics of Basidiomycota

Basidiomycota are unicellular or multicellular, sexual or asexual, and terrestrial or aquatic. Indeed, Basidiomycota are so variable that it is impossible to identify any morphological characteristics that are both unique to the group and constant in the group. The most diagnostic feature is the production of **basidia** (sing. basidium), which are the cells on which sexual spores are produced, and from which the group takes its name. A long-lived **dikaryon**, in which each cell in the thallus contains two haploid nuclei resulting from a mating event, is another characteristic feature. Finally, **clamp connections** are a kind of hyphal outgrowth that is unique to Basidiomycota, although they are not present in all Basidiomycota. The following description of the characteristics of Basidiomycota traces the life cycle of a "typical" species, beginning at the site of meiosis.

» Basidium and basidiospores are the characteristic features of the group. Unlike the endogenous production of ascospores, basidium always bears basidiospores exogenously and typical number of basidiospores is four per basidium

» The mycelium is well developed, branched and septate. The mycelium is of primary, secondary and tertiary types.

» A typical basidium is a club shaped structure, bearing specially 4 basidiospores on pointed projections called sterigmata.

» Basidiospores are haploid, uninucleate and are the result of plasmogamy, karyogamy and meiosis.

» Dikaryotic phase dominates the life cycle.

» Presence of clamp connections on the mycelium.

» Presence of dolipore septum, except in rusts and smuts.

» Absence of motile spores.

» No specialized sex organs. Sexual reproduction takes place by somatogamy and spermatization.

» Cell wall consists of chitin and glucans.

Types of mycelium in basidiomycota

a. **Primary mycelium**: Monokaryotic mycelium grows from germinating basidiospores for a shorter period.

b. **Secondary mycelium**: It is dikaryotic hyphae consists of two nuclei of opposite mating types brought together during plasmogamy. The whole or greater part of the life cycle of basidiomycetous fungi lives in this stage.

c. **Tertiary mycelium**: It is the specially organized secondary mycelium that forms the basidiocarp or basidioma of more complex tissues (hymenial layer).

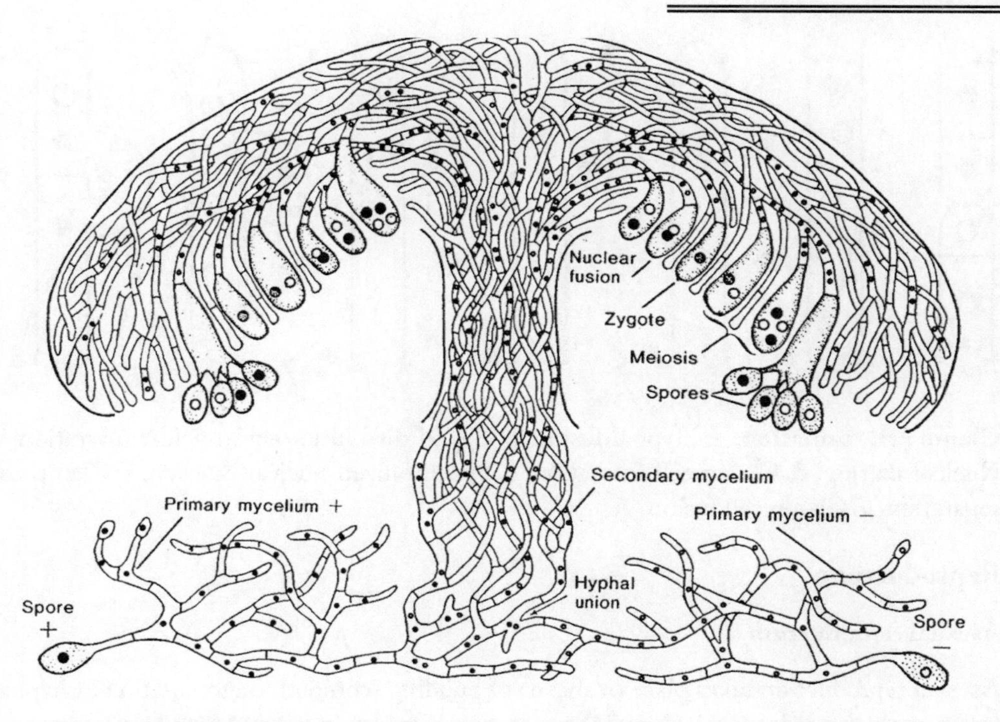

Basidiocarp consists of primary, secondary and tertiary mycelium

Special characters of basidiomycota

Dolipore septum: They are the complex type of septa with barrel shaped central pore with a hemispherical cap (parenthosome / pore cap)

Clamp connection: It is a bridge like hyphal connection formed during cell division in the dikaryotic secondary mycelium having dolipore septum. All fungi that produce clamp connections are members of the Basidiomycota, but not all Basidiomycota produce clamp connections. The main function is to ensure dikaryotic condition.

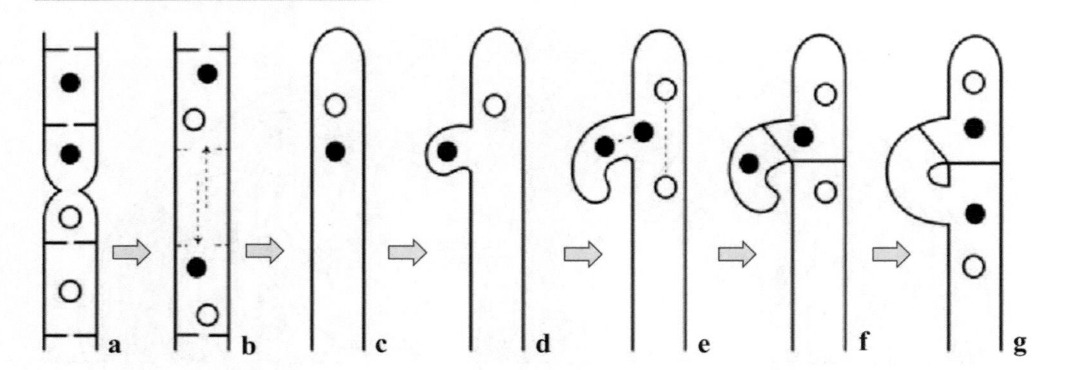

Clamp cell formation: a. Hyphal fusion, b. Septal dissolution and nuclear migration, c. Nuclear pairing, d. Clamp cell formation, e. Synchronized nuclear division, f. Clamp cell separation, g. Clamp cell fusion

Reproduction

Asexual reproduction

Asexual reproduction takes place by means of budding (conidia), fragmentation of hyphae (arthrospores), uredospores. Conidial production is common in smuts while rusts produce uredospores (summer spores) that are conidial in origin and function.

» It is lacking in some members (*Agaricomycotina*)

» But others reproduce by conidia/ oidia (rust, smut and bunt fungi)

» The rust fungi exhibit polymorphism by forming several spores in their life cycle of which uredospores act as asexual spores or repeating spores.

» They are formed successively by the dikaryotic phase.

Sexual reproduction

Sexual reproduction results in the production of basidium bearing haploid basidiospores. Basidiospores are formed as a result of karyogamy and meiosis taking place in basidium. In most of the members, sex organs (gametangia) are not produced and the somatic hyphae or detached somatic cells (arthrospores) undergo sexual process by somatogamy.

» No separate sex organs. In majority of cases, somatogamy occurs.

» Other types of plasmogamy include

a. Conjucation of two basidiospores (sporidia), a common method in smut fungi

b. Fusion of spermatia of one primary mycelium and another primary mycelium of opposite mating type i.e., spermatization, a common method in rust fungi.

Three major clades (subphylum) are strongly supported within the Basidiomycota:

1. Pucciniomycotina includes rusts (Pucciniales) and other taxa;
2. Ustilaginomycotina includes smuts (Ustilaginales) and others; and
3. Agaricomycotina includes mushrooms (Agaricomycetes), jelly fungi (Auriculariales, Dacrymycetales, Tremellales) and others

Basidioma or basidiocarp

Basidiocarp is a fruiting body that bears basidia which may be crust like, gelatinous, papery, spongy, corky, woody in texture. They vary in size from microscopic to a meter or more in diameter. Most Basidiomycotina bear their basidia in basidiocarps except in rusts and smuts.

Basodiocarp producing fungi are mushrooms, shelf fungi/ bracket fungi, coral fungi, puffballs, bracket fungi, birds nest fungi, earth stars etc.

» The open basidioma bears basidia and basidiospores on its under surface in definite layers called hymenium.

» The basidiole or paraphyses (aborted or sterile basidia) and cystidia (bladder like clavate projection beyond basidia) are some other structures present in hymenium.

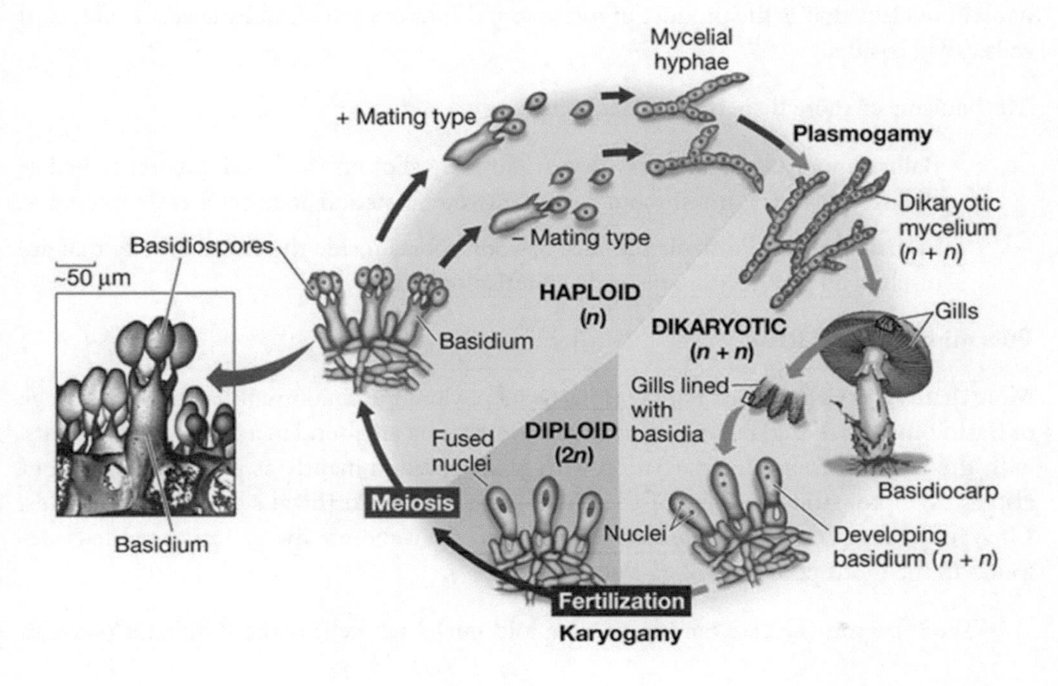

Basidium

The **basidium** is the cell in which karyogamy (nuclear fusion) and meiosis occur, and on which haploid basidiospores are formed (basidia are not produced by asexual Basidiomycota). Unicellular, club shaped structure formed terminally on dikaryotic hyphae. It bears basidiospore exogenously, usually on projections called sterigmata. The point at which the basidiospore is attached to the sterigmata is called hilum. The numbers of basidiospores (haploid meiotic) are generally four per basidium.

The resting spores on germination produce promycelium (metabasidium) into which diploid nucleus moves and after meiosis four haploid nuclei are produced. These nuclei later, result in the formation of haploid basidiospores.

This is divided into two orders based basidia arising from a thick-walled probasidium

1. Basidia becoming septate, bearing 2 to 4 (mostly 4) basidiospores, one at each septum and one nearly terminal - Pucciniales
2. Basidia aseptate or septate, number of basidiospores indefinite – Ustilaginales.

Basidiospores

A basidiospore is a reproductive spore produced by Basidiomycete fungi, a grouping that includes mushrooms, shelf fungi, rusts, and smuts. Basidiospores typically each contain one haploid nucleus that is the product of meiosis, and they are produced by specialized fungal cells called basidia.

Mechanisms of spore dispersal in Basidiomycota

- » Ballistospores: Basidiospores that are forcibly shot off the basidium are termed as Ballistospores. E.g., mushrooms, basidiomycete yeasts and pathogenic rusts and smuts.
- » Statismospores: Puffballs produce basiodiospores inside the fruiting body that are discharged passively are termed as Statismospores.

Pucciniomycotina: Rust

More than 8% of all described Fungi belong to subphylum Pucciniomycotina, the basal lineage of Basidiomycota. Fungi belonging to Pucciniomycotina are found in a diversity of habitats, with the majority found in association with plants, predominantly as phytopathogens but also as asymptomatic members of the phylloplane and mycorrhizal symbionts of orchids. Life cycles range from simple teliosporic yeasts to the elaborate five spore-stage life cycles found in the biotrophic rust fungi, Pucciniales.

The dikaryon (i.e., containing two haploid nuclei per cell) is the dominant phase in

Pucciniomycetes and only one of the orders, Septobasidiales, is known to have a monokaryotic (E.g., containing a single haploid nucleus per cell) yeast stage, a feature that is otherwise prominent in Pucciniomycotina. Unlike many of the other basidiomycetes, including some Pucciniomycotina species, the members of Pucciniomycetes lack clamp connections on their hyphae. All members of the subphylum thus far studied have simple septal pores lacking dolipores (septal pore swellings) and septal pore caps. Production of asexual spores is often well developed, especially among rusts where one species has been observed to produce five morphologically distinct types of asexually produced spores.

» The popular name for the Uredinales is the rust fungi, which relates to the reddish brown colour of some of the spores. All are obligate parasites of crop plants.

» The mycelium is primary in the early stage and in the later stages secondary. The mycelium is intercellular and produce haustoria that penetrate the host cells and obtain nourishment. There is no tertiary mycelium and hence there is no basidiocarp.

» Clamp connections are rare or absent. Dikaryotisation takes place either through somatogamy or spermatisation.

» Teleospores originate from the apical cells of dikaryotic hyphae. They may be uni or multi cellular. The structure of teleospores forms the basis for identification of the rust genera. The teleospore acts as an encysted basidium in which karyogamy occurs. It germinates by producing a promycelium (metabasidium) in which meiosis takes place.

» The rusts have polymorphic life cycle. Production of many spore forms in the life cycle is called polymorphism. Generally, 5 types of spores are seen during the life cycle *viz.*, spermatia (uninucleate) in spermagonium, aeciospores (binucleate) in aecium, uredospores (binucleate) in uredium, teleospores (binucleate) in telium and basidiospores (uninucleate) on promycelium or metabasidium. The spermagonium represents gametic stage (male gamete- spermatium, female sex organ- receptive hypha), aecia represent the stage in which dikaryotisation occurs, uredia represent conidial or repeating asexual stage, telia represent sexual stage and act as encysted basidium in which karyogamy occurs and subsequently giving rise to basidiospores from promycelium or metabasidium.

Types of rust

1. Macrocyclic rust or long cycled rust: Five spore stages are produced in the life cycle of fungi.

a. **Autoecious rust**: Five spore stages are formed on a single host. E.g., Sunflower rust - *Puccinia helianthi*, Bean rust - *Uromyces phaseoli-typica.*

b. **Heteroecious rust**: Two different hosts (*viz.*, primary host and alternate hosts) are required for completion of its life cycle. E.g., Wheat stem rust - *Puccinia graminis* var. *tritici* (Wheat and Barberry).

2. Demicyclic rust: Uredial stage is absent. E.g., Cedar apple rust - *Gymnosporangium juniperi virginianae* (heteroecious rust); Rubus orange rust - *Gymnoconia pikiana* (autoecious rust).

3. Microcyclic rust or Short cycled rust: Fungi which lack pycniospores and aeciospores are called micro cyclic rust. The uredial stage may or may not occur. All microcyclic rusts are autoecious in nature. E.g., Jamine rust - *Uromyces hobsonii.*

Types of teliospores

Teliospores formed in a sorus called telium. As host plants mature, uredia gradually develop into telia, producing teliospores. Teliospores are blackish-brown, oblong, diploid, two–celled spores.

There are four families in Uredinales

a. Teliospores sessile

- Teliospores in single, palisade-like layers or solitary, deep sub-epidermal crusts, germinating to produce a septate promycelium; mostly the spores are unicellular – Melampsoraceae.
- Teliospores in waxy crusts of one or two layers, becoming septate during germination without forming an external promycelium, forming as horn like columns or telial horns – Coleosporiaceae.
- Teliospores in chains - Cronartianceae.

b. Teliospores pedicellate, germinating to form a promycelium, which become septate; spores uni -or multicellular, free, erumpent cusions – Pucciniaceae.

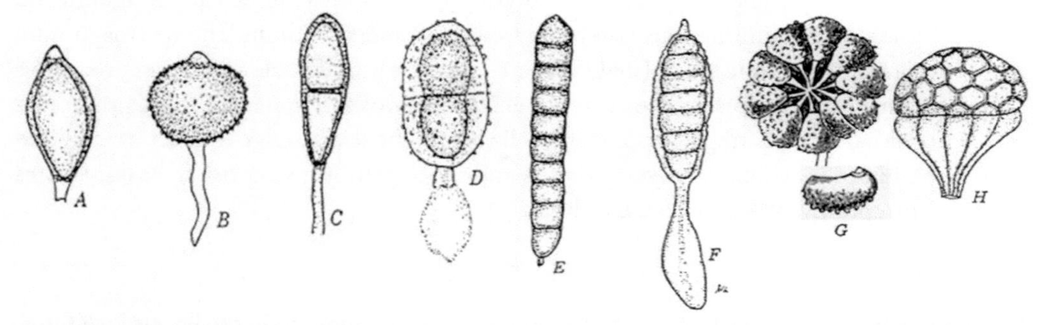

Variations in teliospore morphology: A. *Uromyces*, B. *Pileolaria*, C. *Puccinia*, D. *Uropyxis*, E. *Xenodochus*, F. *Phragmidium*, G. *Hemileia*, H. *Ravenelia*

Genus: *Puccinia*

Obligate parasites. Teleutospores are two celled and stalked. They lie free in the sorus.

Disease: *Puccinia graminis* var. *tritici* - Black or Stem rust of wheat

Symptoms

Reddish brown rust pustules (uredia) are noticed mainly on stem than leaf sheaths and leaves. At maturity a mixture of brown and black lesions can be seen (telia). Alternate host is *Berberis* sp. and *Mohania* sp.

Five stages of rust fungi

Stage 0: Pycnia - Spermagonia with spermatia (pycniospores) and receptive hyphae.

Stage I: Aecia with aeciospores

Stage II: Uredia with uredospores

Stage III: Telia with teleutospores

Stage IV: Basidia with basidiospores

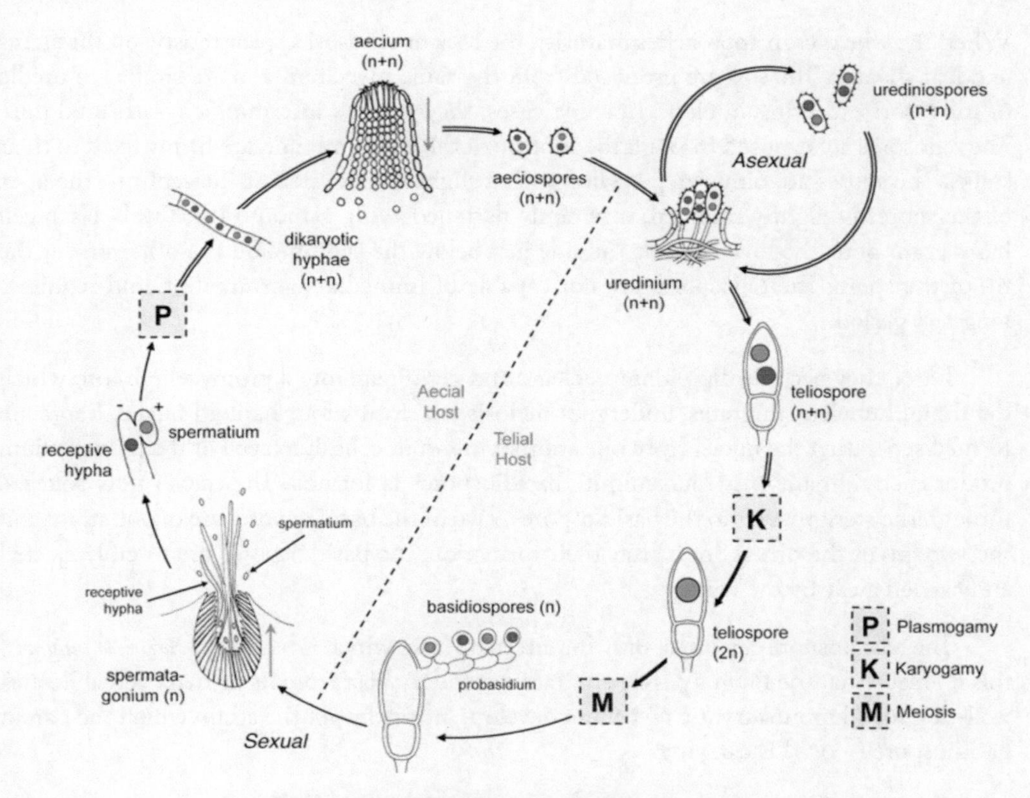

Life-cycle of *Puccinia graminis* var. *tritici*

0. Pycnium: Flask shaped, ostiolate (spermagonia); Pycniospores are spherical, thin-walled, hyaline, unicellular, uninucleate

1. Aecium: Cup like and thick walled; Aeciospores are unicellular, yellow, hexagonal, thick walled, binucleate, echinulated.

2. Uredospore/ Uredinospore: Oval, single celled, binucleate, golden brown, thick walled, echinulated, pedicillate (repeating spores).

3. Teliospores/ Teleutospore: Two celled, dark brown, thick walled apex with pointed at the tip with slight constriction at the septum (resting spore), Young teleutospores are binucleate at maturity it has a single diploid nucleus.

4. Basidium or promycelium: Long thin walled, hyaline and 4 celled basidiospore, unicellular, round and uninucleate.

Life cycle

When the wheat crop approaches maturity, the telia or teliosori appear mostly on the stems and leaf sheaths. The sori are produced from the same mycelium and are similar to uredia or uredosori excepting in clour. In some cases, they may be intermingled with uredosori. The teliosores are exposed through the rupture of the epidermis and are firmly fixed to their stalks. The spores are binuclete, 2 celled with a slight constriction at the septum. The apex of the spores is slightly rounded, very rarely flattened, with a smooth thick wall. Each cell has a gram pore, the lower one at the side just below the septum and the other one at the tip of the spore. The teliospores are not capable of immediate germination and require a long rest period.

Later, they become diploid uninucleate and germinate into a promycelium into which the diploid uncleus migrates, undergoes meiosis and forms four haploid nuclei. Septa are formed separating the nuclei from one another into four cells. Each cell of the promycelium produces sterigmata on which minute basidiospores is formed. The nuclei now squeeze through the sterigmata into the basidiospores. Two of the basidiospores are of one strain and the two are of the other. Soon after their formation, the basidiospores are forcibly ejected and carried away by the wind.

The basidiospore can infect only the alternate host, which is barberry (*Barberis vulgaris*) this it is seen that the fungus passes on to the secondary host from its primary cereal host. A well-developed monokaryotic mycelium develops, its nuclei get the strain which the parent basidiospore (+ or -) is carrying.

The hyphae of the fungus nearest the upper epidermis of the host develop spermagonia. Each spermagonium contains numerous spermatiophores that cut off into succession of minute spermatia. The spermatia (pycniospores) are uninucleate, haploid and are exuded from the ostiole in small droplets of nectar at the mouth of the spermagonium (pycnium). Several periphyses are also formed in the upper part of the spermagonium. Each spermatium contain a large number of nucleus carrying the + or – strain, depending on the strain of mycelium that produced the spermagonium. All permatia from a single spermagonium carry the same factor. The same mycelium that produces the spermatia also gives rise to receptive hyphae with the same genetic makeup as the spermatia. The spermagonial periphyses may also change to receptive hyphae. Spermatization now takes place through the agency of insects. They are attracted by the fragrance of the spermagonial mass and feed on the nectar. During the process, spermatia in the nectar adhere to the mouth parts of the insect and are subsequently brushed off by the respective hyphae and periphyses of the next spermagonium. If + spermatia thus happen to be transferred to –receptive hyphae spermatization is effected, and the spermatial contents pass into the walls at the point of contact. Meanwhile the mycelium had formed a number of aecial primordial. The spermatial nuclei which pass from the spermatia into the receptive hyphae travel down the hyphae apparently pass through the septal perforations of the mycelium, and reach the cells of the aecial primoridia, rendering them binucleate.

Dikaryotization is followed by the formation of aecia and aeciospores. These aeciospores are the first binucleate spores produced in the life cycle of the rust. The aecium is a yellow cup-shaped structure. It consists of an outer wall (memberanous) peridium enclosing chains of yellow aeciospores arising from the base of the cup. They are binuleate, one celled with germ pores. The spore is hexagonal while inside the cup and angular when free. The aeciospores chain eventually breaks through the lower epidermis of the barberry permitting the spores to escape and they get disseminated by wind and germinate on susceptible grass host. The aeciospores produce the binucleate mycelium and masses of uredinial cells from which binucleate stalked urediniospores arise.

The urediniospores are one celled, oval brown structure with a thick wall having numerous spiny projections called echinulations which are borne on the surface of the wall and used for anchorage. Elongated uredinial pustules burst open liberating the urediospores. The urediniospores are the spores perpetuating the fungus throughout the growing season and spread from plant to plant and from field to field. Upon germination they produce binucleate mycelium. Latter in the season the uredinia start producing teliospores are known as teliosori of telia. The teliospores germinate and produce the promycelium and thus the cycle goes on.

Important rust diseases affecting crop plants by *Puccinia*

Brown or leaf rust of wheat - *P. triticina*	Wheat black rust - *P. graminis* ar. *tritici*
Yellow or strips rust of wheat - *P. striiformis*	Oat rust - *P. graminis avenae*
Barley rust - *P. graminis hordei*	Rey rust - *P. graminis secalis*

Pearl millet rust - *P. penniseti*	Sorghum rust - *P. sorghi*
Orange rust of sugarcane - *P. kuehnii*	Groundnut rust - *P. arachidis*
Sunflower rust - *P. helianthi*	Brown rust of sugarcane - *P. melanocephala*

Genus: *Uromyces*

Uromyces is the second largest rust genus with about 600 species. They mostly cause diseases on legumes and carnation and are characterized by unicellular teliospore with thickened apex. Uredial, aecial and spermagonial characters are similar to *Puccinia*. Teleutospores are single celled with a thick apex (papillum) and stalked. The stalks are fragile and short. The species may be heteroecious or autocious and the life cycle may be macro or microcyclic.

Disease: *Uromyces appendiculatus* - Bean rust

Symptoms

Reddish brown rusty pustules (Uredosori/uredia) are produced on the lower surface of the leaves. Teliosori / telia are formed later, which are linear and are black or dark brown. Pycnia appears as yellow spot on the upper surface and aecia appear as orange pustules on the lower surface of the same leaf. Intense pustule formation causes drying and shedding of leaves.

Life cycle

It is autoecious macrocyclic rust. The surviving teliospore found in crop debris / continuously grown bean plants and other species of *Phaseolus* germinates at 10-15°C and initiate the disease cycle. Pycniospores produced on the upper surface of the leaf inside yellow colour pycnia are hyaline and spherical. Cupulate 0.25-0.3mm diameter aecia produced on the underside of the leaf contains hyline and elliptical aeciospores with minute verrucose markings in chain. Uredia are brownish and powdery. Uredospores are golden brown, pedicellate, globose, one celled, echinulated with 2 equatorial germ pores. During the crop season the disease spreads through wind dispersed uredospores. Telia develop in the Uredia and appear dark brown or black. Teliospores are dark brown. Smooth, one celled, pedicellate with warty apical papilla.

Genus: *Hemileia*

So far 40 *Hemileia* species known, most of them parasitize members of family *Rubiaceae* especially all species of coffee. The uredia are easily identifiable by their reniform uredospores and are formed in the sub-stomatal chamber. Telia are formed like the uredia. The species of *Hemileia* infect coffea Arabica are *H. vastatrix and H. coffeicola*. Of which, *H. vastatrix* is commonly reported throughout the world and *H. coffeicola* is restricted to west and central Africa as a minor disease and it occasionally become serious under very wet condition. *H.*

vastatrix froms both uredo and teliospores in the same sorus. It is generally agreed that coffee rust does not complete is life cycle on coffee tree, but no alternate host reported.

Disease: Coffee rust - *Hemileia vastatrix*

Symptoms

Leaves show pale chlorotic lesions on upper surface with orange yellow uredosori on the underside. Lesions may coalescs to involve the whole leaf becoming necrotic. Severe rust attack causes die-back of young shoots. Young leaves are generally more susceptible to infection than older ones.

Life cycle

It is autoecious microcyclic rust; pycnia and aecia are lacking. The fungus persists as mycelium, uredia and urediospore on the infected leaves. The spore germinates only in the presence of water and enters leaves through stomato of the lower surface in less than 12 hours. Hyaline septate mycelium grows intercellularly and sends haustoria into the cells of young leaves. The mycelium gathers and produces new orange yellow uredia within 10-25 days from infection. Uredospores are cuniform or reniform or orange segment-like in shape. Convex side of the spores is echinulated and other side is smooth. Once uredia are formed, premature falling of infected leaves may occur at any time. Even one pustule may cause the leaf to fall off. The fungus occasionally produces are unicellular, thick walled, smooth and turnip-shaped teliospores mixed up with uredospores in the same sorus. They germinate in site without any resting period and therefore, termed as leptospore.

Basidiospore formed by the germination of teliospores does not infect coffee and no alternate host has so far been found the repeatedly formed uredospore continuously and successfully infects the coffee and keeps the life cycle going.

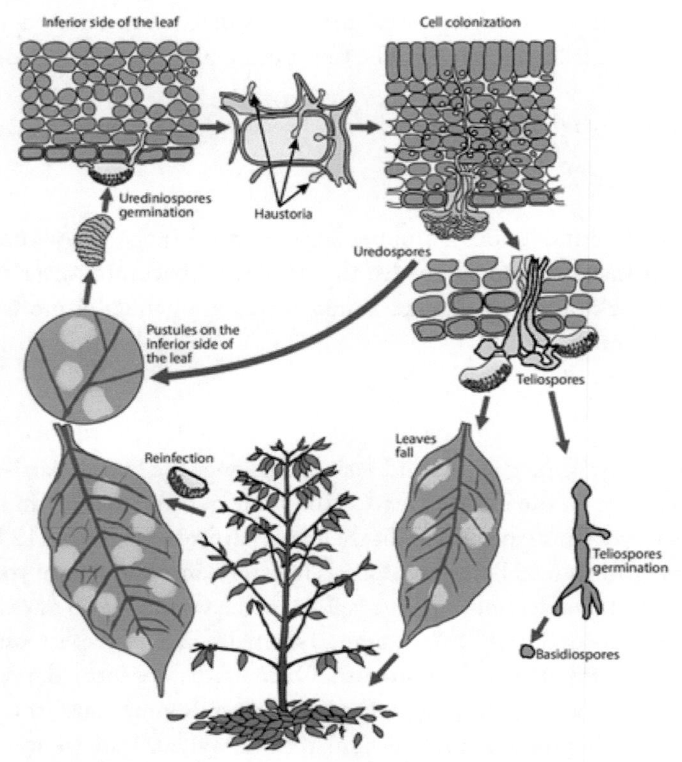

Life cycle of *Hemelia vastatrix*

Ustilaginomycotina: Smut, Bunt, and Blight

Most of the species in this sub-phylum are plant parasites with a biphasic life cycle and largely strong host preference, some species are recognized solely from their anamorphic phase. Smut fungi have earned their sordid name from the masses of black powdery teliospores that many of them produce. All are parasites of flowering plants and are common throughout the world. Although they are obligate plant parasites in the dikaryotic parts of their life cycles most can grow as saprotrophs in the monokaryotic state.

The most extensively studied members of Ustilaginomycotina are species of *Tilletia* and *Ustilago*. *Ustilago maydis*, which causes corn smut, produces conspicuous tumor-like growths on infected host plants. These structures eventually become filled with dark teliospores and are considered a delicacy in Mexico called "cuitlacoche." The most economically important species of *Tilletia* are the wheat bunts (*T. caries* and *T. laevis* common bunt; *T. contraversa* dwarf bunt; and *T. indica* Karnal bunt), plant pathogens that convert host ovaries into masses of dark, thick-walled teliospores that smell like rotting fish.

General characters

- The fungi included in this order are popularly called as smut fungi and the diseases they cause are called smut diseases. The smuts are so called because they form, black dusty spore masses (chlamydospores/ teleutospores/ smut spores) which resemble soot formed in the sori on the leaves, stem and flowers of host plants.
- This order also includes a group of fungi referred to as Basidiomycetes yeasts.
- All smut fungi are facultative saprophytes.
- The mycelium is inter-cellular, except in *Ustilago maydis* (intra cellular) and form haustoria. The primary mycelium originates from germination of basidiospore or conidia later it becomes dikaryotic. No tertiary mycelium. Mycelium generally does not develop profusely in artificial media. Yeast like cells which reproduce by budding is produced.
- Clamp connections are present while dolipore septum absent.
- Chlamydospores are formed from intercalary cells.
- Basidiospores are not produced on sterigmata and are not produced in definite numbers and are not discharged violently.
- Asexual reproduction is by means of budding of sporidia producing secondary sporidia (also called as conidia).
- No sex organs are produced.
- Plasmogamy takes place by fusion of any two basidiospores or two conidia. Somatic hyphae may also fuse resulting in dikaryotic condition.
- The basidium (promycelium) may be septate or aseptate and may bear 4 or many sporidia.
- Formation of teleutospore: Teleutospores are the characteristic structures of the order. These are formed in masses in sori developing in various places like flowers, leaves stem etc. At the time of formation of teleutospores, the dikaryotic mycelium develop profusely in certain portions of the host, lays down many septa and forms masses of hyphae composed of short cells. The protoplast of each hyphal cell round up and hyphal wall gelatinize. Each protoplat then secretes around itself a thick wall that eventually converts the protoplast into a round teleutospore. Since, these spores are formed in the manner of chlamydospore formation; theses are also called as chlamydospores. Teleutosores are of considerable importance in taxonomy of Ustilaginales. Surface ornamentation is the most important character. They may be reticulate, spiny, tuberculate, or smooth. The spores are typically globose, sessile, one celled and variously coloured like brown, black or yellow. In some species, the spores are free from one another and in some they held together to form spore balls. The

spore balls have fertile and sterile cells with fertile cells capable of germination.

» Cause a group of diseases called as smuts or bunts.

Smuts

» Smut spores are called teleutospores, chlamydospores or ustilospores.

» Smut fungi are more dangerous than rust fungi. They are facultative saprophytes.

» The family Ustilaginaceae produces a septate promycelium bearing terminal and lateral basidiospores. E.g., *Ustilago, Sporisorium* and *Tolyposporium.*

Host-parasite relationship

» Smut spores (teleutospore / chlamydospore) are binucleate at the beginning but become diploid, uninucleate later. Outer wall is pigmented, smooth, variously ornamented and its inner wall is thin and hyaline.

» Spores survive in the plant debris, soil, seed, etc. and serve as primary inoculum. They germinate by forming germ tube of determinate growth called promycelium / metabasidium. Septa formed later in the promycelium and give rise to 4 hyaline, uninucleate spores called sporidia.

Genus: *Ustilago*

» Teleospores are singles.

» Sori dusty at maturity.

» Sori covered peridium (membrane) of host origin. E.g., *U. nuda tritici*- loose smut of wheat.

Genus: *Spacelotheca*

» Teleospores are singles.

» Sori dusty at maturity.

» Sori covered by membrane (peridium) made up of fungal cells.

» Central columella present. E.g., *Sphacelotheca sorghi* - short or grain or kernal smut of jowar; *S. cruenta*- loose smut of jowar; *S. reliana*- head smut of jowar

Genus: *Tolyposporium*

» Spores in balls

» Spore balls permanent, spores adhering by thickenings of exospore. E.g., *Tolyposporium ehrenbergii*- long smut of jowar; *T. penicillariae*- smut of bajra

Sporidia infect seedlings in the following ways

1. Embryo infection. E.g., Loose smut of wheat (*Ustilago tritici*)
2. Floral infection. E.g., Cumbu smut (*Tolyposporium penicillariae*)
3. Seedling infection. E.g., Flag smut of wheat (*Urocystis agropyri*)
4. Shoot infection. E.g., Whip smut of sugarcane (*Ustilago scitaminea*)

Loose smut of wheat - *Ustilago tritici*

Symptoms

This disease is found on wheat crop in the plains as well as in the hill. Loose smut manifests itself only when the ears emerge from the boot leaf, usually all the spikelets are affected, having been transformed in to a mass to black powdery spores. In the beginning each smutted spikelete is covered by a thin silvery membrane which breaks while ear emerges. The powdery mass of spores is blown off by wind. Leaving behind only the central rachis.

Life cycle of *Ustilago tritici*

The innumerable smut spores develop on the mycelium in the intercellular spaces of the host. The smut spores are minute, pale olive-brown, lighter on one side, spherical or occasionally oval. The outer wall is thick and echinulate, whereas the inner wall of smut uredospore is thin and smooth. On maturity the two opposite strained (+ and -) nuclei of each smut spore are fused together and a diploid (2n) nucleus is formed. Wind, insects and other agencies carry the smut spores to the stigma of the flowers where they germinate producing basidia or promycelia. The exosporium ruptures and the tubular basidium come out. The diploid nucleus of smut spore divides mitotically forming 4 haploid nuclei. The basidium becomes segmented and the four uninucleate cells are produced, two of + and two of – strains. From each segment single infection thread develops. The two opposite + and – strained infection threads unite and a dikaryotic cell is produced. The dikaryotic hypha develops from the dikaryotic hypha enters the style and reaches the embryo where it develops into thick irregular and branched mycelium. The mycelium remains dormant in the embryo. The seeds, with dormant mycelium in their embryos look quite healthy in outer appearance.

The dormant mycelium established in the embryo of the seeds becomes active at the time of the germination of the infected seed and grows along with the apex of coleoptiles. The optimum temperature for the growth of the mycelium ranges from 20° to 25°C. the fungus is systemic and the mycelium reaches the floral parts and becomes on the emergence of the ears, mycelium reaches the floral parts and becomes activated. Thousands of smut spores are produced in smut sori on the floral axis.

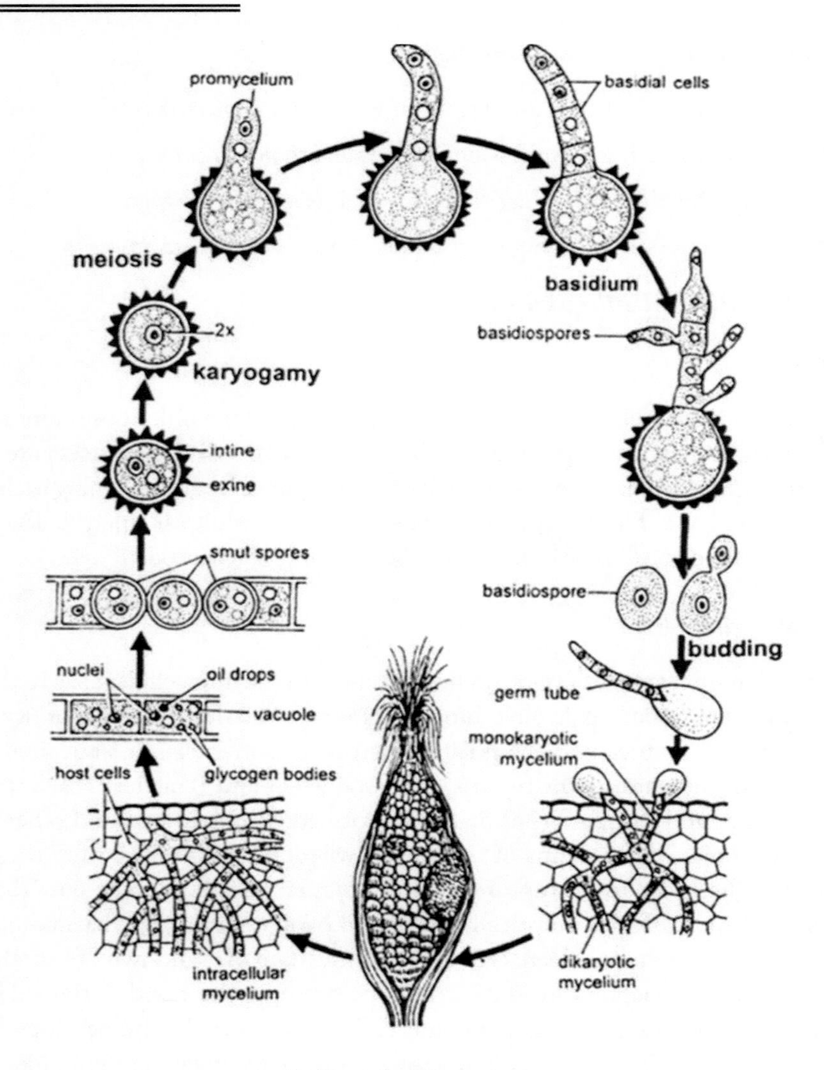

Life cycle of *Ustilago*

Life cycle of *Sporisorium sorghi*

The hypha is intercellular at the time of sporulation, the mycelium develops profusely in certain portions of the host, lays down several septa and forms masses of hyphae. These hyphal masses are converted into large dusty sori.

The grains are replaced by oval or cylindrical dirty gray sacs called sori. Each sorus consists of a wall made up of fungus tissue called peridium. The interior of the sorus is filled with smooth walled, dark olive brown consisting of chlamydospores (spores of the fungus.)

In the center, three is a slender column of hard tissue called columella. Chlamydospores are globose oval smooth walled. The spores are librated by the rupture of the sorus wall either at the time of harvest or subsequently. Spores are round to oval, smooth walled and often united to form loose balls, which often breakup into indivividual spores. Under favourable conditions of moisture, spores germinate usually producing hyaline hyphae which is divided into four cells by the formation of septa. This tubular four celled structure is called promycelium or basidium. Chlamydospore does the function of reproduction. There are two phases in the life cycle viz., haploid and diploid phase. The chlamydospores are produced after diploidization only. A small outgrowth develops just near a cross wall. It bends and touches the cell on the other side of cross wall. The place of contact gets dissolved and regular passage between the two cells occurs. In the meanwhile, the nucleus or the cell gets divided into two and one migrates into the protruberance and fuses with the nucleus in the other cell and thus diploidization takes place.

Each cell of the promycelium gives rise to a spindle shaped spore (hyaline) called sporidium, which arise from the sides of the lower cells and from the tip of the terminal cells. These sporidia when mature are librated and new ones may be formed until the protoplasm within the promycelia cells get exhausted. The sporidia are also capable of budding like yeast cells to give out similar cells or germinate by producing a branching germ tube instead of the sporidia on the promycelium. Fungus is systemic and it can infect the host only when two compatible hypha of opposite strain. Monosporidial lines are not-pathogenic.

The hyphae either from the sporidia or from the promycelium penetrate the young tissues of the shoot, and entry seems to be chiefly through region. The hypha having entered the seedling occupies the growing point of the shoot and keeps growing along with the host in a systemic manner. Only when two compatible strains of opposite unite the fungus can infect the host. Mono spotidial lines are not pathogenic.

Smut diseases infecting crop plants

Loose smut of barley - *Ustilago tritici*	Common smut of corn - *Ustilago maydis*
Covered smut of barley - *Ustilago hordei*	Loose smut of sorghum - *Sporisorium cruenta*
Smut of pearl millet - *Moesziomyces bullatus*	Grain smut of sorghum - *Sporisorium sorghi*
Long smut of sorghum - *Sporisorium ehrenbergii*	Head smut of sorghum - *Sphacelotheca reiliana*
Sugarcane smut - *Ustilago scitaminea*	Onion smut: *Urocystis cepulae*
Flag smut of wheat - *Urocystis agropyri*	Leek smut: *Urocystis colchici*

Differences between Pucciniomycetes (rusts) and Ustilaginomycetes (smuts)

Characters	Rust	Smut
Plant parts affected	Leaves, petiole and stem	Mostly floral parts
Symptoms	Reddish brown pustules	Ovaries turn into black powdery masses
Parasitism	Obligate parasite	Facultative saprophyte
Polymorphism	Polymorphic	Absent
Teleutospores	Terminal, 1-2 celled, stalked or sessile	Intercalary, 1 celled, sessile
Basidiospores	Produced on sterigmata, 4 in number	Sessile, 4 to many (indefinite in number)
Basidiospores discharge	Discharge violently	Passive discharge
Sex organs	Spermatia (male) and receptive hyphae (female)	No sex organs
Clamp connection	Rare	Present
Heteroecium	Common	Absent
Dikaryotization	Spermatization	Fusion of two basidia or basidiospores

Genus: *Exobasidium*

Disease: *Exobasidium vexans* - Tea blister blight

Exobasidium vexans is an obligate pathogen of *Camellia sinensis*

Symptoms: *E. vexans* prefers to attack the young leaves on the lower half of hosts and presents as small yellow translucent leaf spots which progress to lesions. These lesions give the lower surface of the leaf the characteristic blister like appearance.

Pathogen: The mature two-celled basidiospores are very easily dislodged from the sterigmata. Basidiospore is normally one-septate, as many as three septa have been seen in germinated spores.

Genus: *Tilletia*

- Teleospores singles, globose to sub-globose, pale orange-brown to dark reddish brown, mature spores black to opaque.
- Spores dusty and escaping at maturity.
- Sporidia are fused to form H shaped structures.

E.g., *Tilletia indica* - Karnal bunt/ Partial bunt of wheat

Tilletia caries and *T. foetida*- Hill bunt (stinking smut) of wheat.

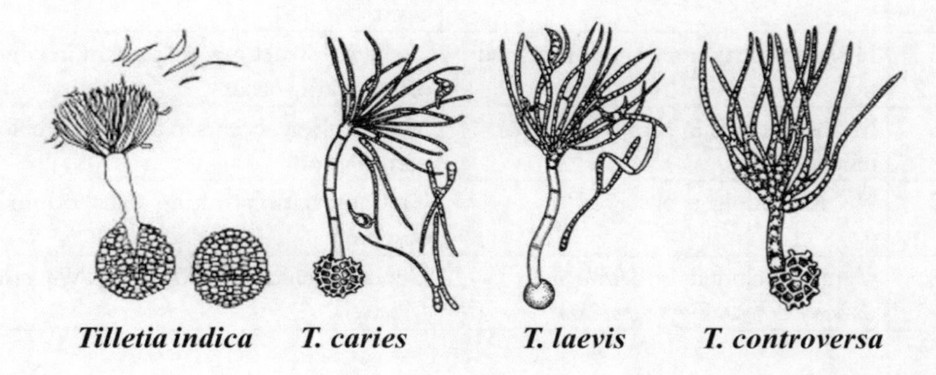

Genus: *Neovossia*

- Teleospores singles
- Spores dusty and escaping at maturity
- Sporidia do not fuse, no H shaped structures

E.g., *Neovossia indica* - Karnal bunt/ Partial bunt of wheat
Neovossia horrida - Bunt of paddy

Genus: *Urocystis*

- Teleospores in balls
- Sori dusty, spore balls surrounded by an adhering layer of hyaline sterile cells.
- Spore balls escape from sorus.

E.g., *Urocystis cepulae*- onion smut

Difference between smut and bunt

S.No.	Smuts	Bunts
1	Belongs to family *Ustilaginaceae*	Belongs to family *Ustilaginaceae*
2	Promycelium is septate	Promycelium is non-septate and hollow tube like
3	Basidiospores are formed laterally from each cell of the promycelium	Basidiospores are formed at the tip of the promycelium
4	Basidiospores are usually four	Basidiospores are more than four usually eight
5	H shaped structures are not formed	H shaped structures are present in which plasmogamy occurs.
6	Meiosis occurs in promycelium or teliospores	Meiosis always occurs in teliospores before germination
7	No fishy odour is observed	Characteristic stinking fishy odour is observed
8	Genera included are *Ustilago, Sphaecelotheca, Tolyposporium*	Genera included are *Tilletia, Neovossia, Urocystis*

Agaricomycotina:

This group includes the fungi commonly known as mushrooms, puffballs, shelf fungi, stinkhorns, jelly fungi and bird's nest fungi. Many species are saprotrophic, utilizing dead plant material including woody substrates. Some of these saprotrophic species are cultivated for food, for example, the common button mushroom (*Agaricus bisporus*), oyster mushrooms (*Pleurotus ostreatus*), and shiitake (*Lentinula edodes*). Other members of this group are important ectomycorrhizal fungi, forming mutualistic associations with the roots of a wide range of trees. Some fruiting bodies produced by ectomycorrhizae are considered choice edibles, for example, chanterelles (*Cantharellus cibarius* and other species), porcini (*Boletus edulis*), and the American matsutake (*Tricholoma magnivelare*). A few members of this group are economically important plant parasites, e. g., species of *Armillaria* and *Rhizoctonia*.

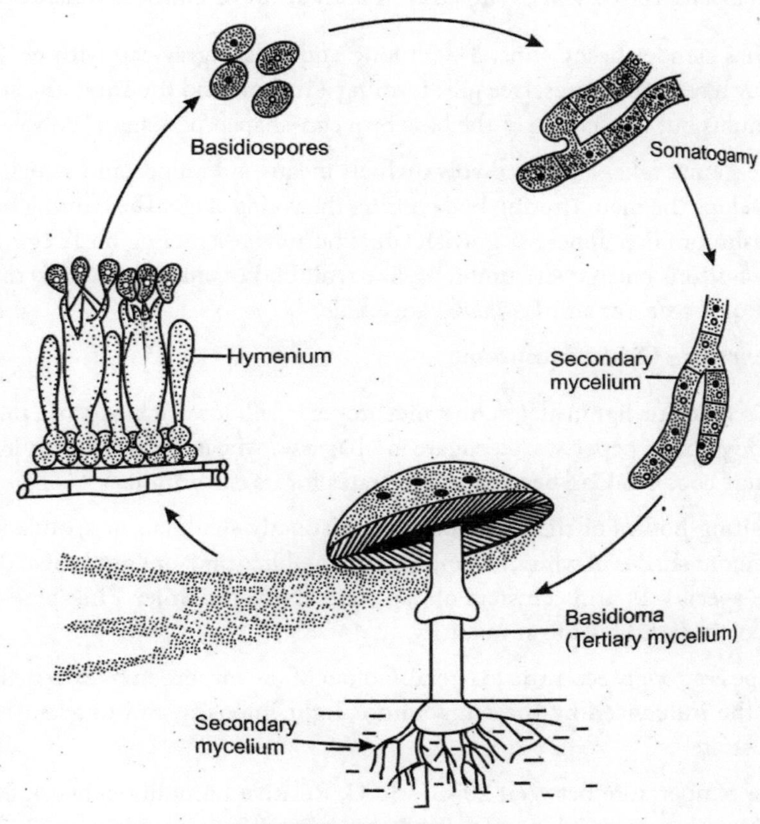

Life cycle of mushrooms

Genus: *Calocybe indica* – Milky mushroom

- Commony called "Dhuth chatta" or milky mushroom because of its milky white colour and robust nature. Mushrooms large with long, thick fibrous stalk, wide cap with incurved margin, abnata gills and whitespores.
- It can be cultivated more successfully during summer and hence they are designated as "**summer mushroom**". Excellent shelf life and no browning with high fibre content.
- Similar to oyster mushroom, it can be cultivated on wide range of cellulosic substrates namely paddy straw, maize stalk, sorghum stalk, pearl millet stalk, palmorosa stalk, vetiver grass, sugarcane bagasse, soybean hay, groundnut hulm ect.
- The mushroom is umbrella shaped and resembles button mushroom. The sporophores are roboust, attractive and milky white in colour. Calocybe require 25-35°C and >85% RH along with good light and aeration for good sporophore production. It needs low intensitiesof 800 lux for spawn running and 1600-3200 lux light intensity for cropping.

Genus: ***Volvariella volvariella / V. diplasia*** **- Paddy straw or Chinese mushroom**

» It has slender fleshy stipe, 3-8cm long and a dark gray cap with 6-12c diameter. They have pink spores, free gills forming a ring around the stipe. The stipe bears no annulus but is enclosed at the base by a cup-shaped persistent 'Volva'.

» The genus takes its name volva which means a wrapper and which completely envelops the main fruiting body during the young stage. The fruiting bodies of this mushroom first appear as grayish white buttons resembling bird's egg. At maturity the buttons enlarge and umbrella-like fruit bodies enlarge after the rupture of the volva. *V. volvacea* and *V. diplasia* are edible.

Genus: ***Pleurotus*** **- Oyster mushroom**

» It is efficient lignin-degrading mushroom. Cellulose rich organic substrates like paddy straw, paper waste, sugarcane bagasse, wheat straw, banana leaves, hulled maize cobs ect. Can be used as substrates for its cultivation.

» Fruiting bodies of the mushroom are distinctly shell, fan or spatula shaped with different shades of white, cream, grey, yellow, blue, pink or light brown depend upon the species. It firms clusters of caps one above the other Pink/blue mushrooms become slightly white at maturity.

» Stipe is strongly eccentric to lateral. Colour of the mushroom is extremely influenced by the influenced by the temperature, light intensity and nutrient status of the substrate.

» The temperature between 20°C - 30°C, Relative humidity between 80-85% with light and cross ventilation for 2-3hrs is required for its cultivation. Among thus, *P. sajor-caju* is the tropical species, which can tolerate the temperature up to 30°C. *P. ostreatus* is the so-called low temperature. *Pleurotus*, fruits mostly at 12-30°C. species of *Pleurotus* are *P. citrinopileatus, P. djamor, P. eous, P. ostreatus, P. sajor-caju, P. florida* and *hypsizycus ulmarius.*

Life cycle

The life cycle of *Pleurotus* spp. has been shown in on maturity, mushroom produces a number of spores which are released at intervals. The spores are dispersed by wind and germinate on a suitable medium to produce a primary mycelium. The cells of homokaryotic mycelium have only one type of nucleus, all being genetically identical. Mating between two compatible homokaryotic mycelia through hyphal fusion i.e plasmogamy results in the formation of a heterokaryon and hence a fertile mycelium occurs through clamp connections. Under appropriate environmental conditions the fertile mycelium produces fruit body. The spore bearing tissues of the fruit body produces columnar layer of club shaped binucleate cells called basidia. The paired nuclei in the casidia fuse (Karyogamy) to form the diploid nucleus.

Immediately meiosis occurs and four haploid nuclei nuclei are formed which move to the basidiospore through stalk like structure called sterigmata. The basidiospores, so formed are discharged and again germination is initiated.

Genus: *Thanatephorus*

» Exist as saprophyte in the soil (Facultative parasite). Under suitable conditions causes damping off and root rots.

» The cells of the hyphae are barrel shaped, branching more or less at right angles, and pale brown to brown in colour.

» Formation of sclerotia of irregular size and shape, brown or black in colour, more or less loosely packed.

» Basidia are barrel shaped, clavate and have four sterigmata with basidiospores which are hyaline and ellipsoid.

Disease: Sheath blight of rice - *Thanatephorus cucumeris* (Anamorph: *Rhizoctonia solani*)

Symptoms

Large irregular straw coloured centre lesions with reddish brown margin appear on the outermost leaf sheath near water level and on leaves. Matured plants and seedlings may become blighted. Small mustard like sclerotia are present on the diseased portion.

Life cycle

It produces characteristic coarse multinucleate septate hyphae and whole lateral branches which are constricted at the point of origin sclerotia are irregular, brown to black. It produces terminal and intercalary barrel shaped chlamydospores. The sclerotia germinate under favourable condition and produce the mycelium

Basidia produced during sexual reproduction are barrel shaped, clavate and have four sterigmata with basidiospores which are hyline and ellipsoid. Produces metabasidium little wider than the pedicels, ellipsoid with one side flattened rarely obpyriform to obovate spores.

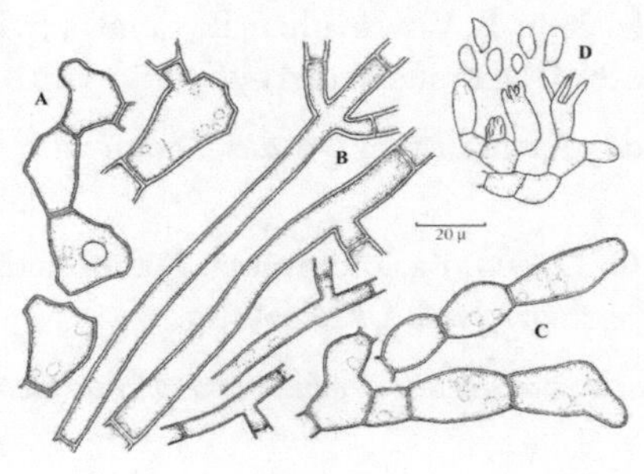

Thanatephorus cucumeris: A. Sclerotia, B. Mycelium, C. monilioid cells of *Rhizoctonia solani*, and D. basidia and basidiospores of its teleomorph

Genus: *Corticium*

Disease: Pink disease - *Corticium salmonicolor* (Syn. *Pellicularia salmonicolor*)

Symptoms

Mango: Dieback of mature mangoes. Gum is seen exuding from the branches and trunk, and branches die.

Rubber: On rubber, pink disease occurs mostly on branch junctions, and sometimes a little further up. The early symptoms are the whitish growth on the bark, which turn pink as disease advances.

Practice Questions

1. The black rust of wheat is a fungal disease caused by...........
2. .... is an alternate host for black stem rust of wheat
3. .... causes jasmine rust
4. .... is resting spore of the genus *Puccinia*
5. Coffee rust is caused by............
6. Explain the clamp connection formation
7. Explain the types of mycelium in the basidioma
8. Explain the polymorphic life-cycle of rust
9. Illustrate the features of different types of teliospores
10. Explain the differences between rust and smut

Suggested Readings

Sundar, A. R., Barnabas, E. L., Malathi, P., Viswanathan, R., Sundar, A. R., & Barnabas, E. L. (2012). A mini-review on smut disease of sugarcane caused by Sporisorium scitamineum. *Botany*, *2014*, 226.

Roelfs, A. P. (1985). Wheat and rye stem rust. In *Diseases, Distribution, Epidemiology, and Control* (pp. 3-37). Academic Press.

Hibbett, D. S., & Thorn, R. G. (2001). Basidiomycota: homobasidiomycetes. In *Systematics and evolution* (pp. 121-168). Springer, Berlin, Heidelberg.

Wilcoxson, R. D. (1996). *Bunt and smut diseases of wheat: concepts and methods of disease management* (Vol. 4). CIMMYT.

Chapter - 14

Prokaryotic Bacterial Plant Pathogens

General Characteristics

Definition: Bacteria are extremely minute, rigid, essentially unicellular organisms (actinomycetes are filamentous), devoid of chlorophyll, most commonly reproduce by transverse binary fission and the resulting cells are identical in size and morphology.

Important characters

1. Microscopic, unicellular prokaryotes (except filamentous actinomycetes) which lack chlorophyll.
2. Contain primitive nucleus lacking a clearly defined membrane.
3. Genetic material relating to pathogenicity, tumour formation and resistance to chemicals are located on plasmids. Plasmids can pass through cells with ease.
4. The shapes of bacteria are round (coccus), rod (bacillus) and spirals (spirillum).
5. All plant pathogenic bacteria are rod shaped.
6. Reproduction by binary fission. They reside space in between cells or in vascular system.
7. Cannot penetrate host tissue but enter through natural openings.
8. Oozing of slimy materials are the signs of bacterial diseases.
9. They survive in plants, seeds, storage organs, plant debris, and equipment and in bodies of insect vectors.
10. Cell walls (no cellulose) surrounded by slime layer (thin) or capsid (thick) comprised of viscous, gummy material.

11. Straight to curved rods with rigid cell walls (except filamentous bacteria). Some bacteria assume irregular shapes like V, Y, L etc., in different stages of their growth. E.g., V form of *Corynebacterium* (*Clavibacter*) and L-forms of *Agrobacterium* and *Erwinia*.
12. Carbohydrate decomposition is mostly aerobic or oxidative (except *Erwinia*, which is a facultative anaerobe).
13. Mostly gram negative, rarely gram positive (Gr+ve genera: *Streptomyces, Corynebacterium, Clavibacter, Curtobacterium*).
14. Plant pathogenic bacteria can be cultured on artificial media. However, pathogenic bacteria grow slowly compared to saprophytes.
15. Majority are flagellate.
16. Plant pathogenic bacteria can be identified based on flagellation, carbohydrate metabolism and pigment production.
17. These are passive invaders, i.e., enter plants through wounds or natural openings.
18. Survive, in/on the seed and in plant debris and spread by means of water, rain, insects, and agricultural implements.
19. All are susceptible to phages.
20. All are non-spore formers except *Bacillus*.
21. None of them cause human and animal diseases.
22. Cell wall rigid made up of pepdidoglycon.
23. Aerobes/ facultative anaerobes.
24. Slow growth compared to other saprophytic bacteria.
25. Incubation period: 36- 48 hrs at 25°C.
26. Majority are flagellate and hence motile.

Bacterial Cell Structure

» A bacterium has a thin, relatively tough, rigid cell wall, and a distinct three layered but thin cytoplasmic membrane.

» Most bacteria have a slime layer made up of viscous gummy material. Slime layer has bacterial immunological property.

» When the layer is thick and firm, it is called capsule.

» Generally, plant pathogenic bacteria lack capsule but some of them like *Pseudomonas* and *Xanthomonas* produce slime.

» Slime layer is mostly composed of polysaccharides but may rarely contain amino sugars, sugar acids, etc.

Flagella

Flagella is a hair like helical structures that protrude through cell wall and are used for locomotion.

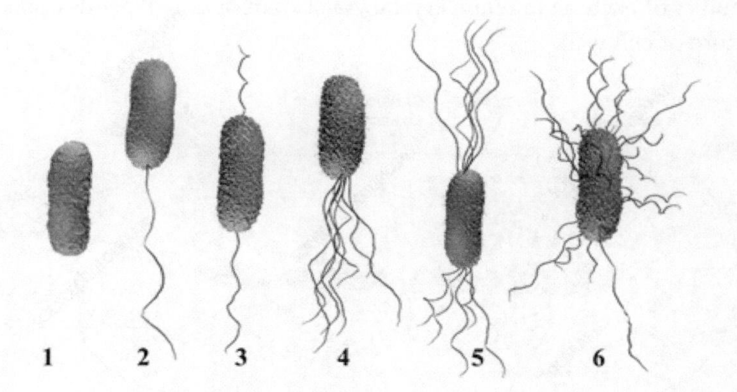

1. Atrichous (without flagella) – *Xylella* sp.
2. Monotrichous (single flagella at one end) - *Xanthomonas* sp.
3. Amphitrichous (at least one flagellum at each pole) – *Pseudomonas* spp.
4. Cephalotrichous (several flagella at one end) - *Pseudomonas fluorescens*
5. Lophotrichous (two or more flagella at both the poles of the bacterium) - Most of the Spirillum types
6. Peritrichous (protruding from all sides) - *Pectobacterium*, *Erwinia*

Gram staining

The Gram staining is one of the most crucial staining techniques in microbiology. It gets its name from the Danish bacteriologist Hans Christian Gram who first introduced it in 1882, mainly to identify organisms causing pneumonia. Often the first test performed, gram staining involves the use of crystal violet or methylene blue as the primary color. The term for organisms that retain the primary color and appear purple-brown under a microscope is Gram-positive organisms. The organisms that do not take up primary stain appear red under a microscope and are Gram-negative organisms.

The basic principle of gram staining involves the ability of the bacterial cell wall to retain the crystal violet dye during solvent treatment. Gram-positive microorganisms have higher peptidoglycan content, whereas gram-negative organisms have higher lipid content.

» In this process, a bacterial smear is heat fixed on glass slide, stained with crystal violet and mordanted with iodine and finally rinsed with ethanol.

- When the bacteria retain the crystal violet stain after rinsing, the bacteria are called gram positive; and those which do not retain the stain are called gram negative.
- The later are then counter stained with pink colour safranin.
- The ability of bacteria to retain crystal violet stain or not, depends upon fundamental structure of cell wall.

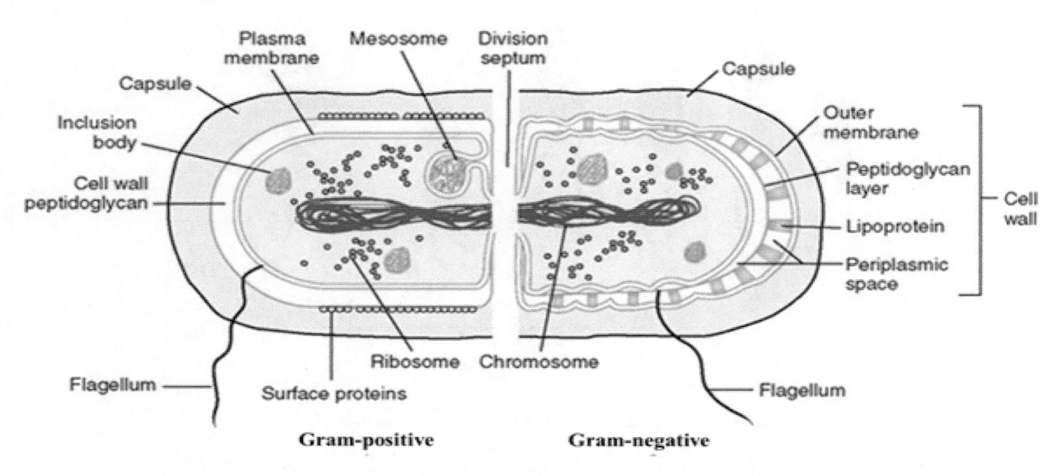

Gram positive *vs.* Gram negative bacteria

Gram positive bacteria	Gram negative bacteria
» Cell wall is thicker and homogemous. » Contains lower content of lipids (5-10%). » Peptidoglycan comprises up to 90% of the cell wall and hence maximum lipid. » Techoic acid present. » Cell wall has higher amino sugar content (10-20%). » Cell wall is simple in shape and is single layered. » Mesosomes more prominent. » Retains violet dye. » Examples: *Bacillus*, *Clavibacter*, *Streptomyces.*	» Cell wall is thinner and usually thin layered. » Contains higher content of lipids (up to 40%). » Peptidoglycan comprises only 10%. » Techoic acid absent. » Low content of amino sugars » Varying cell wall shape and is tripartite (3-layered). » Mesosomes less prominent. » Retains red dye » Examples: *Erwinia, Pseudomonas, Xanthomonas, Agrobacterium, Xylella*

Cell wall of bacteria

The main component that forms the bacterial cell wall is peptidoglycan, also known as murein. Peptidoglycan contains polysaccharide chains that are cross-linked with peptides containing D-amino acids.

- The unique structure of the bacterial cell wall is due to the composition of disaccharide-pentapeptide subunits.
- N-acetylglucosamine and N-acetylmuramic acid are the disaccharides that are linked to N-acetylmuramic acid molecules.
- Polymer subunits like these cross-link to each other via peptide bridges to form peptidoglycan sheets, and these peptidoglycan sheets are present around the whole cell.

Bacterial cell wall is further divided into two categories, the gram-positive and gram-negative bacteria. The main difference between these two layers is the thickness level of the peptidoglycan layer. Gram-positive bacteria have a thicker peptidoglycan layer than gram-negative ones.

- The cell wall of Gram-negative bacteria consists of a thin layer of peptidoglycan in the periplasmic space between the inner and outer lipid membranes. The outer membrane contains lipopolysaccharides on its outer leaflet and facilitates non-vesicle-mediated transport through channels such as porins or specialized transporters.
- The cell wall of Gram-positive bacteria have a single lipid membrane surrounded by a cell wall composed of a thick layer of peptidoglycan and lipoteichoic acid, which is anchored to the cell membrane by diacylglycerol.

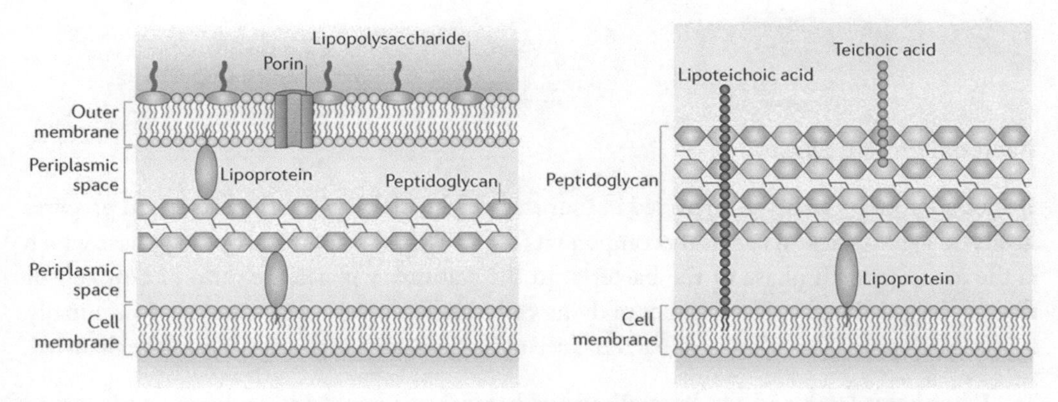

Cell wall of gram negative (left view) and gram positive (right view)

Asexual Reproduction: Bacteria multiply at a phenomenal rate by the process of fission or binary fission.

- As the cytoplasm and cell wall undergo division into two, the nuclear material is organized into a circular chromosome like structure which ultimately duplicates itself and gets distributed equally into 2 newly formed cells.
- Similarly, plasmids also duplicate and come into 2 daughter cells.
- The duplication occurs rapidly, once every 20 minutes.
- However, this number is not reached because of gradual limitations of nutrients and toxic metabolites. Still what is achieved normally is phenomenonal.
- Such prolificacy in multiplication must be of great advantage both in survival of bacterial pathogen, and also for successive plant infections.

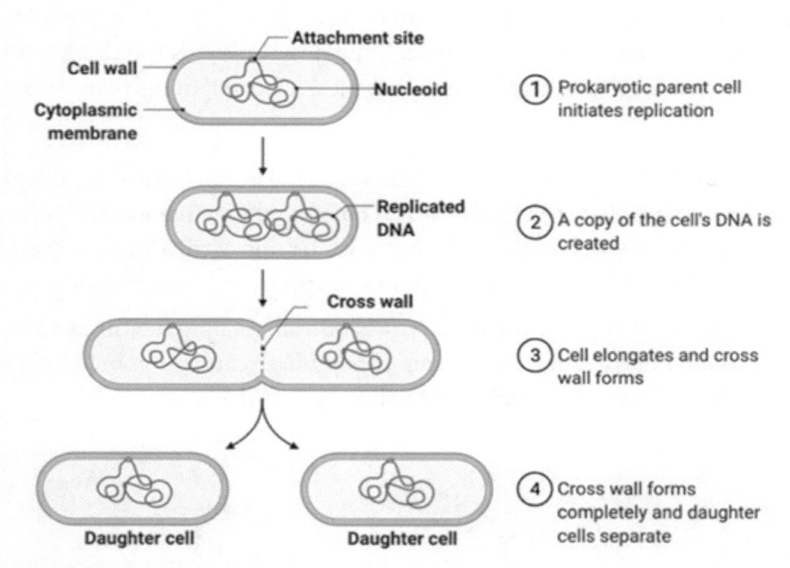

Bacterial growth phases

Bacterial growth can be summarized in four stages. Lag phase when the bacterium prepares itself and synthesizes all necessary components for growth. It is followed by Log phase which is the active growth phase of the bacteria. In the stationary phase the ratio of new cells is directly proportional to the number of dying cells due to depleting nutrient sources. Finally, in the death phase the number of bacterial cell dying increases as the nutrients get exhausted.

1. **Lag phase:** In this phase, bacteria adapt themselves to growth conditions and start to mature. Synthesis of RNA, enzymes and other molecules occurs, there is almost no cell division.

2. **Log phase/Exponential phase:** This is where the number of cells increase proportionally to the existing population. The number of new bacteria appearing per unit time is proportional to the present population. Doubling happens at a rate where both the number of cells and the rate of population increase doubles with each consecutive time period.
3. **Stationary phase:** After a certain period of growth, due to the unavailability of a sufficient quantity of a necessary factor, enough to sustain a huge population as produced in the phase of exponential growth, the bacteria enter the stationary phase. In this phase, the rate of growth and death are more or less equal as the rate of growth of bacteria is limited by the death rate.
4. **Death phase:** This is the final phase of bacterial growth due to depletion of nutrients and accumulation of toxins.

Recombination in bacteria has been noticed by the following sexual-like processes:

1. **Conjugation:** Conjugation occurs when two compatible bacteria come into contact and part of the chromosomal or non-chromosomal genetic material of one is transferred to the other and incorporated into the genome of later through conjugal zygote formation and breakage and reunion. It was first observed by Lederberg and Tatum (1956) in *E. coli*.
2. **Transformation:** It occurs when the bacterium is genetically transformed by absorption of genetic material of another compatible bacterium, secreated by or released in a culture during the rupture, and its incorporation into the genome of the former. It was first observed by Griffith (1928) in *Enterococcus pneumoniae*.
3. **Transduction:** When genetic material from one bacterium is carried by its phage (virus) to another bacterium that it visits next and the later is genetically transformed. It was first discovered by Zinder and Lederberg (1952) in *Salmonella*.

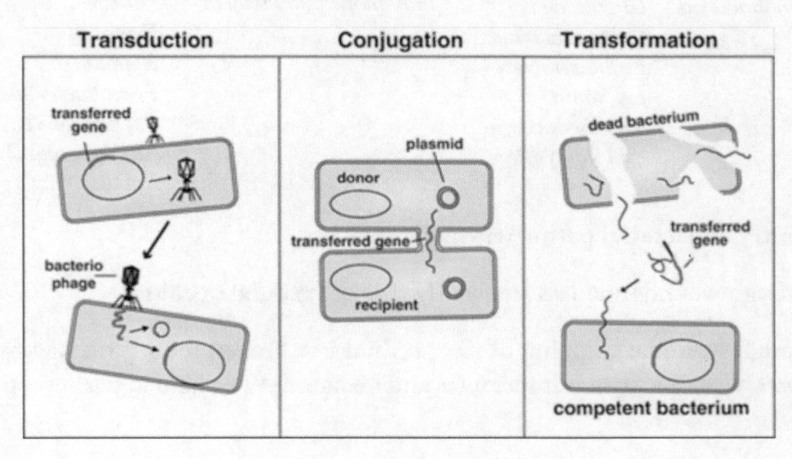

Phylogenetic classification of bacteria

Based on cell wall composition, fatty acid and protein profiling, comparison of DNA and RNA composition and sequencing.

Molecular characterization of 16S ribosomal RNA distinguishes bacteria from one another.

There are 21 phyla within the domain *Bacteria* and the plant pathogens are found in 3 phyla.

1. Phylum: *Firmicutes – Bacillus, Phystoplasma, Spiroplasma*
2. Phylum: *Actinobacteria* (Gram positive bacteria) - *Clavibacter, Leifsonia, Curtobacterium, Rathayibacter* and *Sterptomyces*
3. Phylum: *Proteobacteria* (Gram negative bacteria) – *Agrobacterium* (*Rhizobium*), *Erwinia, Pectobacterium, Pseudomonas, Xanthomonas* and *Xylella*

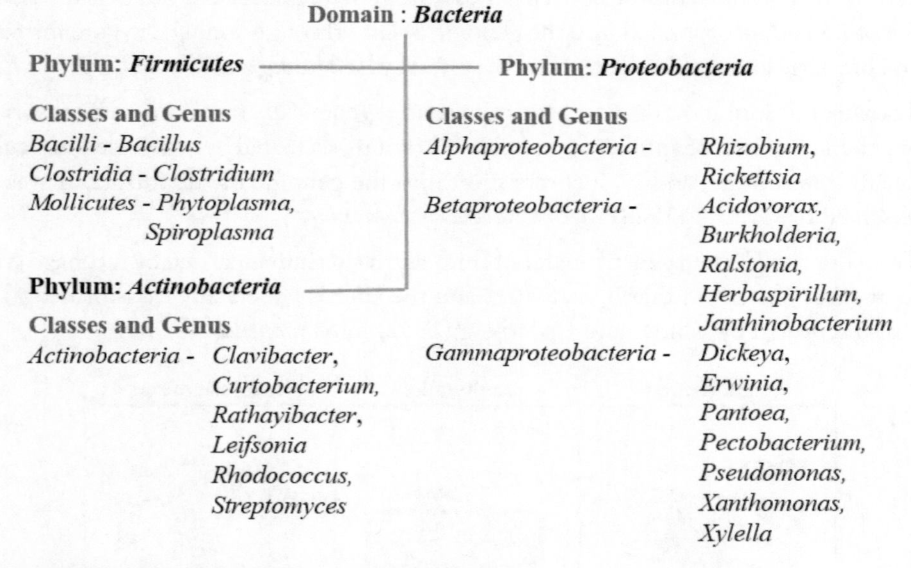

Mode of entry of bacterial pathogens in plants

Mostly through wounds and less frequently through natural openings

Entry through wounds: Clipping of rice seedlings or broken roots pave way for entry of *Xanthomonas oryzae* pv. *oryzae*. Insect bites and nematodes also help in the entry of bacterial pathogens.

Entry through natural openings

» Stomata - *Erwinia amylovora* (fire blight) in apple and pear; *Pseudomonas tabaci* (wild fire) in tobacco.
» Nectarthodes - *Erwinia amylovora* in apple and pear flowers
» Hydathodes - *Erwinia amylovora* and *Xanthomonas axonopodis* pv. *campestris*
» Trichomes - *Clavibacter michiganensis* sub sp. *michiganensis* (bacterial canker) enters through trichomes of tomato.

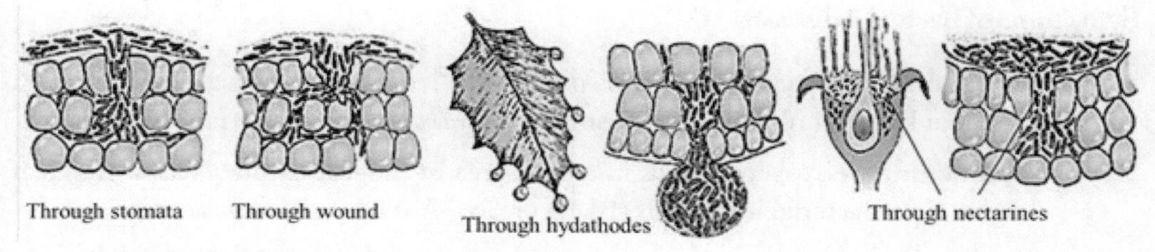

Survival of bacterial pathogens

1. Survival with seed (internally seed-borne & externally seed-borne)

   E.g., *Xanthomonas* causing bacterial blight of rice and cotton

2. Survival in plant residue

   E.g., *Clavibacter*, *Pectobacterium carotovorum* subsp. *carotovorum*

3. Survival in soil

   Soil invaders E.g., *Xanthomonas, Erwinia, Clavibacter*

   Soil inhabitants E.g., *Rhizobium radiobacter, Streptomyces*

   Soil saprophyte E.g., Fluorescent pseudomonads, *Bacillus*

4. Survival in perennial host

   E.g., *Erwinia amylovora, Xanthomonas axonopodis* pv. *citri*

5. Survival as epiphyte

   E.g., *Erwinia amylovora* on apple buds

6. Survival with insects

   E.g., *Pantoea stewartii* in cucuber beetle & corn flea beetle

7. Survival with non-host materials

   E.g., *Xanthomonas axonopodis* pv. *malvacearum*

Mode of spread

Plant pathogenic bacteria spread by different means.

They are soil, wind, irrigation water and rainwater, seed, insects (Beetles: cucumber wilt by striped and spotted beetles), nematodes – *Anguina tritici* transmitting tundu disease of wheat caused by *Rathyibacter tritici,* collateral hosts, activities of man (fire blight from USA to Europe; BLB from Japan to India), self-sown and volunteer crops, crop debris and contaminated tools.

Symptoms of bacterial diseases

» **Leaf spot:** Water soaked or brown spots on the leaves are known as leaf spot. E.g., Bacterial leaf spot of chillies and tomato - *Xanthomonas vesicatoria*

» **Leaf blight:** Necrosis of tissues in a large area in the leaf lamina is known as leaf blight. E.g., Bacterial leaf blight (BLB) of rice -*Xanthomonas oryzae* pv. *oryzae*

» **Canker:** Brown, corky, cankerous growth on leaves, stems and fruits are known as canker. E.g., Canker of citrus - *Xanthomonas axonopodis* pv. *citri*

» **Scab:** A roughened crust-like diseased area on the surface of a plant organ is known as scab. E.g., Common scab of potato - *Streptomyces scabies*

» **Crown gall:** A swelling or outgrowth produced on the stem and roots of plants as a result of infection by certain bacteria like *Agrobacterium.* E.g., Crown gall of apple - *Agrobacterium tumefaciens.*

» **Wilt:** Yellowing of leaves and death of the plant with bacterial ooze from the vascular bundles is seen in bacterial wilt. E.g., Moko wilt of banana - *Burkholderia solanacearum*

» **Soft rot:** Disintegration and softening of plant tissues or fleshy fruits, vegetables due to the enzymatic action of the pathogen is seen in soft rot caused by bacteria. E.g., Soft rot of vegetables - *Erwinia carotovora* pv. *carotovora*

Important plant pathogenic bacteria

» ***Pseudomonas*** is straight to curved rods, 0.5 to 1 × 1.5 to 4 µm. Motile by means of one or many polar flagella. It is inhabitants of soil or of fresh water and marine environments.

» ***Xanthomonas*** is straight rods 0.4 to 1.0 × 1.2 to 3 µm motile by means of a polar flagellum. Growth an agar media usually yellow. The important pathogenic sps under these genera are *X. oryzae* pv. *oryzae* (Bacterial leaf blight in paddy), *X. o.* pv. *oryzicola* (Bacterial leaf streak in paddy), *X. campestris* pv. *campestris* (Black rot of crucifers),

X. c. pv. *malvacearum* (Black arm/angular leaf spot of cotton), *X. c.* pv. *vesicatoria* (Bacterial leaf spot of tomato and chillies), *X. c.* pv. *translucens* (bacterial blight of cereals), *X. c.* pv. *vasculorum* (gummosis in sugarcane) and *X. axonopodis* - (*X. c.* pv. *citri*) - Citrus canker.

» ***Agrobacterium*** is rod shaped bacteria with 0.8 × 1.5 to 3 μm size. They are motile by means of 1 to 4 peritrichous flagella. When only one flagellum is present it is more often lateral than polar. The colonies are non-pigmented and usually smooth. These bacteria are rhizosphere and soil inhabitants.

» ***Clavibacter*** is Gram-positive, non-acid fast, pleomorphic rods, often arranged at an angle to give v-formation as a result of snapping or bending type of cell division; no coccoid cells are seen. It is non-endospore forming bacterium and non-motile.

» ***Curtobacterium*** is cells small, short rods, coccoid cells found in old cultures. It is weakly Gram-positive and motile by lateral flagella.

» ***Rhodococcus*** cells in young culture measure 0.5 to 0.9 × 1.5 to 4.0 μm. They are slightly curved rods and are arranged singly, at angles or as palisade. Gram positive and non-motile.

» ***Erwinia*** is rods shaped, 0.5 to 1.0 × 1.0 to 3.0 μm in size. They are motile by means of several many peritrichous flagella. Erwinia is the only plant- pathogenic genus, which is facultative anaerobe.

» ***Streptomyces*** is slender, branched hyphae without cross walls, 0.5 to 2 micrometre in dia. At maturity the aerial mycelium forms chains of three to many spores on nutrient media. All species are Gram-positive and soil

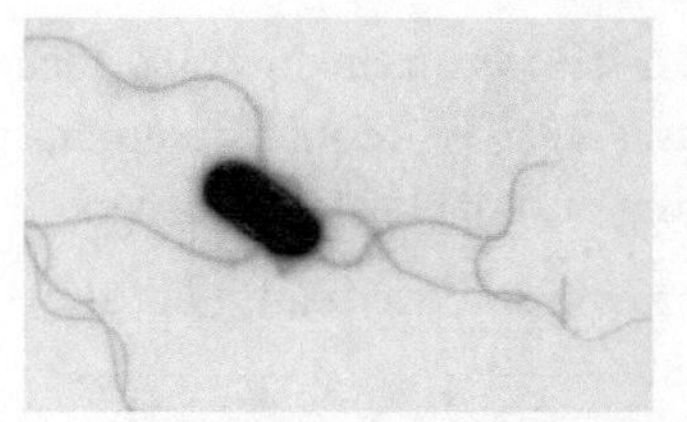

Agrobacterium tumefaciens
Courtesy: Ozyigit 2012

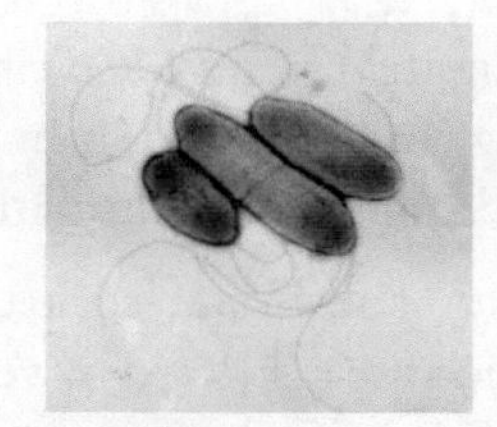

Burkholderia pseudomallei
Courtesy: R.G. Hans, 2014

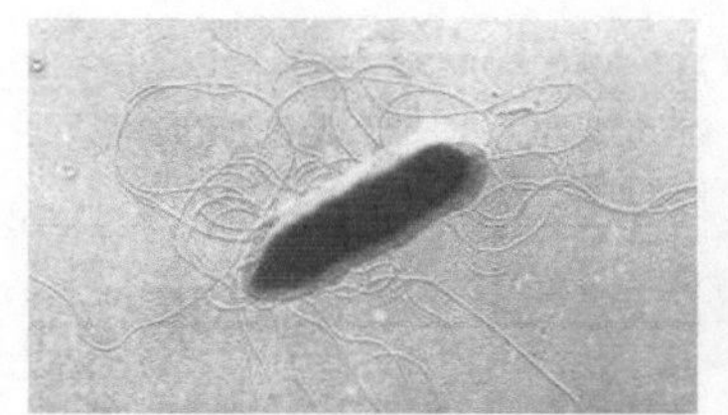

Erwinia amylovora
Courtesy: Guerrero-Prieto et al. 2006

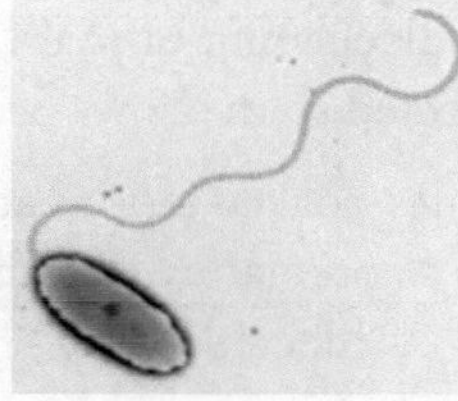

Xanthomonas campestris
Courtesy: Vicente & Holub, 2012

Important plant bacterial diseases

Bacterial leaf spot of chilli: ***Xanthomonas vesicatoria***

- Small, circular or irregular, dark brown or black greasy spots appear on leaves. The spots coalesce to form irregular lesions.
- Severely infected leaves become chlorotic and fall off. Stem infection leads to the formation of cankerous growth and wilting of branches.
- On the fruits, raised water soaked spots with a pale yellow border are produced.

Bacterial blight of rice: ***Xanthomonas oryzae*** **pv.** ***oryzae***

- Bacterial leaf blight of rice has three phases *viz.,* leaf blight, kresek or wilt and yellow leaf.
- Leaf blight - Small, water soaked spots on the margin of the lamina near the tip, later enlarge and coalesce; wavy margin of the dried areas, drying from tip downwards and from margin towards midrib, leaving a small green patch in the centre.
- Kresek stage - Systemic infection in the seedling leads to death of seedlings in one or two weeks after transplanting.
- Yellow leaf - Yellowing of young leaves and withering.

Bacterial blight / Black arm of cotton: ***Xanthomonas axonopodis*** **pv.** ***malvacearum***

- It causes symptoms like seedling blight, angular leaf spot, vein blight, black arm and boll rot.
- Seedling blight: Water-soaked circular lesions on the lower surface of cotyledons of germinating seeds are seen. It increases in size become irregular, spreads to leaves, petioles and stem and results withering and drying of seedlings.
- Angular leaf spot: Water-soaked, brown coloured, angular spots appear on leaf lamina, which are restricted by veinlets.
- Vein blight: Browning of veins and veinlets which cause crinkling and inward twisting and drying up of leaves.
- Black arm: Blackening of main stems and branches, which are linear and sunken. Heavily infected stem shows deep cracking and gummosis.
- Boll rot: Water-soaked lesions appear on the bolls. They turn dark black and sunken, which leads to premature shedding of bolls. The lint is stained yellow and the seeds inside also get infected.

Citrus canker: *Xanthomonas axonopodis* pv. *citri*

» Round yellowish spots initially, later corky brown outgrowth with raised margin mainly on leaves, petioles, stems and fruits

» crater-like depressions in the centre of the corky outgrowths

» yellow halo surrounding the corky outgrowths are seen especially on leaves and not on fruits

Moko wilt of banana: *Ralstonia solancearum* race 2

» Yellow discolouration of lower leaves and quickly followed by the appearance of similar colour change in all the other leaves; collapse of the lamina near the junction of lamina with petiole; pseudostem turn to yellow or brown

» When the pseudostem or sucker is cut across, the cut surface becomes covered with opaque, greyish or dirty white slimy coat of bacterial ooze and vascular bundles discoloured to pale yellow or dark brown or bluish black

» Cross section of fruit shows central black discoloured patch; side suckers show necrotic symptoms.

Crown gall of rosaceae: *Agrobacterium tumefaciens*

» Small outgrowth appears on the stem and roots. At the young stage, the galls are soft, spherical, and white or flesh coloured.

» The galls are hard and corky on woody stems and vary in size from 7 to 10 mm in diameter. The affected plants are stunted with chlorotic leaves.

Soft rot of vegetables: *Erwinia carotovora* pv. *carotovora*

» The bacterium attacks the vegetables both in the field and in storage, water-soaked region appears on the tissue which enlarge quickly in diameter and depth

» affected area become soft and mushing; affected surface is discoloured, depressed or appear blistered or wrinkled; cracks develop and slimy mass exudes to the surface; whole fruit is converted into a soft, watery, colourless decayed mass within three to five days.

Common scab of potato: *Streptomyces scabies*

» Circular lesions around the lenticels of tubers; skin turns brown and necrotic; lesions increase in size (3-4 mm deep in to the tubers), coalesce and cover larger areas on the tuber.

Practice Questions

1. ……….is a causative of citrus canker
2. ………is a causative of bacterial blight of rice
3. ………is main compoenent of gram-positive cell wall
4. Crown gall of rose is caused by……..
5. ……..bacterium produces peritrichous flagella
6. Write the differences between gram positive and gram negative bacteria
7. Write the symptoms and causative of citrus canker
8. Explain the five different symptoms of bacterial blight of cotton
9. Illustrate the flagellar arrangement of bacteria with examples
10. Ellaborate the phyllogenitic classification of bacteria

Suggested Readings

Borkar, S. G. (2017). *Laboratory techniques in plant bacteriology*. CRC Press.

Kado, C. I. (2010). *Plant bacteriology* (No. PA 581.2 K33 2010.). St. Paul, MN: American Phytopathological Society.

Goto, M. (2012). *Fundamentals of bacterial plant pathology*. Academic Press.

Breed, R. S., Murray, E. G. D., & Kitchens, A. P. (1948). Sergey's manual of determinative bacteriology. *Sergey's manual of determinative bacteriology. Sixth edition.*

Gnanamanickam, S. S. (Ed.). (2006). *Plant-associated bacteria* (Vol. 1). Dordrecht: Springer.

Chapter - 15

Prokaryotic Mollicutes

Important Characteristics of *Candidatus Phytoplasma* (Phytoplasma or MLO (Yellow disease agents)

Generally called as Mollicutes (*Mollis* = soft or pliable; *cutis* = skin) and are the smallest known free-living organisms. Prokaryotic, pleomorphic, resemble mycoplasma, wall less, self- replicating, pass through bacterial filters, sensitive to tetracyclines, transmission by leaf hoppers, found in sieve elements of plants.

General characters

» Differ from true bacteria in the absence of typical cell wall.

» The cells are bounded by a triple layered membrane.

» Unicellular, Gram-positive, having both RNA and DNA, non-motile, pleomorphic, facultative saprophytic and their size ranging from small spherical -large irregular bodies (175-250 nm). Some species produce myceloid structure (7.5 to 10 nm in width).

» Free living, both parasitic and saprophytic and reproduces by budding and binary fission.

» Cells contain cytoplasm, ribosomes and strands of nuclear material devoid of nuclear membrane. Both DNA and RNA are present. Can be cultured artificially (some species) on cell free agar medium with sterol and form fried egg colonies.

» Generally present in sap of phloem sieve tubes.

» Resistant to penicillin and sensitive to tetracycline.

- » Production of virescent flowers (green flower parts). It changes the angle of growth of the branches. As a result, the branches become stiff and become upright.
- » Affected leaves become pale or yellow. It causes stunting, chlorosis, brooms, distortion of flowers and necrosis.
- » Transmission mostly by leaf hoppers.

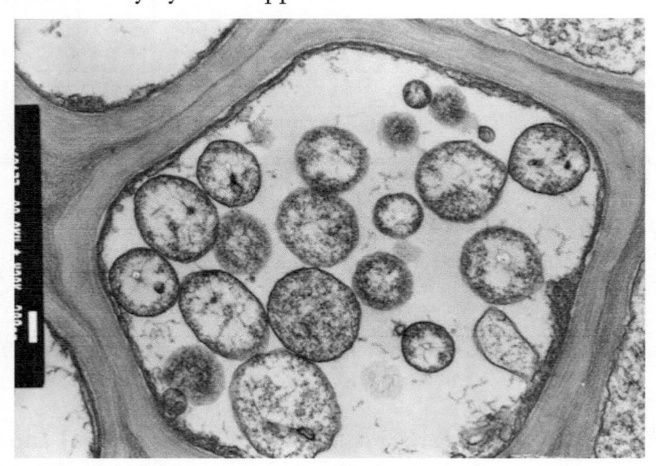

Phytoplasma in a phloem cell of a plant

General symptoms

- » Characterized by yellowing, chlorosis or bronzing of foliage, shortening of internodes, reduction in leaf size, proliferation of auxiliary buds (often resulting in witches' broom effect), phyllody, proliferation of secondary roots and abnormal fruits and seeds.
- » The symptoms vary with host species, time and mode of infection and environmental factors.
- » The yellowing, stunting and wilting may be due to excessive callose formation and disintegration of phloem tissues.
- » It also disrupts the growth regulators level in infected plants which lead to development of witches' broom, phyllody, etc.

Phytoplasma infection cycle

When a leafhopper inserts its piercing stylet into the vein of a plant infected with phytoplasmas and draws up the sap from the phloem, the phytoplasmas are transferred into the leafhopper's body, where they multiply. Next, when the leafhopper feeds on sap from a healthy plant, phytoplasmas are introduced into the plant and the plant becomes infected.

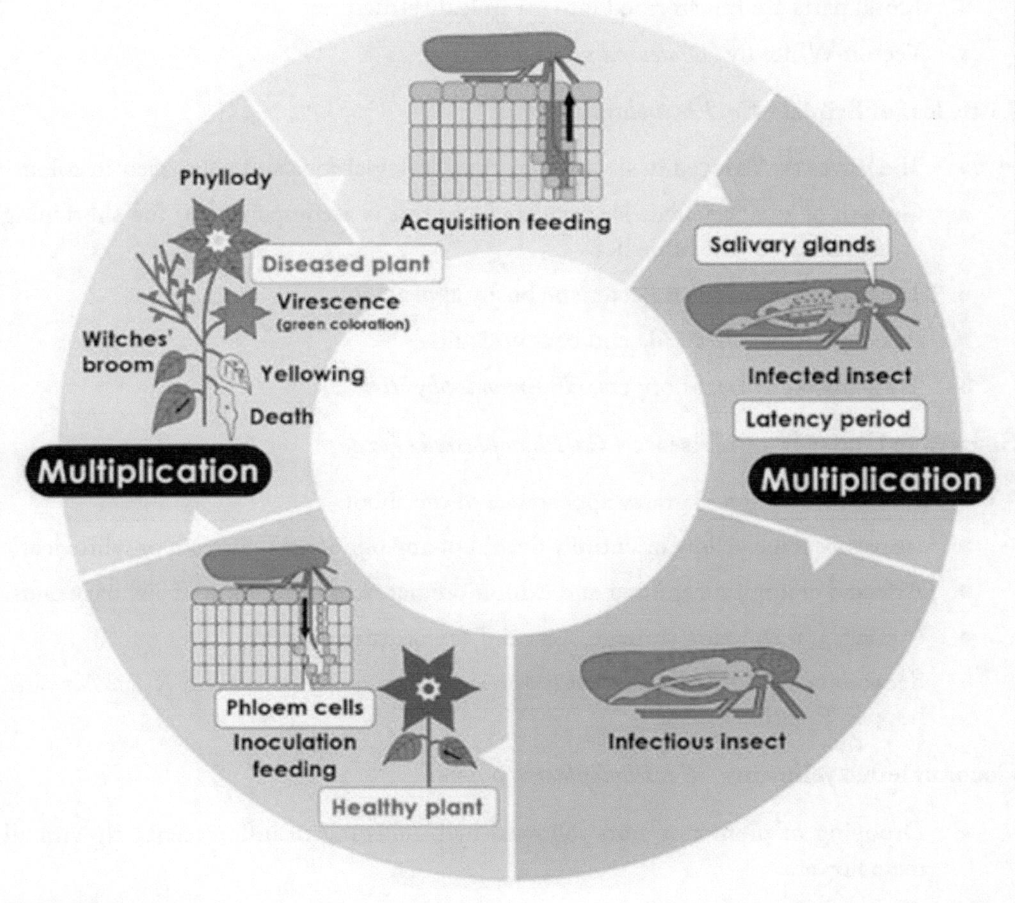

Major diseases

Sesame Phyllody – *Ca. Phytoplasma asteris*

- This is also known as green flowering.
- Reduction in leaf size, shortening of internodes, stimulation of auxillary buds, transformation of floral parts into green leafy structures.
- Diseased plants are completely sterile.
- Vector – Leaf hopper, *Orosius albicinctus*

Jasmine Phyllody – *Ca. Phytoplasma aurantifolia*

- Clustering of leaves and bushy appearance of plants
- Leaves are linear, small, closely arranged

- Floral parts are transformed into green leafy structures
- Vector: White fly (*Dialeurodes kirkaldii*)

Little leaf of Brinjal - *Ca. Phytoplasma trifolii*

- The leaves are reduced in size, sessile, thin, soft, glabrous and pale green in colour.
- Growth of auxiliary buds is stimulated and this is accompanied by the shortening of internodes of the branches.
- The plant presents a characteristic bushy appearance.
- Affected plants are sterile and bear no fruits.
- **Transmission:** Leaf hoppers, *Hishimonas phycitis*

Sugarcane Grassy Shoot Disease - *Ca. Phytoplasma oryzae*

- Profuse tillering and grassy appearance of the shoot.
- Leaves become yellow or entirely devoid of any pigments (albinism or white leaf)
- Affected clumps are stunted and exhibit premature proliferation of auxiliary buds.
- Primary transmission through setts and cane cutting knives.
- Secondary transmission through aphid, *Rhaphalosiphum maidis*, *R. sacchari* and hopper, *Proutista moesta*

Coconut lethal yellowing - *Ca. Phytoplasma palmae*

- Drooping of premature nuts followed by blackening of inflorescence tip and all male flowers.
- The lower leaves exhibit yellowing which progress upward to the young one.
- Later the lower leaves die prematurely, turn brown and cling to the tree.
- Finally, all the leaves and vegetative buds die, fall away leaving the trunk alone.
- Transmission: Leaf hopper, *Myndus crudus*

Rice yellow dwarf - *Ca. Phytoplasma oryzae*

- General chlorosis, stunting, and excessive tillering.
- Plants infected early may die prematurely, but usually they survive until maturity.
- They produce very poor panicles or none at all.
- Transmitted by Green leaf hoppers, *Nephotettix cincticeps, N. virescens, N. nigropictus.*

Sandal Spike

» Rosette spike: Reduction in leaf size, shortening of internodes branches become stiff and pointed spike- like structure

» Pendulous spike: Rare. Individual infected shoots show continuous apical growth and assume a drooping habit.

» Transmission by grafting, dodder and insect, hopper - *Nephotettix virescens*

Important Characteristics of Spiroplasma

» Helical, wall less prokaryotes, seen in phloem of diseased plants, a kind of mycoplasmas, but can be cultured.

» Pleomorphic, spherical to ovoid, helical and branched structures.

» Helical filaments are motile.

» Lack rigid cell wall and can be cultured on artificial media

» Reproduction by binary fission.

» Cholesterol is required for its growth.

» It is transmitted mostly by leaf hoppers.

» Spiroplasmas are resistant to penicillin but inhibited by erythrocin and tetracycline.

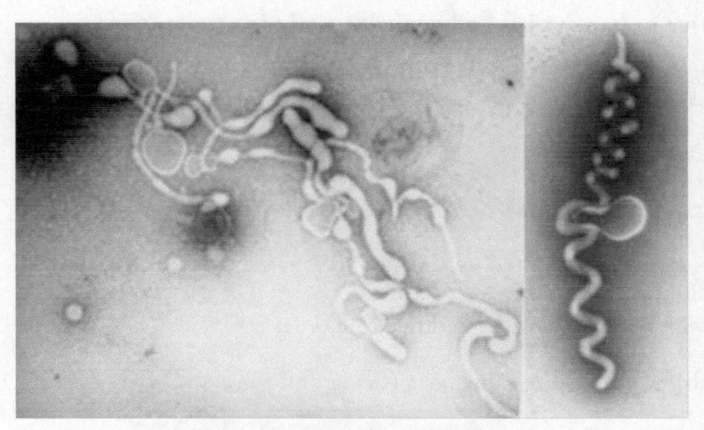

Spiroplasma in a phloem cell of a plant

Major diseases

Citrus Stubborn - *Spiroplasma citri*

» Affected trees show a bunchy upright growth of twigs and branches.

» Internodes are shorter. Multiple buds and sprouts are common.

» Diseased plants exhibit excessive defoliation and produce small lopsided fruits with thin rind. Fruits are usually with sour taste and have an unpleasant odour.

» Fruits – mummified; discoloured with aborted seeds.

» Gram positive, wall less organism transmitted by budding, grafting and by leaf hoppers, *Circulifer tenellus*

Corn Stunt disease – *Spiroplasma kunkelii*

» Infected plants are stunted bushy with chlorotic leaves.

» The cobs are smaller, little or no seed.

» Tassels of infected plants are usually sterile.

» Leaf hoppers *viz., Dalbulus elimatus, D. maidis, Graminella nigrifrons.*

Important Characteristics of Fastidious Vascular Bateria (FVB)

» All are Gram negative (except sugarcane ratoon stunt which is Gram positive XLB)

» Non motile rods (No flagella) / pleomorphic in shape

» Cell wall is thin and undulating

» Intra cellular parasites, non spore former found in xylem and phloem vessels of plants

» Transmission by leaf hoppers, bugs, psyllids, nematodes and mechanical

» Sensitive to tetracycline and penicillin and high temperatures

» Xylem limiting FVB can be cultured artificially than phloem limiting FVB. E.g., Pierce's disease of grapevine.

FVB can be divided into

1. Phloem-inhabiting **Fastidious** bacteria (G+ve)

Ca. Liberobacter asiaticum (Citrus Greening, Zibra chip of Potato)

Serratia marcescens (Cucurbit Yellow Vine)

2. Xylem-inhabiting **Fastidious** bacteria (G+ve and -ve)

3. Laticifer- inhabiting **Fastidious** bacteria (*Rickettsia* – Bunchy top of Papaya – leaf hopper *Empoasca papaya, E. stevensi*)

Phloem inhabiting fastidious vascular G+ bacteria

Citrus greening caused by Candidatus *liberobacter asiaticus* (transmission by citrus psyllid, *Diaphorina citri*)

General Symptoms

- Stunting and yellowing of young leaves
- Virescence of floral parts
- Premature flowering and fruit drop
- Transmission leafhoppers, dodder and grafting

Xylem inhabiting fastidious vascular bacteria

- All are Gram negative (except sugarcane ratoon stunting caused by *Leifsonia xyli* subsp. *xyli*, which is Gram positive)
- They are non-spore forming, non-motile with well-defined cell wall
- Transmission by xylem feeding insect vectors (leaf hoppers and spittle bugs)

General symptoms

- Marginal chlorosis and necrosis of leaves
- Stunting of plants
- Decline in vigour and reduction in yield

Gram positive - Xylem inhabiting fastidious vascular bacteria

- Sugarcane Ratoon Stunting (*Leifsonia xyli* subsp. *xyli*) mechanical transmission (syn. *Clavibacter xyli* subsp. *xyli*)

Gram negative - Xylem inhabiting fastidious vascular bacteria

- Pierce's disease of grapevine (*Xylella fastidiosa*)

***Xylella* - Fastidious vascular bacteria**

- Previously known as Rickettsia-like organisms or RLO.
- Similar to bacteria in all respects found inhabiting in plant vascular system (Phloem limited and xylem limited).
- Single straight rods, non-motile. These are parasitic bacteria that cannot easily be grown on simple culture media.
- Causing wilt diseases transmitted by glassy-winged sharpshooter *Homalodisca vitripennis.*

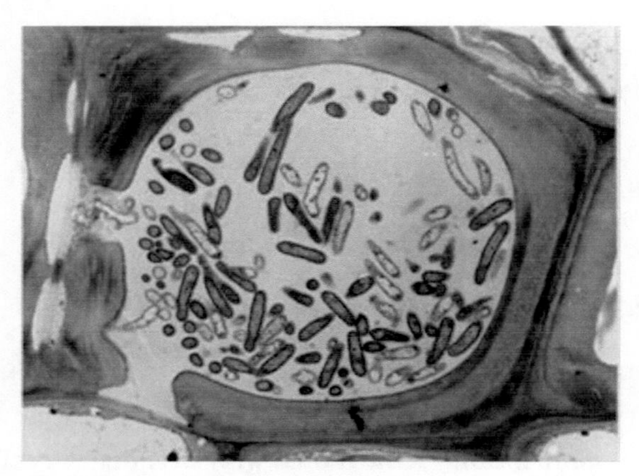

Xylella fastidiosa (Source K. Matsuoka)

Practice Questions

1. Little leaf of brinjal is transmitted by.......
2. Sesame phyllody is caused by.........
3. Reproduction of spiroplasma is mainly by............
4. Spiroplasmas are resistant to penicillin but inhibited by.......... and..........
5. Candidatus *liberobacter asiaticus* causing citrus greening transmission by..........
6. Write the differences between phloem limited and xylem limited bacteria
7. Explain the symptoms of phytoplasmal diseases in crop plants
8. Write the differences between phytoplasma and spiroplasma

Suggested Readings

Bertaccini, A., & Duduk, B. (2009). Phytoplasma and phytoplasma diseases: a review of recent research. *Phytopathologia mediterranea*, *48*(3), 355-378.

Lee, I. M., Davis, R. E., & Gundersen-Rindal, D. E. (2000). Phytoplasma: phytopathogenic mollicutes. *Annual Reviews in Microbiology*, *54*(1), 221-255.

Lee, I. M., Gundersen-Rindal, D. E., & Bertaccini, A. (1998). Phytoplasma: ecology and genomic diversity. *Phytopathology*, *88*(12), 1359-1366.

Williamson, D. L., & Whitcomb, R. F. (1975). Plant mycoplasmas: a cultivable spiroplasma causes corn stunt disease. *Science*, *188*(4192), 1018-1020.

Cisak, E., Wójcik-Fatla, A., Zajac, V., Sawczyn, A., Sroka, J., & Dutkiewicz, J. (2015). Spiroplasma-an emerging arthropod-borne pathogen?. *Annals of Agricultural and Environmental Medicine*, *22*(4).

Chapter - 16

Virus

Virus: Submicroscopic, infectious intracellular obligate parasite, do not have its own metabolism but depend upon other living cells for their multiplication, Disease producing entities, no sexual reproduction and resting spores, it has nucleic acid and protein coat, Either DNA or RNA and never both, Nucleic acid carries its genome, it may be single stranded or double stranded, Protein coat present as an envelope around the nucleic acid.

Viruses are **living** because

- They have ability to assimilate (metabolism) with the release of energy
- They are able to multiply and mutate
- They exhibit response to environment like temperature, chemicals, etc.,
- They have genetic materials like RNA or DNA
- They have the ability to infect

Viruses are **non-living** because

- Viruses can be crystallized. Viruses can be purified a concentrated from the extract of infected cells and show uniform size, shape and chemical composition even after crystallization by chemical treatments.
- Stanley (1935) considered virus as a molecule and a molecule is not capable of self-replication
- Viruses are inert outside the living host
- Viruses do not have cell wall or cell membrane of any type
- They do not show functional autonomy

» They do not respire or excrete
» They lack any energy producing system.

No scientific report on plant virus disease until late 19th century

Difference between bacteria and viruses

Bacteria	Viruses
Prokaryotic	Mesobiotic
Possess cellular organization	Do not possess cellular organization
Grow on inanimate media	Do not grow on inanimate media
Multiply by binary fission	Do not multiply by binary fission
Possess RNA and DNA	Possess either RNA or DNA
Possess ribosome	Do not possess ribosome
Sensitive to antibiotics	Insensitive to antibiotics
Insensitive to interferon	Sensitive to interferon

Composition and structure of virus

A virus is a set of one or more genomic nucleic acid molecules, normally encased in a protective coat or coats of protein or lipoprotein, which is able to mediate its own replication only within suitable host cell. Within such cells, virus replication is

i. Dependent on the hosts' protein – synthesizing machinery
ii. Derived from pools of the required materials rather than from binary fission; and
iii. Located at sites not separated from the host cell content by a lipoprotein bilayer membrane - Mathews, 1992

Genetic material DNA or RNA, not both, consist of nucleic acid core enclosed in a protein or protein-lipid shell Direct an infected cell to make 'virus parts and ultimately complete virions; Have no ATP generating system; Use host ribosomes for their replication, Viruses are reproduced entirely from their nucleic acid; do not undergo binary fission, in order to multiply viruses, have to get into a cell in which they can replicate - a host cell. Viruses take advantage of the host cell's metabolic machinery encoded by the host cell's chromosome

Concepts: Virus - a capsid-encoding organism that is composed of proteins and nucleic acids, that self-assembles in a nucleocapsid, and that uses a ribosome-encoding organism for the completion of its life cycle (Raoul & Forterre, 2008)

Terms: Virus: Viruses are obligate intracellular parasites at the genetic level

Virion: The entire infectious unit or the complete virus particles (syn. Virus; virus particle (Latin: Poison))

Protein coat: Coat Protein, Capsid, Protein shell

Nucleoprotein: Nucleocapsid, NP

Protein sub unit: Structural unit

Morphologic unit: Capsomere

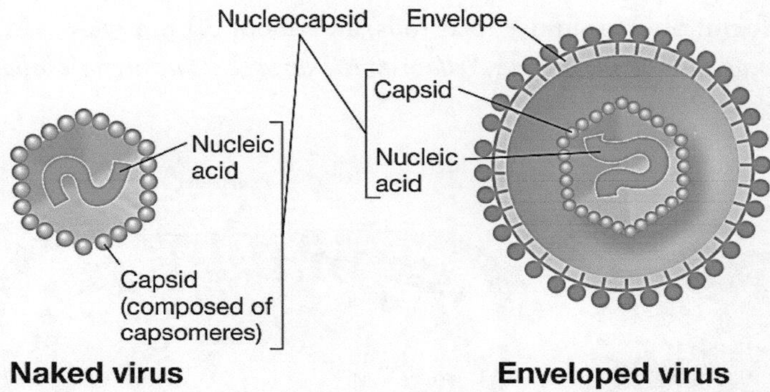

» Virus-like agents classified and studied with viruses

» **Viroid:** Viroids consist solely of a single molecule of circular RNA without a protein coat or envelope

» **Satellite Virus** (Kasanis, 1966)**:** The plant viruses, which are unable to replicate without the help of another virus are called satellite virus or defective virus. Satellite virus has a non-functional protein coat of its own but a small RNA genome. It needs another virus to supply enzymes. It acts as a parasite of plant parasitic virus. A good example is *Tobacco necrosis satellite virus* (sTNV), which has a small piece of ssRNA which codes only for a capsid protein, and depends for its replication on the presence of TNV.

» **Satellite Nucleic Acids** (RNA): Certain viruses have associated with the nucleic acids that are dispensable in that they are not part of the genome, which have no (or very little) sequence similarity with the viral genome, yet depend on the virus (helper virus) for replication, and are encapsidated by the virus. These are mainly associated with plant viruses and are generally ssRNA, both linear and circular. E.g., *Cucumber mosaic virus, Tobacco ring spot virus*

» **Prions:** Prion is infectious protein particle that are composed solely of protein, i.e. they contain no detectable nucleic acid.

Nucleoprotein has two basic structure types

Helical: Rod shaped, varying widths and specific architectures; no theoretical limit to the amount of nucleic acid that can be packaged. E.g., *Tobacco mosaic virus*

a. **Rod shaped:** About 20-25 nm in diameter and 100 to 300 nm long. Appear rigid and often have a clear central canal. Some viruses have two or more different lengths of particle and these contain different genome components. E.g., *Tobacco mosaic virus*, *Barley stripe mosaic virus*, *Papaya mosaic virus*, *Barley yellow mosaic virus*, *Wheat streak mosaic virus*.

b. **Bacilliform:** Short round-ended rods, and about 30 nm wide and 300 nm long. E.g., *Cocoa swollen shoot virus*, *Yellow mottle virus*, *Lettuce necrotic yellow virus*, *Potato yellow dwarf virus*.

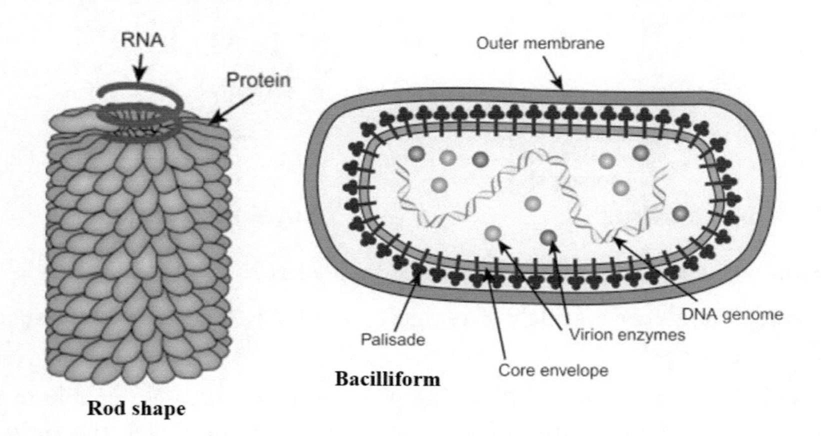

Rod shape **Bacilliform**

Cubical (Icosahedral): Spherical, amount of nucleic acid that can be packaged is limited by the particle. E.g., *Tomato bushy stunt virus*

a. **Isometric:** Spherical (depending on the species) from about 18 nm in diameter. E.g., *Tobacco necrosis virus*, *Tomato spotted wilt virus*, *Maize chlorotic dwarf virus*, *Turnip yellow mosaic virus*, *Cucumber necrosis virus*, *Potato leaf roll virus*, *Barley yellow dwarf virus*.

b. **Filamentous:** Usually about 12 nm in diameter and more flexuous than the rod-shaped particles. They can be up to 1000 nm long, or even longer in some instances. E.g., *Potato virus Y*, genus *Potyvirus* with particles 740 nm long.

c. **Geminate:** Twinned isometric particles about 30 × 18 nm. These particles are diagnostic for viruses in the family *Geminiviridae* which are widespread in many crops. E.g., *Maize streak virus*, *Bhendi yellow mosaic virus*.

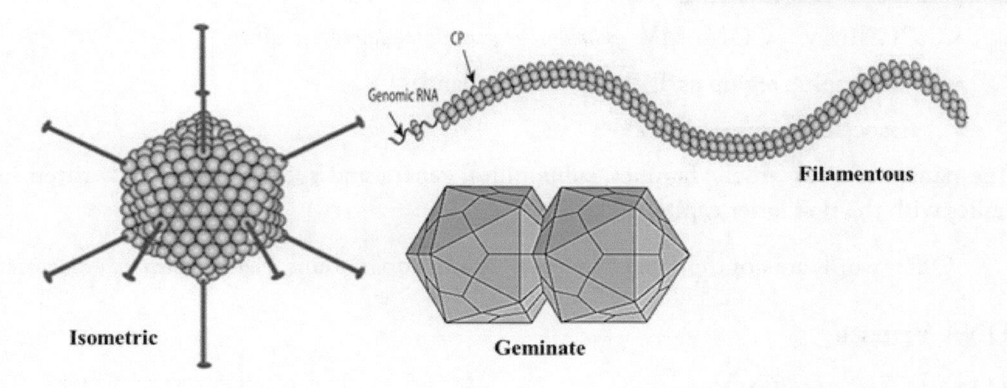

Classification of plant viruses

- According to **ICTV (International Committee on Taxonomy of viruses)** system of classification, basically viruses can be classified based on the nature of their genetic material, i.e., DNA or RNA.
- Presently in plant viruses, 27 groups have been identified.
- Among these, two groups (families), *viz.*, Rhabdoviridae and Reoviridae, can infect plants as well as animals and arthropods.

ICTV Recommendations & Guidelines

The designation of abbreviations is based on

- Abbreviation should be as simple as possible
- It must not be duplicate
- The word 'virus' in a name is abbreviated as 'V'
- The word 'viroid' as 'Vd'
- Mosaic – M; Mottle – Mo - E.g., *Cowpea mosaic virus* CPMV; *Cowpea mottle virus* – CPMoV
- Ringspot – 'RS' in many cases, sometime 'R'
- Symptomless – 'SL' in many cases, sometime 'S'
- The second / third lr in lower case represent host: CdMV – *Cardamom mosaic virus*
- Sometime as two capitals as exception: CPRMV – *Cowpea rugose mosaic virus*
- Abbreviation for single word should not exceed 2 lrs
- Secondary lr in abbreviation are omitted when their use would make the abbreviation excessively long, generally more than 5 lrs

- CGMMV – CGMoMV - *Cucumber green mottle mosaic virus*
- Geographic region as TYLCV-Th (Thailand)
- Associated virus – GLRaV-3

The names of virus orders, families, subfamilies, genera and species should be written in italics with the first letter capitalized.

Other words are not capitalized unless they are proper nouns, E.g., *Tobacco mosaic virus.*

DNA Viruses

1. Double stranded DNA viruses

   - Isometric particles - E.g., *Caulimovirus (Cauliflower mosaic virus)*

2. Single stranded DNA particles

   - Geminate (twin) particles - E.g., *Geminivirus (Maize streak virus)*
   - Single isometric particles - E.g., *Banana bunchy top virus*

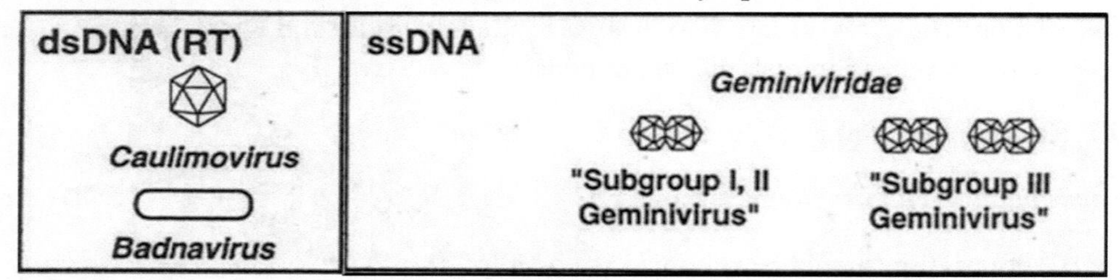

RNA Viruses

1. Double stranded RNA viruses

E.g., *Rice dwarf virus, Rice ragged stunt virus,* and *Sugarcane fiji virus*

2. Single stranded RNA viruses

   - Rod shaped particles - E.g., *Tobamovirus (Tobacco mosaic virus)* and *Furovirus (Potato mop top virus)*
   - Filamentous particles - E.g., *Potexvirus (Potato virus - X)* and *Potyvirus (Potato virus - Y)*
   - Isometric particles - E.g., *Waikavirus (Rice tungro virus), Comovirus (Cowpea mosaic virus), Tospovirus* (*Tomato spotted wilt virus*), *Nepovirus (Grape fan leaf virus)* and *Potato leaf roll virus.*

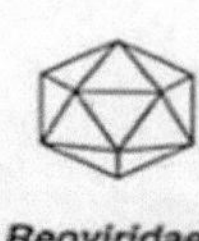
dsRNA
Reoviridae
Phytoreovirus
Fijivirus
Oryzavirus

Partitiviridae
Alphacryptovirus
Betacryptovirus

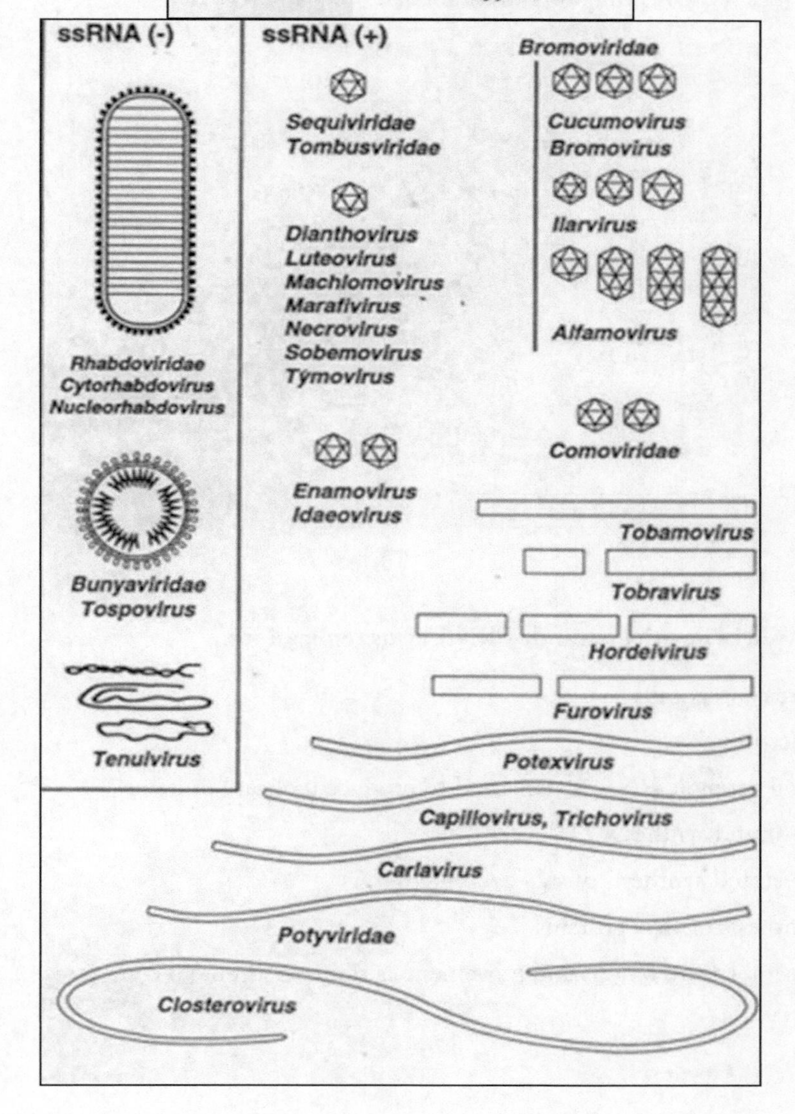
ssRNA (-)
Rhabdoviridae
Cytorhabdovirus
Nucleorhabdovirus
Bunyaviridae
Tospovirus
Tenuivirus
ssRNA (+)
Sequiviridae
Tombusviridae
Dianthovirus
Luteovirus
Machlomovirus
Marafivirus
Necrovirus
Sobemovirus
Tymovirus
Bromoviridae
Cucumovirus
Bromovirus
Ilarvirus
Alfamovirus
Comoviridae
Enamovirus
Idaeovirus
Tobamovirus
Tobravirus
Hordeivirus
Furovirus
Potexvirus
Capillovirus, Trichovirus
Carlavirus
Potyviridae
Closterovirus

Replication

Virus infection

Virus enters into the plant cell through wounds made mechanically or by vectors or naturally by deposition into an ovule by an infected pollen grain. Once the virus enters inside the host, it becomes naked by shedding its capsid or coat.

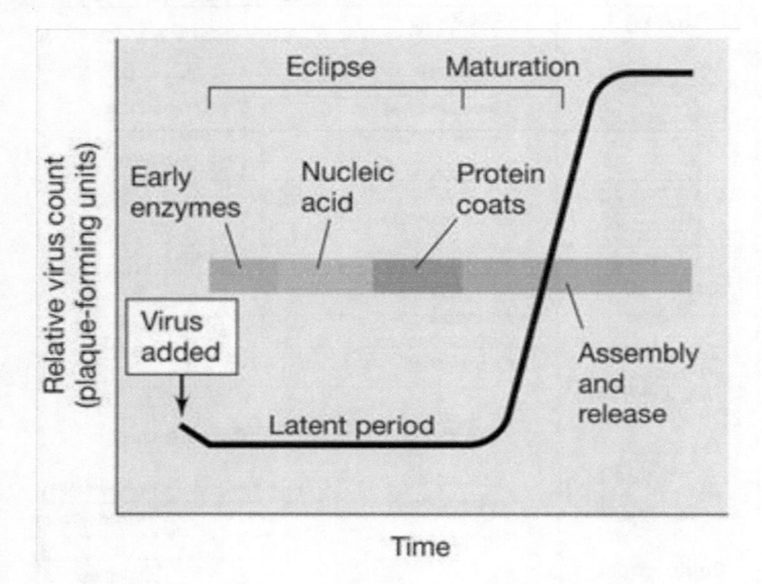

Steps involved in positive-stranded RNA virus replication:

1. Virus enters cells
2. Uncoating
3. Viral genomic RNA is translated to produce replicase proteins
4. (-)-strand synthesis
5. (+)-strand synthesis of sub-genomic RNAs
6. Synthesis of viral proteins
7. Assembly into virions and movement as ribonucleoprotein complexes.

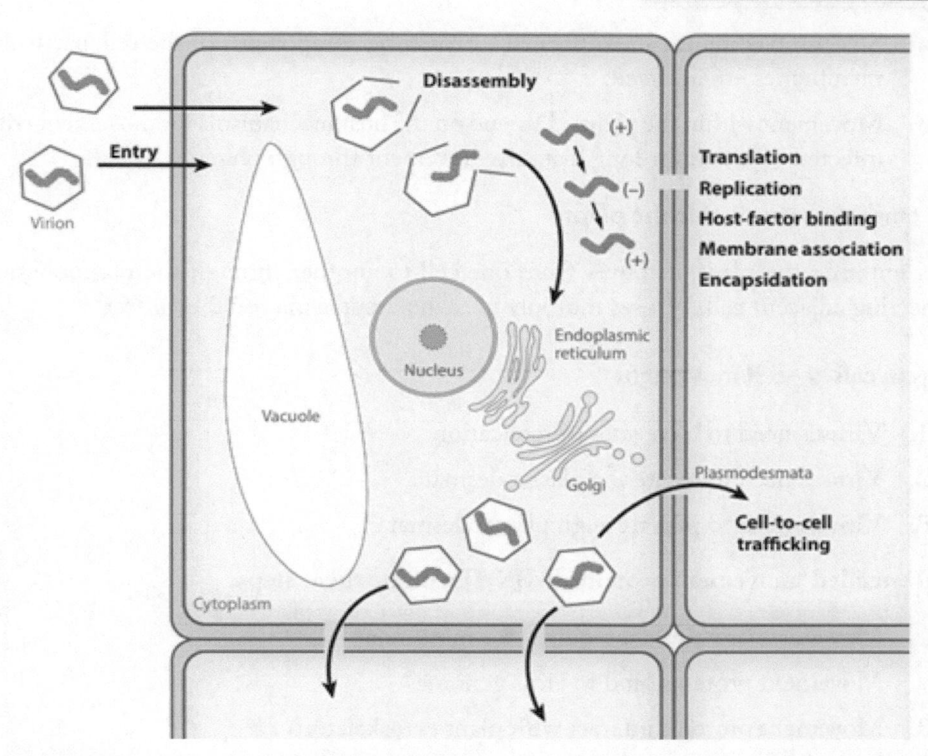

Replication of positive-sense ssRNA viruses in plants: The hexagon represents an icosahedral virion with a genome of one ssRNA. Positive-sense RNA is represented by (+), and the negative-strand synthesized during replication is represented by (–).

Host Function used by virus

- Plant viruses intimately dependent on the activities of the host cell for many aspects of replication.
- Components for viral synthesis - Use amino acids and nucleotides synthesized by host cell metabolism to build viral proteins and nucleic acids.
- Energy - The energy for polymerization process in viral protein and RNA synthesis is provided by host cell (nucleoside triphosphate).
- Protein synthesis - Viruses use the ribosomes, tRNAs and enzymes & factors of host cells' protein synthesizing system.
- Nucleic acid synthesis - Viruses code for an enzyme / enzyme involved in synthesis of their nucleic acids. But they may not contribute all of the polypeptides involved in genome replication.

» Structural components of the cell - Structural components of the cell, particularly membranes are involved.

» Movement within the plant - Depend on the host mechanisms for movement within infected cells and for long distance movement through phloem.

Movement of virus inside the plants

After entering the plant, it moves from one cell to another through the plasmodesmata connecting adjacent cell. Viruses multiply in each parenchyma cell they infect.

Steps in cell-to-cell movement

1. Viruses need to leave sites of replication
2. Viruses need to locate the plasmodesmata
3. Viruses need to pass through plasmodesmata

Viral encoded "movement proteins (MPs)" facilitate these steps.

1. Movement proteins are required for movement
2. Movement proteins bind to virus genomes
3. Movement proteins interact with plant cytoskeleton
4. Movement proteins localize to plasmodesmata
5. Movement proteins gate plasmodesmata

Transmission of plant viruses

The viruses are infectious and capable of being transmitted from diseased to susceptible healthy plants.

The viruses may be transmitted in the following methods.

» Transmission through seeds or seed materials

» Transmission by budding or grafting

» Transmission by sap or mechanical transmission

» Transmission by insects

» Transmission by mites

» Transmission by fungi

» Transmission by nematodes

» Transmission by phanerogamic parasites

1. Transmission through seeds or seed materials

- **Seed:** *Urdbean leaf crinkle virus, Tobacco ring spot, TMV on tomato, Soyabean mosaic virus* etc.,
- **Seed material:** *Potato viruses, Sugarcane mosaic virus, Citrus triszteza virus, Cassava mosaic virus, Banana bunchy top virus* etc.,

2. Budding or grafting

- Most of the viruses are being transmitted either by budding or grafting. Only grafting can transmit viruses like Potato witches' broom.

3. Sap or mechanical transmission

- Generally, mosaic group of viruses are readily transmitted by sap inoculation. E.g., TMV

4. Transmission through insects

- More than 80 per cent of the viral diseases are spread by different types of insects. The insect, which act as specific carriers in disseminating the diseases, are called insect vector.

A. Aphids: Aphids are the most important insect vectors of plant viruses and transmit the great majority of all stylet-borne viruses.

- *Aphis gossypii: Bean mosaic virus, Cucumber mosaic virus*
- *A. craccivora*: *Bean common mosaic virus*
- *A. maydis: Sugarcane mosaic virus*
- *Pentalonia nigronervosa: Banana bunchy top virus*, Katte disease of cardamom
- *Toxoptera citricidus*: *Citrus tristeza virus*
- *Myzus persicae*: *Potato virus Y*
- *Raphalosiphum maidis*: *Sugarcane mosaic virus*

B. Leafhoppers: Leafhoppers are phloem feeders and acquire the virus from the phloem region. All leafhopper transmitted viruses are circulatory. Several of these viruses multiply in the vector (propagative) and some persist through the moult and are transmitted through the egg stage of the vector.

- *Nephotettix cincticeps: Rice dwarf virus*
- *N. virescens, N. nigropictus: Rice tungro virus*
- *Nilaparvata lugens*: *Rice grassy stunt virus, Rice ragged stunt virus* etc.,

C. Plant hoppers: *Maize mosaic virus, Rice hoja blanca virus, Sugarcane Fiji virus*

D. White flies (*Bemesia tabaci*): *Bhendi yellow vein mosaic virus, Cassava mosaic virus, Chilli leaf curl virus, Papaya leaf curl virus, Tobacco leaf curl virus* etc.,

E. Thrips: *Frankliniella schultzei, Scirtothrips dorsalis, Thrips tabaci: Tomato spotted wilt virus* and *Tobacco streak virus*

F. Mealybugs: *Planococcoides citri*: *Cocoa swollen shoot virus* and *Banana streak virus*

G. Grasshoppers: *Potato virus X, Tobacco mosaic virus*

H. Lady bird beetle: *Epilachna* sp.: *Cowpea mosaic virus*

5. Transmission by Mites: *Aceria cajani*: *Pigeon pea sterility mosaic virus*

6. Transmission by Fungi:

- » *Olpidium brassicae*: *Tobacco necrosis virus, Lettuce big vein virus*
- » *Synchytrium endobioticum*: *Potato virus X*

7. Transmission by Protozoa

- » *Spongospora subterraneae*: *Potato mop top virus*
- » *Polymyxa graminis: Soil borne wheat mosaic virus*
- » *Polymyxa betae: Beet necrotic yellow vein virus*

8. Nematode transmission

a. NEPO – Polyhedral virus

- » *Xiphinema index - Grapevine fan leaf virus*
- » *Longidorus* sp.- *Tobacco black ring virus*

b. NETU – Tubular virus

- » *Trichodorus, Paratrichodorus – Tobacco rattle virus, Pea early browning virus*

9. Transmission by parasitic phanerogams

- » The parasitic dodder *Cuscuta californica, C. subinclusa, C. campestris* have been reported to transmit virus diseases.
- » The virus passes through food material drawn by the parasite and reaches the healthy plants and infects them.

E.g., *Sugarbeet curly top virus* and *Cucumber mosaic virus* have been transmitted by this method.

Vector-borne viruses

S.No	Name of Disease	Vector
Aphids		
1	Potato virus Y (Poty virus)	*Myzus persicae*
2	Potato leaf roll virus (PLRV)	*Myzus persicae*
3.	Bean common mosaic virus	*Aphis craccivora*
4.	French bean mosaic virus	*Aphis craccivora, A. gossypii, Myzus persicae*
5.	Chilli mosaic virus	*Aphis craccivora, A. gossypii*
6.	Cucumber mosaic virus	*Aphis craccivora, A. gossypii*
7.	Papaya ring spot virus (PRSV)	*A. gossypii, Myzus persicae*
8.	Katte disease of cardamom	*Pentalonia nigronervosa*
9.	Banana bunchy top virus	*Pentalonia nigronervosa*
10.	Alfaalfa mosaic virus	*Acryrthosiphon pisum*
11.	Lettuce mosaic virus	*Myzus persicae*
12.	Soybean mosaic virus	*Aphis gossypii*
13.	Sugarcane mosaic virus	*Rhopalosiphum maidis*
14.	Watermelon mosaic virus	> 38 sp of Aphids
15.	Zucchini yellow mosaic	> 4 sp of Aphids
16.	Citrus tristeaza virus (Citrus quick decline)	*Toxoptera citicidus*
17.	Groundnut rosette virus	*Aphis craccivora*
18.	Pea enation mosaic virus	*Acrythosphum pisum*
19.	Cauliflower mosaic virus	*Myzus persicae*
20.	Grassy shoot of sugarcane (Phytoplasma)	*Rhopalosiphum maidis*
Whitefly		
21	Cotton leaf curl virus	*Bemisia tabaci*
22.	Yellow mosaic of legumes	*Bemisia tabaci*
23.	Yellow vein mosaic of bhendi	*Bemisia tabaci*
24	Leaf curl of cotton, Tobacco, Tomato, Chilli, Papaya, etc	*Bemisia tabaci*
25.	Cassava mosaic virus	*Bemisia tabaci*
26.	Cowpea mild mottle virus	*Bemisia tabaci*
27	Tomato yellow leaf curl virus	*Bemisia tabaci*
28.	Squash leaf curl virus	*Bemisia tabaci*

29.	Bean golden mosaic virus	*Bemisia tabaci*
30.	Horse gram yellow mosaic	*Bemisia tabaci*
Leaf hoppers and plant hoppers		
31.	Rice tungro virus	*Nephotettix virescens*
32	Rice yellow dwarf (Phytoplasma)	*Nephotettix virescens*
33.	Rice stripe virus	*Loadelphax striatellus*
34.	Rice dwarf virus	*Nephotettix impecticeps*
35.	Rice grassy stunt virus	*Nilaparvatha lugens*
36.	Rice ragged stunt virus	*Nilaparvatha lugens*
37.	Sandal spike (Phytoplasma)	*Nephotettix virescens, Moonia albimaculata*
38.	Little leaf of brinjal (Phytoplasma)	*Hishimonas phycitis*
39.	Sesamum phyllody	*Orosius albicinctus*
40.	Grassy shoot of sugarcane (Phytoplasma)	*Proutista moesta*
41.	Coconut root wilt / Kerala wilt (Phyotplasma)	*Proutista moesta*
42.	Citrus stubborn (Spiroplasma)	*Circulifer tenellus*
43.	Corn stunt (Spiroplasma)	*Dulbulus maidis*
Beetles		
44.	Potato virus X	Epilachna ocellata
45.	Urdbean leaf crinkle virus	Henosepilachna & Epilachna
46.	Bottle gourd mosaic virus	Red pumpkin beetle
47.	Squash mosaic virus	*Diabrotica undecimpunctata*
48.	Bud rot of coconut (*Phytophthora palmivora*)	Rhinoceros beetle (*Oryctes rhinoceros*)
49.	Bacterial wilt of cucumber	Cucumber beetles
50	Fire blight of apple (*Erwinia amylovora*)	Bees, wasps, flies, beetles, etc
51.	Dutch elm disease	*Scolytus scolytus*
Mealy Bug		
52.	Cacao swollen shoot virus	*Planococcoides njalensis*
53.	Grapevine leaf roll virus	Planococcoides njalensis
54	Banana streak virus	Planococcoides njalensis
Psyllids		
55.	Citrus greening (Fastidious vascular bacteria)	Diaphorina citri

Thrips		
56.	Tomato spotted wilt virus, Groundnut bud necrosis virus	*Thrips tabaci*
57.	Tobacco streak virus, Sunflower necrosis virus	*Franklinella occidentalis*
58.	Tobacco ring spot virus	*Thrips tabaci*
Mites		
59.	Pigeon pea sterility mosaic virus	*Aceria cajani*
60.	Wheat streak mosaic virus	*Aceria tulipae*
Fungus		
61.	Lettuce big vein mosaic virus	*Olpidium brassicae*
62.	Wheat mosaic virus	*Polymyxa graminis*
63	Potato virus X	*Synchytrium endobioticum*
64.	Potato mop top virus	*Spongospora subterranean*
Nematode		
65	Grapevine fan leaf virus	*Xiphnema index*
66.	Tobacco rattle virus (Tobra / NETU)	*Trichodorus*
67.	Tobacco ring spot virus	*Longidorus*
68.	Tundu disease	*Anguina tritici*

Virus -vector relationship

Viruses achieve transmission by interacting directly with their insect vectors and benefitting from their feeding behaviors/activities (vector feeding) via a continuum of three inter-related transmission processes such as acquisition, retention and inoculation. Virus, vector, and plant undergo bilateral interactions and trilateral interactions involving all three partners (represented by the entire triangle) to influence various aspects of the transmission processes.

Bi-lateral and trilateral interactions can also trigger plant cues and/or vector responses that indirectly influence virus transmission by eliciting specific vector responses that is, attraction, deterrence and dispersal (indicated outside the triangle). Specific virus and/ or vector triggered plant resistance/defense responses may also indirectly influence virus transmission by targeting/inhibiting the virus and/or the vector.

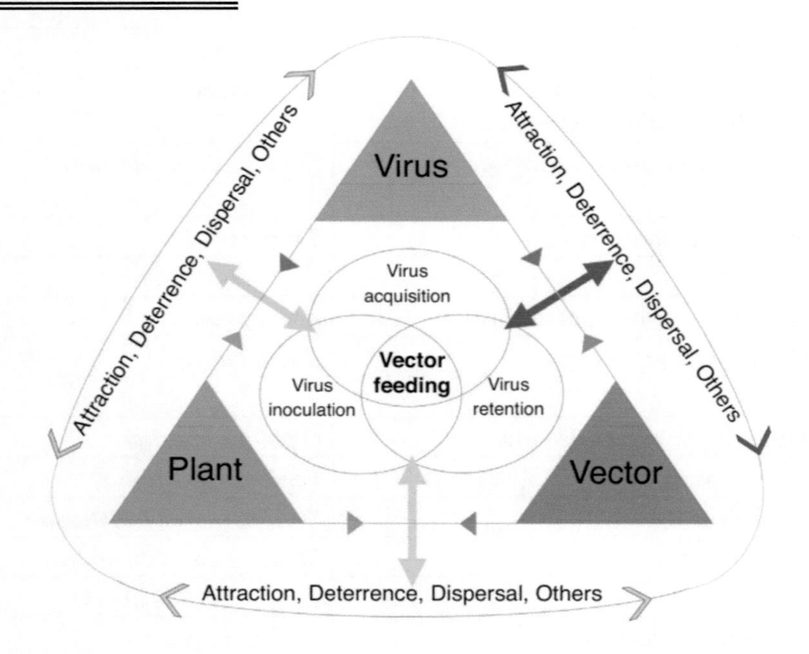

- » Acquisition feeding period: It is the time from which a virus free vector actually feeds on a virus infected plant to acquire the virus.
- » Inoculation feeding period: The actual period of feeding to inoculate the virus is called as inoculation feeding period.
- » Viruliferous vector: Ability of the vector to transmit the virus. Any organism that carries disease causing pathogens is called vector.
- » Latent period: The period from acquisition of virus by the vector till the time when the vector becomes capable of infecting healthy plant with the virus.
- » Incubation period: The time between penetration of a host by a pathogen and the first appearance of disease symptoms.
- » Indexing: Testing of a plant for infection, often by mechanical transmission or by grafting tissue from it to an indicator plant.

On the basis of length of time for which the insect vector remains viruliferous, they are classified into

1. Persistent viruses (circulative viruses / propagative viruses)
2. Semi-persistent viruses (non-circulative / non propagative)
3. Non-persistent viruses (non-circulative)

1. Persistent viruses (Circulative/ propagative viruses)

- These viruses are highly specific and have intimate biological relationship with the vector.
- Immediately after acquiring, these viruses are circulated through alimentary canal, gut wall and body fluid of the insect and hence called as circulative viruses. Sometimes they multiply inside the vector body and hence named as circulative propagative viruses. Certain viruses are carried to the progeny of vector such as eggs and are called as transovarial transmission. E.g., *Rice dwarf virus* is transmitted to eggs of leaf hopper, *Nephotettix cincticeps*.
- They have a latent period in vector body and persist for long.
- Moulting has no effect on persistence of such viruses.
- They have long acquisition and inoculation feeding period.
- Usually present in phloem cells of host plant.

2. Semi-persistent viruses

- They are intermediary between non-persistent and persistent viruses.
- Basically, these viruses are non-persistent because they do not circulate within the vector. But their vector retains their ability to transmit the virus (persistency) for 3-4 days. The virulence is lost at moulting.
- Like persistent viruses, semi-persistent viruses are usually associated with phloem cells. These viruses also show greater vector specificity than non-persistent viruses.
- They have long acquisition feeding period but become viruliferous without latency.

3. Non-persistent viruses

- Virus is acquired by the insect just after feeding on the infected plant i.e. a few seconds and the virus is transmitted immediately. They are retained only for a short period in the body of vector.
- The acquisition and inoculation feeding period is very short (in seconds) and do not have any latent period in the body of vector.
- Viruses are transmitted mechanically and are lost by the vector during moulting.
- No vector specificity and individual viruses are usually transmitted by several aphid species.
- Most of the non-persistent viruses are transmitted by aphids.
- E.g., Aphids such as *Myzus persicae, Aphis craccivora, A. gossypii* transmits several viruses.

Some features of virus-vector relationships

Features	Types of persistence of virus in vector		
	Non-persistent (Stylet-borne)	Semi-persistent (Stylet-borne)	Persistent circulative/ non propagative
Pre-acquisition fasting	Enhances chances of acquisition	No such effect	No such effect
Tissue of acquisition	Epidermis	Epidermis, mesophyll	Mesophyll, phloem
Mechanical transmission	Yes	No	No
Tissue of virus acquisition	Epidermis	Mesophyll/phloem	Mesophyll/phloem
Minimum acquisition time	Seconds to minutes	Minutes to hours	Hours to days
Latent period	Nil	Nil	12 hours or more
Period for which vector remains infective (Retention time)	Usually less than 4 hours but can be up to 10 hours	10-100 hours	More than 100 hours
Effect of moulting	Stops transmission	Stops transmission	Transmission not affected
Multiplication of virus in vector	No	No	Yes, in some vectors
Retention of virus through moulting	No	No	Yes
Casting effect on transmission	Yes	No	No
Vectors	Aphids, mites, chytrids	Aphids, beetles, leaf hoppers, mealy bugs, white flies	Aphids, beetles, bugs, fungi, leafhoppers, mites, nematodes, thrips, white flies
Location of virus in vector	Externally or near mouth parts	Near junction of stylet and tip	Internally in haemolymph
Virus-vector specificity	Mostly low	Intermediate	Mostly high
Common symptoms	Mosaic	Yellowing	Leaf roll, yellowing
Some examples of plant viruses (aphid-transmitted)	*Bean yellow mosaic virus, Soybean mosaic virus* and *Cowpea mosaic virus*	*Cauliflower mosaic virus* and *Citrus tristeza virus.*	*Potato leaf roll virus* and *Sugarbeet curly top virus*

Symptoms of Virus Diseases

- Systemic infection: The virus is present throughout the plant and the symptoms produced are called systemic symptoms.
- Local infection: The virus causes the formation of small, chlorotic or necrotic lesions only at the points of entry and the symptoms are called local lesions.
- Latent viruses: Viruses infect hosts without causing development of visible symptoms on them and the hosts are called symptomless carriers.
- Plants that usually develop symptoms on infection with a certain virus may remain temporarily symptomless under certain environmental conditions (E.g., high or low temperature) and such symptoms are called masked.

External Symptomatology

Mosaic	Leaves show various shades of green and yellow area usually irregularly angular but sharply delimited. E.g., *Tobacco mosaic virus, Cucumber mosaic virus* and *Cowpea mosaic virus*
Vein band	A broader band of chlorotic tissue along the veins. E.g., *Strawberry vein banding virus* and vein banding virus on tobacco.
Mottle	Diffusely bordered variegation. E.g., *Peanut mottle virus*
Ring Spot	Infected leaves show rings or circular spots. E.g., *Tobacco ring spot virus, Tomato ring spot virus*, *Papaya ring spot virus*
Streak	Mosaic in leaves parallel to the veins. E.g., *Maize streak virus, Wheat streak mosaic virus, Banana streak virus*
Stripe	Sharply mosaic parallel to veins looking like stripes E.g., *Barley stripe mosaic virus*
Line pattern	Systemically infected leaves may show line pattern. E.g., *North American Plum line pattern virus*
Chlorosis	Colour may vary and distributed throughout organs or entire plants in which less chlorophyll is produced. E.g., *Raspberry vein chlorosis virus*
Yellows	Yellow pigments may predominate. E.g., *Solanum yellows virus, Rice transitory yellowing virus, Beet yellow virus.*
Vein chlorosis	Chlorosis will be restricted to tissues adjoining the veins. E.g., *Raspberry vein chlorosis virus*
Vein clearing	In systemically infected leaves, clearing or chlorosis of the tissue in or immediately adjacent to the veins occurs. This pattern is called as vein clearing. E.g., *Bhendi vein clearing virus*

Reddening	Red pigments predominate. E.g., *Carrot red leaf virus.*
Browning	Rapid cell death leads to necrosis. Necrosis leads to browning of tissues. E.g., *Pea early browning virus*
Blackening	Intensive necrosis leads to blackening of tissues. E.g., *Potato black ring spot virus.*
Necrosis	Several types of necrosis are seen in virus infected tissues. Necrosis of specific organs is also common. E.g., *Rice necrosis mosaic virus, Muskmelon vein necrosis virus, Cocoa necrosis virus, Sunflower necrosis virus.*
Wilt	Plants wilt due to shortage of water and nutrients. E.g., *Tomato spotted wilt virus*
Etch	Desiccation of superficial tissue (epidermal cells) is called etching. E.g., *Tobacco etch virus*
Leaf rolling	Leaf distortion by rolling, curling and crinkling. E.g., *Tobacco leaf curl virus, Tomato yellow leaf curl virus, Mung bean leaf crinkle virus, Strawberry crinkle virus, Potato leaf roll virus, Cotton leaf curl virus, Carrot thin leaf virus, Beet curly top virus.*
Fan leaf	Grape vine fan leaf virus
Rugose	Retardation of veinal tissue. E.g., *Citrus leaf rugose virus, Bean rugose mosaic virus.*
Enation	Small outgrowth, on leaves, especially on veins. E.g., *Pea enation mosaic virus* and *Cotton leaf curl virus.*
Big vein	Veins become big. E.g., *Lettuce big vein virus.*
Rosette	Rosetting due to impeded internodal expansion at of stem. E.g., *Groundnut rosette virus.*
Tumours	Swellings on stems or roots. E.g., *Cacao swollen shoot virus, Wound tumour virus.*
Bud blight	Necrosis of buds. E.g., *Groundnut bud blight* and *Soybean bud blight*
Colour breaking	Mosaic or variegation of petals of flowers. E.g., *Tulip colour breaking virus*
Dwarf, Stunt	Retardation of plant growth. E.g., *Tobacco stunt virus, Tomato bushy stunt virus, Tomato yellow dwarf virus, Sorghum stunt mosaic virus, Potato yellow dwarf virus,* and *Maize rough dwarf virus.*
Sterility	Plants become sterile with mosaic in leaf. E.g., *Pigeonpea sterility mosaic virus*

Internal Symptomatology

Diagnosis by internal symptoms: Various changes in histology of virus infected plant parts especially leaves, petioles and stems are caused by virus infection. Such histological changes or abnormalities can be categorized in mainly three groups, namely.

» Necrosis or death of cells. E.g., *Tobacco necrosis virus*

» Hyperplasia or excessive growth. E.g., *Lettuce big vein virus*

» Hypoplasia or reduced growth differentiation of cell. E.g., *Rice dwarf virus*

Viroids

Viroids are small naked, circular low molecular weight ssRNA without coat protein capable of causing diseases in plants. They are pathogenic RNAs. These **'mini viruses'** are the smallest known agents of infectious diseases. They are subviral in size. The term **'viroid'** was first introduced by Diener in 1971.

They code for nothing but their own structure, and are presumed to replicate by somehow interacting with host RNA polymerase, and to cause pathogenic effects by interfering with host DNA/RNA metabolism and/or transcription.

General symptoms

» Viroids produce stunting, mottling, leaf distortion and necrosis.

» Degenerative abnormalities have been found in the chloroplasts of viroid-infected cells.

» Viroid infection caused marked change in the amounts of host proteins.

Transmission: Viroids are transmitted by mechanical means in most of their hosts. Transmission in the field may be by contaminated tools and by vegetative means.

Classification of Viroids

Genus	Species	Length (nt)	Natural host(s)
Family: *Pospiviroidae*			
Pospiviroid	Potato spindle tuber (PSTVd)	341-364	Potato, Tomato, Avocado
	Chrysanthemum stunt (CSVd)	348-356	Chrysanthemum
	Citrus exocortis (CEVd)	366-475	Citrus, Tomato
	Tomato apical stunt (TASVd)	360-363	Tomato

Hostuviroid	Hop stunt (HSVd)	294-303	Citrus, grapevine, *Prunus* spp., Hop, Cucumber
Cocadviroid	Coconut cadang-cadang (CCCVd)	246-301	Coconut palm, African oil palm, other monocots
	Hop latent (HLVd)	255-256	Hop
Apscaviroid	Apple scar skin (ASSVd)	329-333	Apple, Pear
	Australian grapevine (AGVd)	369	Grapevine
Family: *Avsunviroidae*			
Avsunviroid	Avocado sun blotch (ASBVd)	239-251	Avocado
Pelamoviroid	Chrysanthemum chlorotic mottle (CChMVd)	397-401	Chrysanthemum
	Peach latent mosaic (PLMVd)	335-351	Peach, Nectarine
Elaviroid	Eggplant latent (ELVd)	332-335	Eggplant

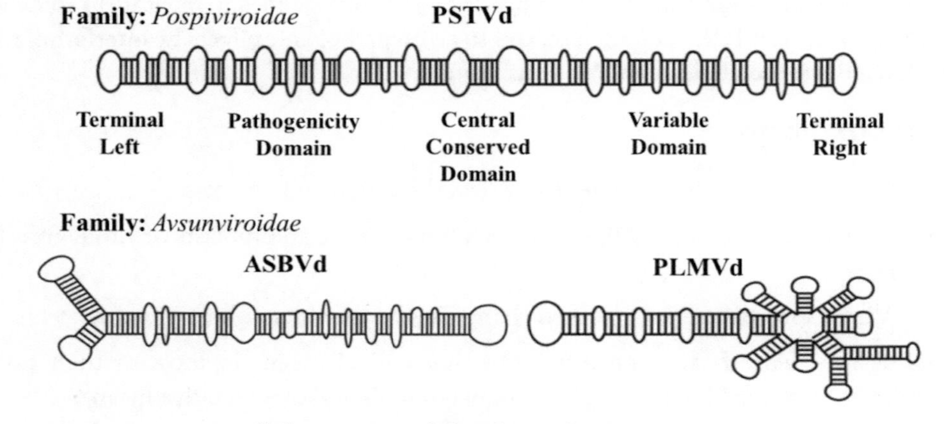

Schematic models of viroid structures

Major diseases

1. Coconut Cadang-cadang

- It was first described in Philippines
- Cadang-cadang is a lethal disease of coconut characterized by yellow-bronze colouration of the lower two-third of fronds in the crown, cessation of nut production, crown diminution and death.
- Leaflets become brittle. The crown of the palm dies. Generally diseased palms die between 8 and 16 years.

» No vector has been identified.

2. Potato spindle tuber

» Potato spindle tuber viroid (PSTVd) is the first recognized viroid disease in plants.

» Diseased plants are less vigorous, erect, and spindly dwarf. Leaves are small, erect, and dark green.

» Leaflets sometimes show rolling and twisting.

» The tubers are elongated with a cylindrical middle and tapering ends.

» PSTVd is mechanically transmitted by knives

Practice Questions

1. Coconut cadang-cadang is caused by………
2. Potato spindle tuber is caused by……
3. The composition of virus particles consists of………
4. Most of the plant viruses transmitted by…………
5. ……….mode of persistence and transmission is most viruliferous at field conditions
6. Write the differences between virus and bacteria
7. What are viruses? How do they differ from other plant pathogens?
8. Viruses do not generally disseminate in regions of high altitude. Explain why this is so and how this feature has been utilized in agriculture.
9. List some important viral diseases and their vectors in a tabular form.
10. Discuss the nature of the development of the concept of viruses.
11. How will you find out whether a disease is due to virus or due to nutritional deficiency?
12. What are viroids? Who discovered viroids?
13. Name some important diseases of viroids.
14. Give one example each of ssRNA (positive RNA), ssRNA (negative RNA), ssDNA and dsDNA viruses.
15. Name a few diseases of international importance, caused by viruses and viroids.
16. How can tissue culture techniques help in controlling some of the diseases caused by viroids?
17. Explain the virus vector relationship in plant disease incidence
18. Define viroid and its infectious nature

19. Difference between virus and viroid
20. Write the replication mechanism of virus
21. Give two example(s) each of plant virus transmission through fungi and nematode.
22. Attempt a brief note on stylet-borne transmission of Plant viruses.
23. Explain 'circulative- propagative transmission' of Plant viruses.
24. Attempt a brief note on transmission of plant viruses through vegetatively propagated plant parts.
25. Discuss approaches to prove multiplication of plant viruses within tissues of its insect vector.

Suggested Readings

Smith, K. M. (2012). *A textbook of plant virus diseases*. Elsevier.

Roossinck, M. J. (2013). Plant virus ecology. *PLoS pathogens*, *9*(5), e1003304.

Sutic, D. D., Ford, R. E., & Tosic, M. T. (1999). *Handbook of plant virus diseases*. CRC Press.

Garcia-Ruiz, H. (2019). Host factors against plant viruses. Molecular Plant Pathology, 20(11), 1588-1601.

Scholthof, K. B. G., Adkins, S., Czosnek, H., Palukaitis, P., Jacquot, E., Hohn, T., & Foster, G. D. (2011). Top 10 plant viruses in molecular plant pathology. *Molecular plant pathology*, *12*(9), 938-954.

Hull, R. (2013). *Plant virology*. Academic press.

Walkey, D. G. (2012). *Applied plant virology*. Springer Science & Business Media.

Wilson, C. R. (2014). *Applied plant virology*. CABI.

Hull, R. (2009). *Comparative plant virology*. Academic press.

Khan, J., & Dijkstra, J. (2006). *Handbook of plant virology*. Food Products Press.

Bowman, C. (2019). *Plant Virology*. Scientific e-Resources.

Mandal, B., Rao, G. P., Baranwal, V. K., & Jain, R. K. (Eds.). (2017). *A Century of Plant Virology in India*. Springer.

Chapter - 17

Phanerogamic Parasites

Certain flowering plants (Phanerogams) also parasitize the crop plants in addition to the microorganisms. They mostly belong to Loranthaceae, Convolvulaceae, Scrophulariaceae, Orabanchaceae, Lauraceae, Santalaceae and Balauophoraceae. They produce flowers and seeds and parasitize their host by drawing nutrition and water. Some phanerogams have green leaves, roots and they have the ability to synthesis food materials but they obtain only the mineral constituents of food from the host, then they are called hemiparasite/waterparasite/partial parasite. Some of the phanerogams which do not have any chlorophyll completely depend on host for water and all minerals. They are called as holoparasite or complete or total parasite.

The parasitic flowering plants of cultivated crops

S.No.	Parasite name	Type of parasitism	Major host plants
1.	Broomrape (*Orobanche* spp.)	Complete root parasite	Tobacco, egg plant, safflower, tomato, mustard, cumin etc.
2.	Witchweed (*Striga* spp.)	Partial root parasite	Sorghum, sugarcane, rice, kodo millet etc.
3.	Dodder (*Cuscuta* spp.)	Complete stem parasite	Onion, alfalfa, (lucerne), flax, clover, ornamentals etc.
4.	Love-vine (*Cassytha* spp.)	Complete stem parasite	Citrus, ornamentals
5.	Giant mistletoe (*Dendrophthoe* spp.)	Partial stem parasite	Mango, citrus, jack, pomegranate, neem, forest trees etc.

6.	Dwarf mistletoes (*Arceuthobium* spp.)	Partial stem parasite	Coniferous trees

The angiospermic parasites can also be classified as holoparasites (total parasites) or hemiparasites (semiparasites).

» The holoparasites lack chlorophyll and are totally dependent on the host for nutrition. Thus, they are obligate parasites.

» The hemiparasites contain chlorophyll and make their own food, and absorb water and minerals from their host. But, in some cases, e.g., Arceuthobium, the photosynthesis is negligible and the parasite draws nutrition from the host. Practically, it is an obligate parasite.

The phaneroganic plants are also divided as root parasites or stem parasites.

1. Stem parasite

Total/ Complete parasite – *Cuscuta campestris*

Partial/ Semi-parasite - *Dendropthae falcata*

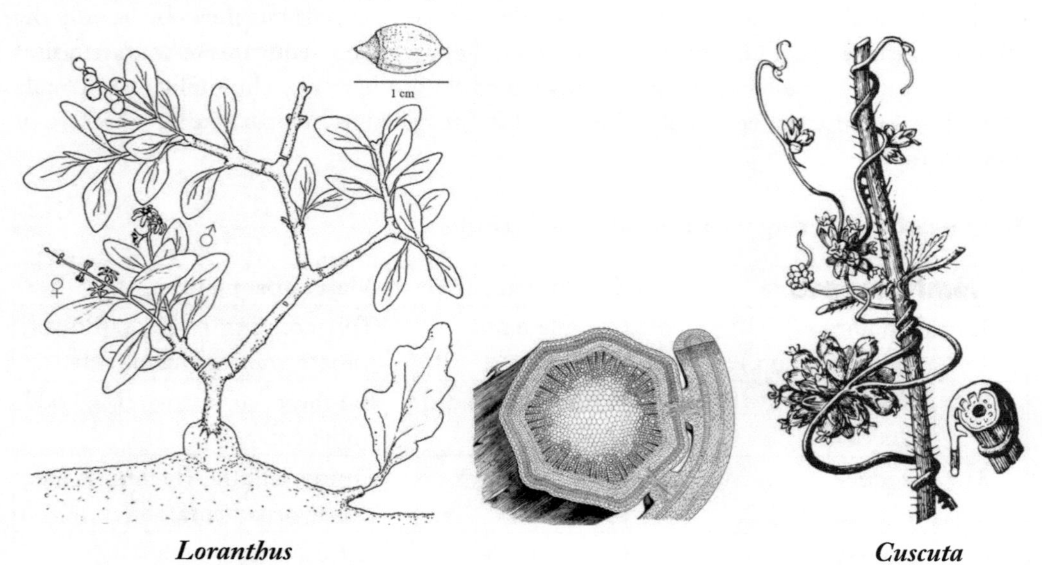

Loranthus ***Cuscuta***

2. Root parasite

Total/ Complete parasite – *Orabanche cernua*

Partial/ Semi-parasite - *Striga asiatica*

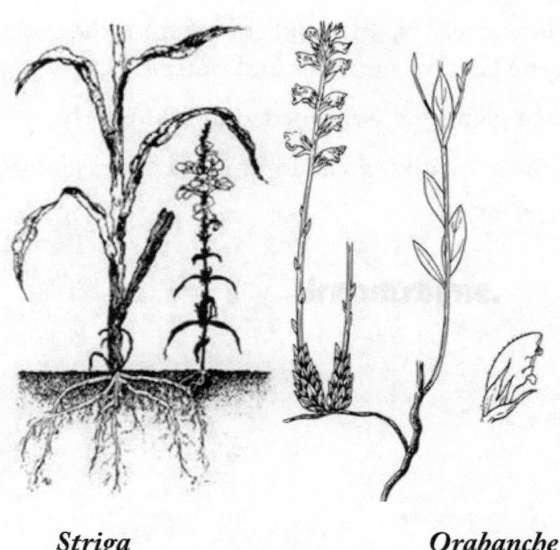

Striga *Orabanche*

Life cycle/ infection cycle of flowering parasite

Phanerogams have haustoria as absorping organ, which are sent deep into the vascular bundle of the host to draw water and nutrients. The haustoria in general secrete some pectolytic and cellulolytic enzymes which soften the host tissue. Haustoria have higher osmotic pressure than that of host tissue which facilitates easy absorption of nutrients. The affected plants show stunting, chlorosis and death.

a) Stem parasite:

i) Complete/ holo/ total parasite: Dodder - *Cuscuta* sp. *C. campestris, C. trifoli, C. planiflora, C. indicora*

» Commonly known as gold thread, hellvine, hair weed, devil's hair and love vine.

» Attacks alfalfa, clover, onion, flax, sugar beet, potato, chillies many ornamentals etc.

» It is a yellow or orange vine strands which grow and twin the plant. They do not have leaves but bear only very minute scale leaves. Dodder produces flowers and fruits. Flowers are white, pink or yellowish, which form seed.

» On severe infection, they form a dense and tangled mat on the crop.

» Seeds of dodder overwinter in the infested soil, germinate to produce a slender yellow shoot, make contact with the susceptible host plant, encircle and send haustoria into the vascular bundle of the host.

» It does not produce any roots. As soon as the dodder is established with the host, base of the dodder shrivels, dries and cut off from the ground. Thus, it completely depends upon the host for nutrients and water.

» Thus the affected plants get weakened and yield poorly.

» Seeds of cuscuta are mainly spread by animals, water and implements.

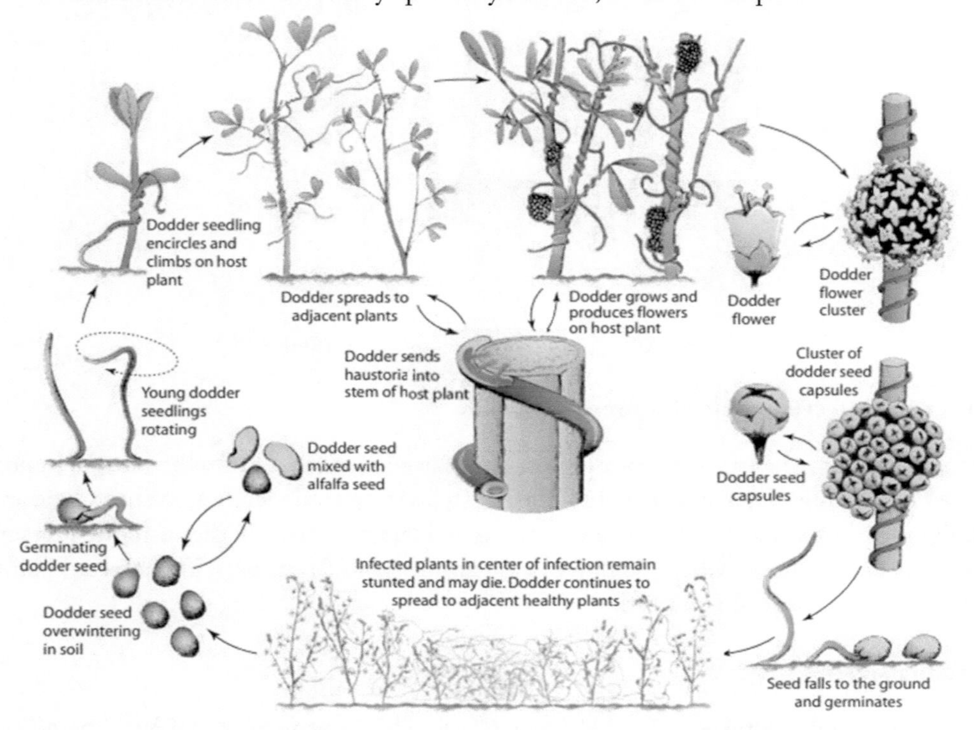

Life cycle of *Cuscuta* (Source: Agrios, 2005)

ii) Partial semi / hemi stem parasite: Commonly known Loranthus, Giant mistletoe, Banda., *Dendrophthae flacata* (Order: Santalales; Family: Loranthaceae).

» Attacks mango, citrus, apple, rubber, guava etc.

» Partial parasite of tree trunks and branches with brown stem, dark green leaves but no roots.

» Stem of the parasite is usually thick, and flattened at the node, appears in clusters at the point of attack which can be easily spotted on the trees.

» At the point of attachment with the tree, it shows swellings or tumourous growth where the haustoria are produced.

» This parasite produces flowers which are long, tubular, greenish white or red and borne in clusters.

- It produces fleshy fruits with single seed. The affected host plant becomes stunted in growth with few small chlorotic leaves.
- Dispersal of the seed is mostly through the birds and to some extend by animals.

Mistletoes are stem holoparasites occurring in three families of the order Santalales as follows: Family Loranthaceae: Showy mistletoes [Loranthus (*Dendrophthoe*)].

Family Santalaceae: sandalwood (*Pyrularia*, *Santalum*).

Family Viscaceae: Dwarf mistletoe (*Arceuthobium*), leafy mistletoe (*Viscum*).

- The showy mistletoes produce large and beautiful flowers that are pollinated by birds.
- The co-evolution of these parasites and the birds is also suggested by the seed dispersal mechanism operating in the birds.

iii) Root parasite / total/ holo/ complete parasite: Commonly known as Broom rape or *Tokra*.: *O. cernuva* var. *dessertorum, O. robancre ramosa, O. minor, O. crenata* (Order: *Orchidales,* Family: *Orabanchaceae)*

- It is a serious parasite in tobacco, tomato, brinjal, cabbage, cauliflower etc.
- It is an annual fleshy flowering plant growing to a height of about 10 - 15" with pale cylindrical stem, thickened at base and covered with brown scaly leaves that end in spikes.
- Plants lack chlorophyll, flowers arise from axils of the scale leaves.
- Flowers have well developed lobed calyx, tubular corolla, superior ovary, numerous ovules and large four lobed stigmata. Fruits are capsules containing small black reticulate and ovoid seed.
- Seeds remain dormant in the soil for many years and they germinate due to a stimulant (benzopyran derivatives) present in the root exudate of susceptible host plant
- Ethylene, gibberellin and coumarins also induce the seed germination.
- In tobacco it appears in clusters of 50 - 100 shoots around the base of a single plant 5 - 6 weeks after transplanting. Affected tobacco plants are stunted, show withering and drooping of leaves leading to wilting.

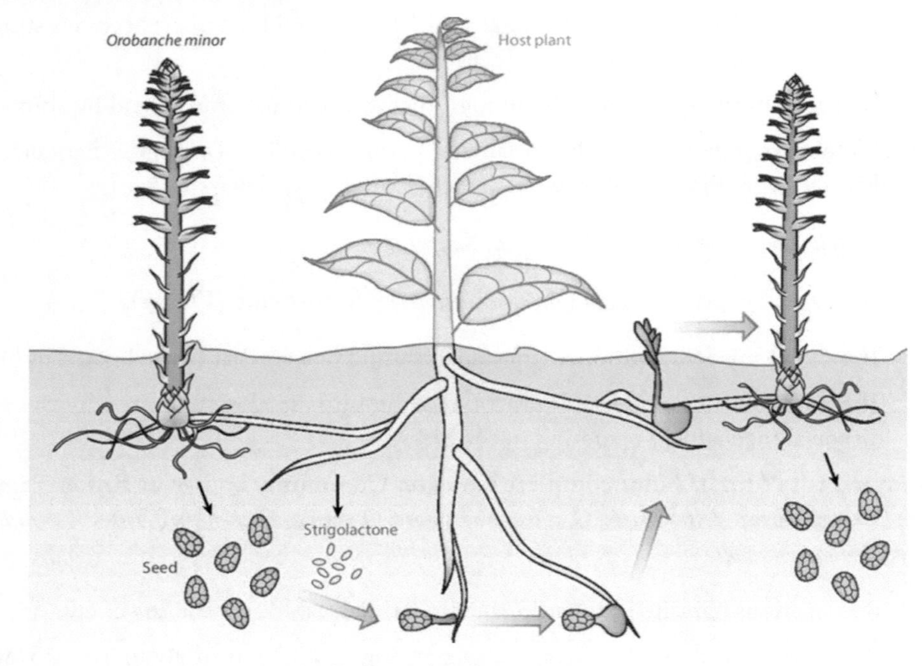

Life cycle of *Orabanche*

iv) Hemi/ partial/ semi parasite commonly known as witch weed or striga.

S. asiatica parasitize sorghum, maize and sugarcane

S. densiflora parasitize sorghum and sugarcane.

- » Mostly affect the monocots
- » It is a small plant with bright green leaves grows upto a height of 15 - 30 cm.
- » It occurs in clusters of 10 - 20/host plant.
- » *S. asiastica* produces pink flowers while *S. densiflora* produces white flower with a pronounced bend in corolla tube.
- » This phanerogam lack typical root hairs and root cap.
- » Fruits contain minute seeds in abundance which survive in soil for many years.
- » Seeds germinate after post-harvest ripening of about two weeks, in response to the host stimulant viz., strigol ethylene, cytokinin, gibberellin and couma in strigol.
- » This parasite slowly attaches to the host root by haustoria, grow below the soil surface and produce underground stem and roots for about 1-2 months. Then it grows
- » faster and appears at the base of the host plant.

» Severe infection of striga causes yellowing and wilting of host leaves. Sometime the host plant may die.

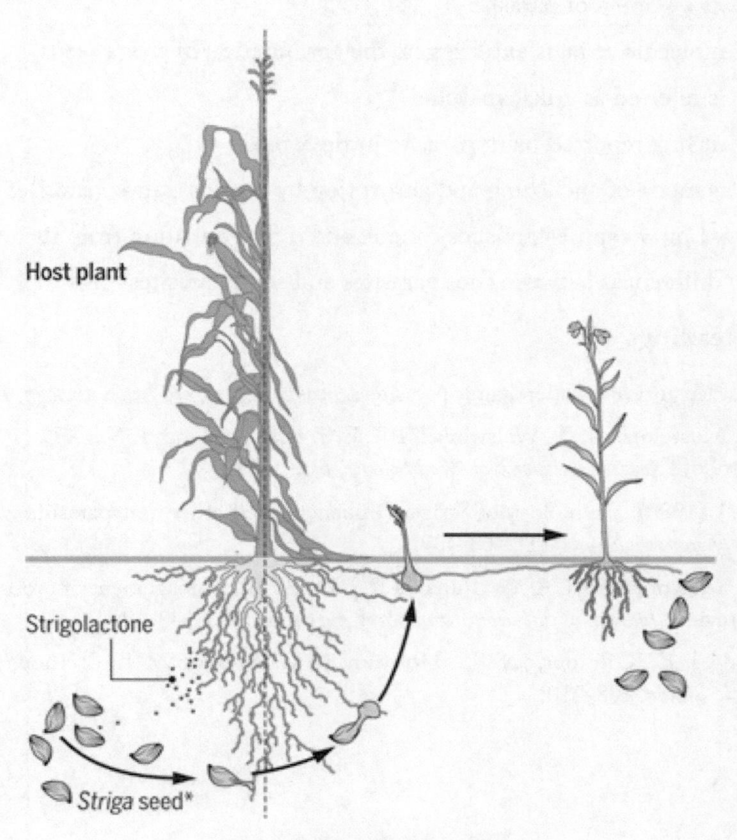

Life cycle of *Striga*

A Phanerogams parasitic plant is a plant that derives some or its entire nutritional requirement from another living plant. All parasitic plants have produce flowers and seeds and parasitize their host by drawing nutrition and water. Some phanerogams have green leaves, roots and they have the ability to synthesis food materials but they obtain only the mineral constituents of food from the host, then they are called hemiparasite/ waterparasite/ partial parasite. The haustoria in general secrete some pectolytic and cellulolytic enzymes which soften the host tissue. Mistletoes cause economic damage to forest and ornamental trees.

Practice Questions

1. ....is a semi-root parasite
2. ....molecule attracts and triggers the germination of striga seeds
3. ....is referred as vrikshabaksha
4. ....is first reported plant parasite in the world
5. ....is mode of anchoring and absorption by phanerogamic parasites
6. Write how *Cuscuta* spp. establishes contact and draws nutrition from its host?
7. Write the differences between root parasites and stem parasites?

Suggested Readings

Kuijt, J. (1977). Haustoria of phanerogamic parasites. *Annual Review of Phytopathology*, *15*(1), 91-118.

Dawson, J. H., Musselman, L. J., Wolswinkel, P. I. E. T. E. R., & Dörr, I. N. G. E. (1994). Biology and control of Cuscuta. *Reviews of Weed Science*, *6*, 265-317.

Musselman, L. J. (1980). The biology of Striga, Orobanche, and other root-parasitic weeds. *Annual review of phytopathology*, *18*(1), 463-489.

Habimana, S., Nduwumuremyi, A., & Chinama R, J. D. (2014). Management of orobanche in field crops: A review. *Journal of soil science and plant nutrition*, *14*(1), 43-62.

Balle, S., Dandy, J. E., Gilmour, J. S. L., Holttum, R. E., Stearn, W. T., & Thoday, D. (1960). Loranthus. *Taxon*, 208-210.

Chapter - 18

Algal Plant Parasite

Characters of Algal Parasite, *Cephaleuros*

» Algae are eucaryotic, unicellular or multicellular organisms and mostly occur in aquatic environments.

» Algal cells may be spherical, rod shaped, barrel shaped, club shaped or spindle shaped and the size ranges from 1.0 mm to many cm in length.

» The cell wall is thin and rigid. It surrounded by a flexible, gelatinous outer matrix secreted through the cell wall.

» They contain chlorophyll and are photosynthetic.

» They reproduce by asexual and sexual processes.

» There are three kinds of photosynthetic pigments in algae: chlorophylls, carotenoids and biloproteins (phycobilins)

» Asexual reproduction by production of motile (zoospores) or non-motile spores (aplanospores)

» Sexual reproduction by fusion of gametes (conjugation) or by oogamy.

Systematic position

Domain : Eukarya

Kingdom : Archaeplastida

Phylum : Chlorphyta

Class : Ulvophyceae

Order	: Trentepohliales
Family	: Trentepohliaceae
Genus	: *Cephaleuros*
Species	: *C. parasiticus, C. virescens*

The following are important species, which are parasitic on plants and cause red rust diseases.

Cephaleuros coffeae - Red rust of coffee

Cephaleuros virescens - Red rust of tea

Cephaleuros parasiticus - Red rust of tea, guava, mango, citrus

Red rust of mango/ guava - *Cephaleuros parasiticus*

Symptoms

It occurs on leaf and on stem as an epiphyte causing cracks and cankers. Fruit infection is also noticed.

Leaf infection

» Circular to irregular, yellowish green to orange, velvety raised hairy pustules appear on upper leaf surface. The rusty colour of the pustule may be due to presence of orange colour pigment (haematochrome) in the sporophore and sterile hairs.
» The matured spores fall off and leave cream to white velvety texture on the leaf surface.
» Affected leaves defoliate prematurely.
» The host cell layers below the algal thallus die and assume a corky appearance. They serve as barrier to further algal penetration.
» The pustules on stem are elongated and lead to the formation of scaly and cracked bark.
» The lesions coalesce often leading to girdling of the stem completely.
» The pustules on the fruit are initially dark grey but later turn brown and which gives ugly appearance to the fruits.

Pathogen

The algae produce a disc-like thallus which is multi-layered in the centre and single layered in the periphery. The cells have chlorophyll a, b and carotenoid as photosynthetic pigments and cell wall contains cellulose, xylan, mannan, etc. It grows between the cuticle and epidermis of the host and very rarely between epidermis and palisade cells. Rhizoids arise from the thallus extend to the underlying cells and act as anchoring as well as absorbing organ.

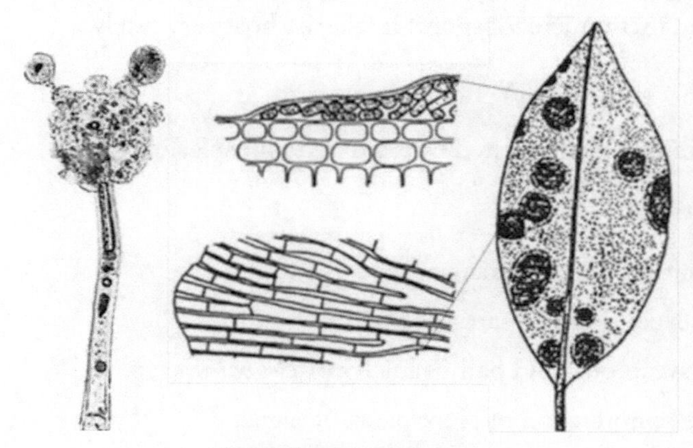

Cephaleuros virescens

Asexual reproduction

- » Algae produce many sterile hairs (setae) and sporangiophores (sporophores) on the thallus after rupturing the cuticle or epidermis.
- » The sporangiophores terminate in a swollen cell called 'vesicle'.
- » Usually four or sometimes eight orange color, unicellular oval to globose zoosporangium arise from vesicle through pedicel.
- » Besides, bigger sessile zoosporangia are also produced on the thallus.
- » The zoosporangia (pedicellate) are disseminated by wind and water and under favourable conditions; sporangia release 8 to 30 zoospores through an ostiole.
- » Zoospores are ovoid or spherical and biflagellate. The zoospores swim actively and when they reach the host cells, lose their flagella and cause infection.

Sexual reproduction

- » It undergoes sexual reproduction by isogametic copulation.
- » Large, sessile, flask shaped gametangia are produced in the thallus.
- » In the presence of water, 8-32 biflagellate zoospores are released from each gametangium. These swarm spores act as gametes.
- » Copulation takes place between two isogametes either from the same gametangium or from different gametangia resulting in the formation of zygote.
- » The zygote produces a dwarf sporophyte consisting of a stalk bearing one or

more cells. Each cell has a small dehiscent microsporangium. After meiosis each microsporangium releases four quadriflagellate 'microzoospores'.

» However, these microzoospores infect the host very rarely.

Mode of spread

» Zoosporangia disseminated by wind and rain splashes

Practice Questions

1. ....is causative of mango algal spot
2. ....is asexual spore of parasitic algae
3. Explain the symptoms and pathogenicity of *Cephaleuros* sp.
4. Describe the reproduction of plant parasitic algae.

Suggested Readings

Ramya, M., Ponmurugan, P., & Saravanan, D. (2013). Management of Cephaleuros parasiticaus Karst (Trentepohliales: Trentepohliaceae), an algal pathogen of tea plant, Camellia sinsensis (L) (O. Kuntze). *Crop Protection*, *44*, 66-74.

Sunpapao, A., Thithuan, N., Bunjongsiri, P., & Arikit, S. (2016). Cephaleuros parasiticus, associated with algal spot disease on Psidium guajava in Thailand. *Australasian Plant Disease Notes*, *11*(1), 1-4.

Jose, G., & Chowdary, Y. B. K. (1980). Studies on the host range of the endophytic alga Cephaleuros Kunze in India. *Revista de Biología Tropical*, *28*(2), 297-304.

Chapter - 19

Nematodes: Morphology and Reproduction

Nematology is an important branch of biological science, which deals with a complex, diverse group of round worms known as Nematodes that occur worldwide in essentially all environments. Nematodes are also known as eelworms in Europe, nemas in the United States and round worms by zoologists, four out of five animals are nematodes in abundance. Many species are important parasites of plants and animals, whereas others are beneficial to agriculture and the environment. Nematodes that are parasites of man and animals are called helminthes and the study is known as Helminthology. The plant parasitic forms are called nematodes and the study is known as Plant Nematology. The name nematode was derived from Greek words nema (thread) and oides (resembling).

Annual crop losses due to these obligate parasites have been estimated to be about $ 78 billion wordwide and $ 8 billion for U.S. growers. The estimated annual crop loss in Tamil Nadu is around Rs. 200 crores. The soils in a hectare of all agroecosystem typically contain billions of plant parasitic as well as beneficial nematodes. The damage to plants caused by nematodes is often overlooked because the associated symptoms, including slow growth, stunting and yellowing, can also be attributed to nutritional and water related disorders.

Importance of Nematodes in Agriculture

In the United States, the nematodes are known to cause six per cent loss in field crops, ($ 100 million / year), 12 per cent loss in fruits and nuts ($ 225 million / year), 11 per cent loss in vegetables ($ 267 million / year) and 10 per cent loss in ornamental ($ 60 million / year).

In India, the cereal cyst nematode, *Heterodera avenae* causes the 'molya' disease of wheat and barley in Rajasthan, Punjab, Haryana, Himachal Pradesh and Jammu and Kashmir. The loss due to this nematode is about 32 million rupees in wheat and 25 million rupees for barley in Rajasthan State alone.

History of Plant Nematology

Vedas occasionally refered to nematode parasites of human beings by the name Krmin or Krmi in Sanskrit (meaning worms) in Rig, Yajur and Atharvavedas written 6000-4000 BC. The vedic people were aware about the krimis, their symptoms, vectors, and their cure.

Later, when indigenous medical science Ayurveda developed following Atharvveda (3000 BC and later), Charak recognized 20 different organsims as krimis in his Samhita, which included nematodes besides arthropods and ticks.

N.A. Cobb (1915) who is considered to be the father of American Nematology, provided a dramatic description of the abundance of nematodes. He stated, "If all the matter in the universe except the nematodes were swept away, our world still would be dimly recognizable we would find is mountaintops, valleys, rivers, lakes and oceans represented by a film of nematodes. The statement "sowed cockle, reaped no corn" in Shakespeare'2 "Love's Labour's List"' act 4, scene 3, as suggested by throne (1961) possibly the first record of plant parasitic nematodes in 1549. The nematode that Throne suspected to be in that reference actually was described by Needham in 1743. Subsequently, discovery of microscope and development in various disciplines of science led to the discovery of plant parasitic nematode and the disease caused by them.

Some of the important milestones on the history of plant nematology are listed below in chronological order.

1743	Needham	First reported wheat gall disease in association with nematode (*Anguina tritici*) in England
1855	Berkeley	Determination of root-knot nematode, *Meloidogyne* spp. to cause root galls on cucumber plants in greenhouse in England.
1857	Kuhn	Reported the stem and bulb nematode, *Ditylenchus dipsaci* infesting the heads of teasel
1892	Atkinson	Report of root-knot nematode and *Fusarium* complex in vascular wilt of cotton.
1907	N.A.Cobb	Joined the USDA in 1907 Proposed the name "nema". Removed the plant-parasitic and free-living nematodes from Helminthology Proposed the science of nematology Considered to be the father of nematology.

History of Nematology in India

Nematology as a separate branch of Agriculture Science in India has been recognized only about 37 years back. The history and development of Nemtology in India have been listed below in chronological order.

1901	Barber	First reported root-knot nematode on tea in Dhovala Estate, Tamil Nadu, South India
1906	Butler	First reported root – knot nematode on black pepper in Kerala.
1913, 1919	Ayyar	Reported reported root – knot nematode infestation on vegetable and other crops in India
1934, 1936	Dastur	Reported white tip disease of rice caused by Aphelenchoides besseyi in Central India
1958	Vasudeva	Molya disease of wheat and barley (caused by Heterodera avenae) recorded in Rajasthan
1959	Prasad, Mathur and Sehgal	Reported cereal cyst nematode for the first time from India.
1961	Jones	First reported golden nematode of potato from Nilgiri Hills in Tamil Nadu.
1961	ICAR & TNAU	Nematology laboratory established at Agricultural College and Research Institute, Coimbatore, with the assistance of Rockfeller Foundation and Indian Council of Agricultural Research.
1961	CPRI	Nematology unit established at the Central Potato Research Institute, Simla.
1963	ICAR	Laboratory for potato cyst nematode research established at Uthagamandalam with the assistance of Indian Council of Agriculture Research
1966	IARI	Division of Nematology established at IARI, New Delhi
1977	AICRP	All India Co-ordinated Research Project (AICRP) on nematode pests of crops and their control started functioning in 14 centres in India with its Project Co-ordinator at IARI, New Delhi.
1979	PG	M.Sc. (Ag.) Plant Nematology course started at Tamil Nadu Agricultural University, Coimbatore.

Economically important nematodes associated with crops

1. Root-knot nematode (*Meloidogyne incognita*, *M. javanica*, *M. arenaria*, *M. graminicola* and *M. indica*)
2. Reniform nematode (*Rotylenchulus reniformis*)
3. Cyst nematodes (*Heterodera zeae* and *H. cajani*)
4. Lesion nematode (*Pratylenchus coffeae*, *P. indicus*, *P. thornei* and *P. zeae*)
5. White tip nematode (*Aphelenchoides besseyi*)
6. Ufra nematode (*Ditylenchus angustus*)
7. Burrowing nematode (*Radopholus similis*)
8. Citrus nematode (*Tylenchulus semipenetrans*)
9. Potato cyst nematodes (*Globodera rostochiensis* and *G. pallida*)
10. Wheat seed gall nematode (*Anguina tritici*)
11. Rice root nematode (*Hirschmanniella* spp.)
12. Stunt nematode (*Tylenchorhynchus brevilineatus*)

Other common plant parasitic nematodes

1. Awl nematode - *Dolichodorus* spp.
2. Dagger nematode - *Xiphinema* spp.
3. Foliar nematode - *Aphelechoides* spp.
4. Lance nematode - *Hoplolaimus* spp.
5. Needle nematode - *Longidorus* spp.
6. Pin nematode - *Paratylenchus* spp.
7. Reni form nematode - *Rotylenchus* spp
8. Ring nematode - *Criconemella* spp.
9. Sheath nematode - *Hemicycliophora* spp
10. Spiral nematode - *Helicotylenchus* spp.
11. Sting nematode - *Belonolaimus* spp.
12. Stubby-root nematode - *Paratrichodorus* spp and *Trichodorus* spp.
13. Stunt nematode - *Tylenchorhynchus* spp.
14. Chrysanthemum foliar nematode - *Aphelechoides ritzemabozi*
15. Bulb nematode - *Ditylenchus dipsasi*

16. Rice stem nematode - *Ditylenchus angustus*
17. Yam nematode - *Scutellonema bradys*

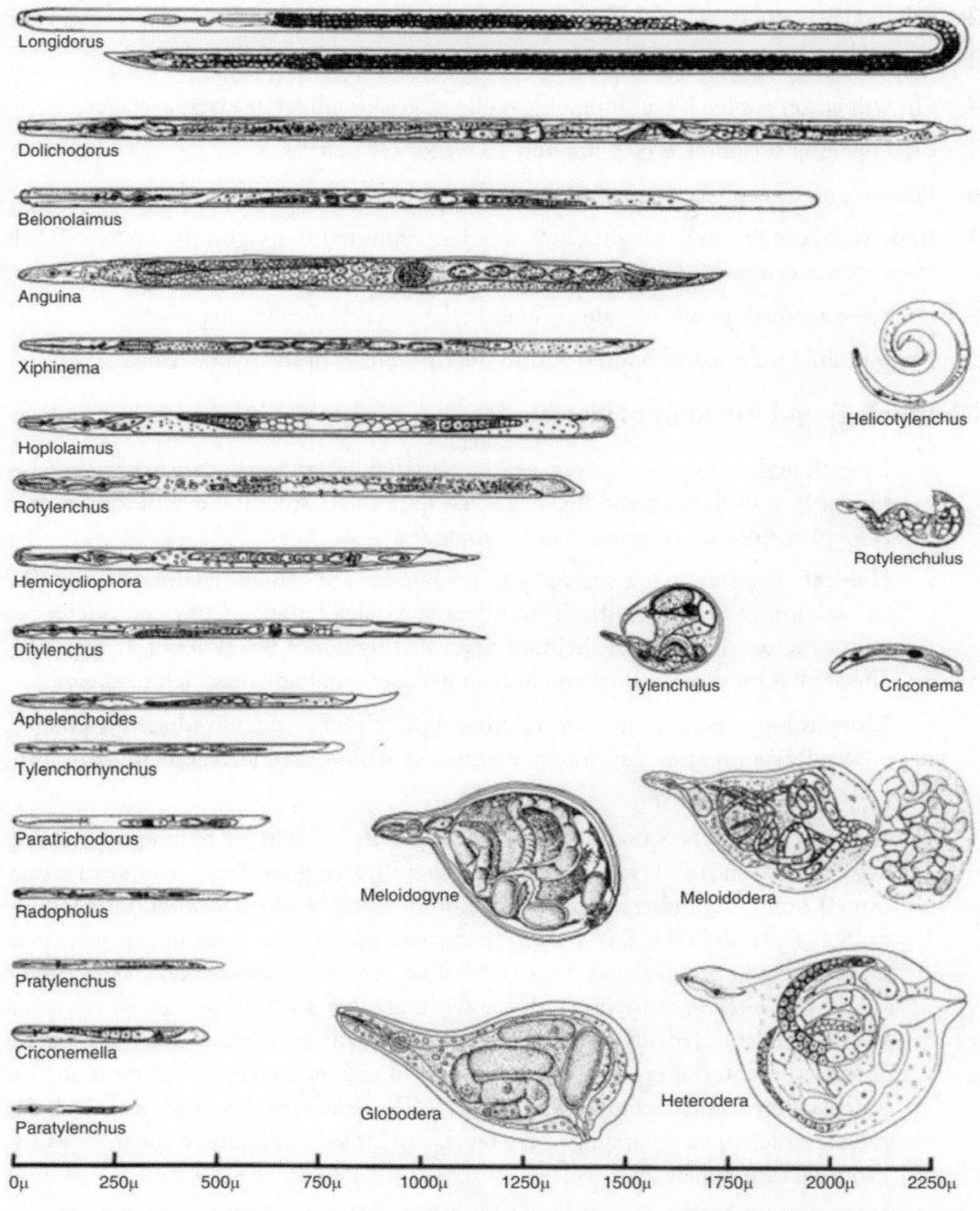

Major genera of phytoparasitic nematodes (Source: Shurtleff and Averre, 2000)

The following are the characteristics of members of the phylum Nemata.

1. Inhabit marine, freshwater and terrestrial environments as free living and parasites.
2. Bilateraly symmetrical, triploblastic, unsegmented and pseudocoelomates.
3. Vermiform, round in cross-section, covered with a three layered cuticle.
4. Growth accompanied by molting of juvenile stages, usually four juvenile stages.
5. Oral opening surrounded by 6 lips and 16 sensory structures.
6. Possess unique cephalic sense organs called amphids.
7. Body wall contains only longitudinal muscles connected to longitudinal nerve chords by processes extending from each muscle.
8. Unique excretory system containing gland cells or a set of collecting tubes.
9. Longitudinal nerve cords housed within the thickening of the hypodermis.

Morphology and Anatomy of Nematodes

» Even though nematodes occupy nearly every habitat on earth, they are remarkably similar in morphology and life stages. Despite their structural complexity, certain basic principles are common to all nematodes.

» **Habitat**: The free living saprophytic nematodes are generally larger in size. The animal and human parasitic helminthes may have length of few centimeters to even a meteer or more. The helminth parasitizing whale fish is about 27 feet long. The studies on these animal and human parasites are known as Helminthology.

» **Morphology**: Nematodes are microscopic, triploblastic, triradiate esophagus, bilaterally symmetrical (mirror-image), unsegmented, pseudocoelomate, vermiform and colourless animals.

» The plant parasitic nematodes are slender elongate, spindle shaped or fusiform, tapering towards both ends and circular in cross section. The length of the nematode may vary from 0.2 mm (*Paratylenchus*) to about 11.0mm (*Paralongidorus maximus*). Their body widths vary from 0.01 to 0.05 mm. In few genera, the females on maturity assume pear shape (*Meloidogyne*), globular shape (*Globodera*), reniform (*Rotylenchulus reniformis*) or saccate (*Tylenchulus semipenetrans*). The swelling increases the reproductive potential of the organism. Radially symmetric traits (triradiate, tetraradiate and hexaradiate) exist in the anterior region. The regions of intestine, excretory and reproductive systems show tendencies towards asymmetry. The nematodes have one or two tubular gonads which open separately in the female and into the rectum in the male which also have the copulatory spicules.

» The nematode body is not divided into definite parts, but certain sub-divisions are given for convenience. The anterior end starts with the head, which consists of mouth and pharynx bearing the cephalic papillae or setae. The portion between the head and the oesophagus is known as the neck. Beginning at the anus and extending to the posterior terminus is the tail. Longitudinally the body is divided into four regions as dorsal, right lateral, left and ventral. All the natural openings like vulva, excretory pore and anus are located in the ventral region. The nematode body is made up of several distinct body systems. They are the body wall, nervous system, secretory - excretory system, and digestive system and reproductive system. Nematodes do not possess a specialized circulatory or respiratory system. The exchange of gases is thought to occur through the cuticle and circulation proceeds through the movement of fluids within the pseudocolelom and by simple diffusion across membranes.

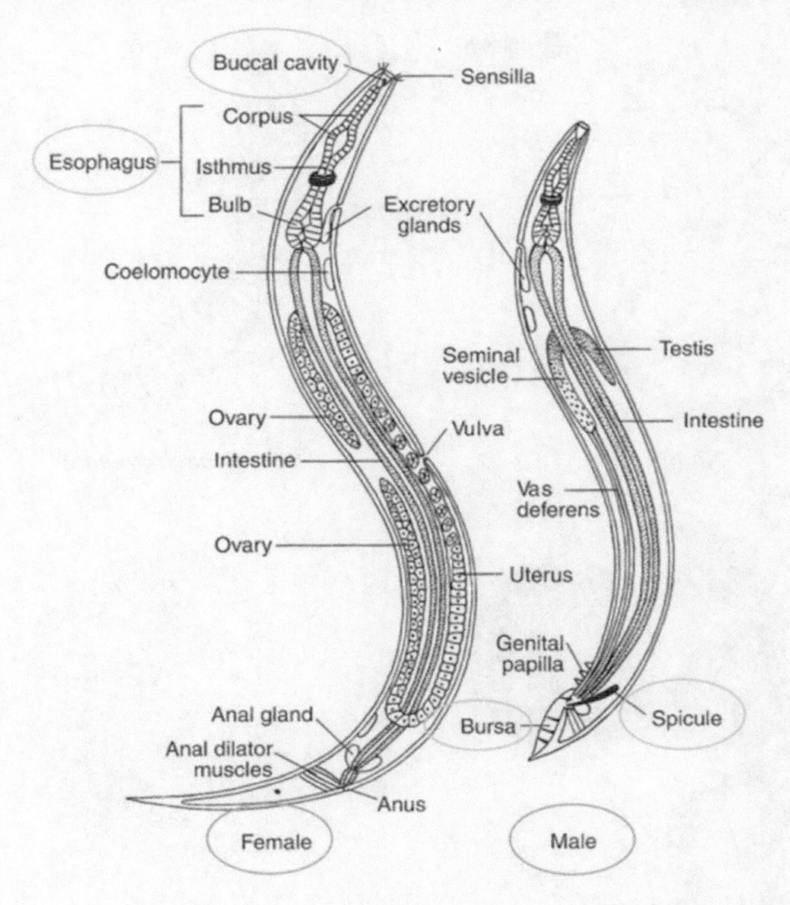

Types of nematodes based on shape

Filiform - More elongated towards the body extremities E.g., *Xiphinema*, *Longidorus*, *Paralongidorus*.

Sausage shaped or plump - When body length is reduced but breadth remains same giving a plump look. E.g., *Criconematids* group

Pyriform or flask shaped - Females of certain Genera swell to acquire saccate, pear shape or flask like structure. E.g., Females of *Meloidogyne* spp. and *Heterodera* cysts.

Kidney shaped or Reniform - E.g., *Rotylenchulus reniformis*

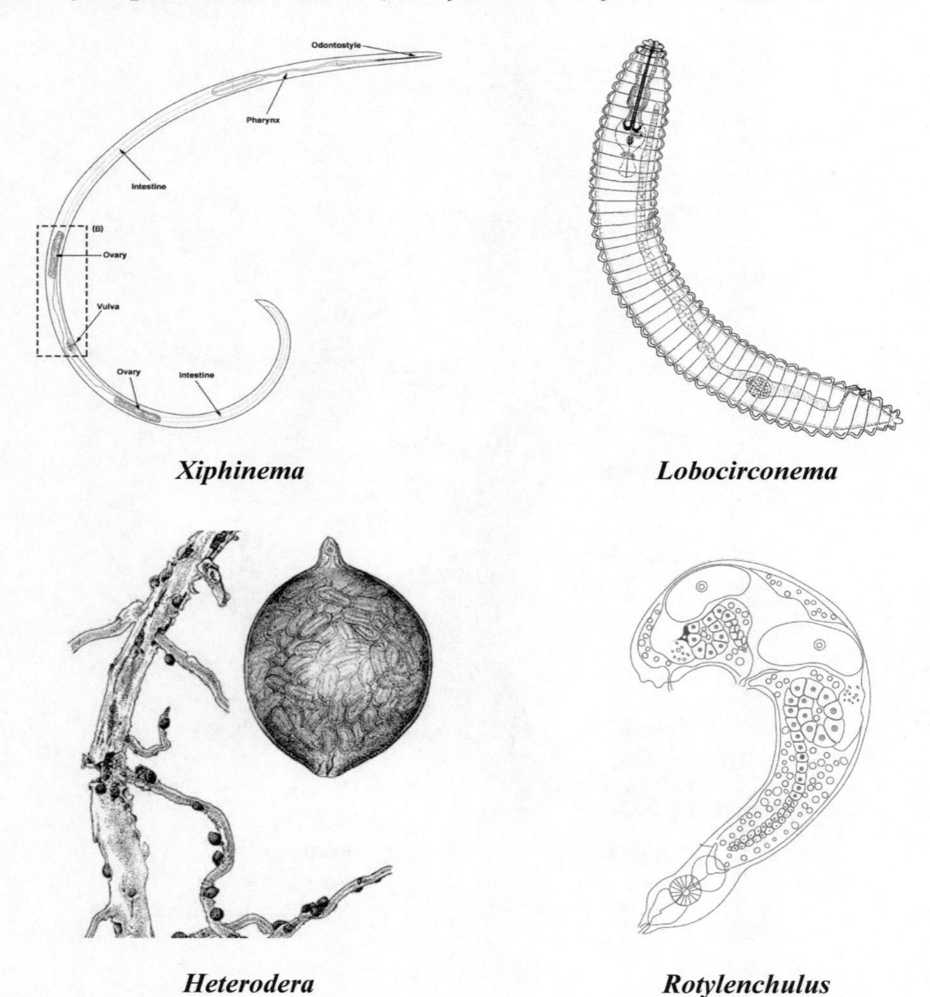

Xiphinema ***Lobocirconema***

Heterodera ***Rotylenchulus***

Sexual Dimorphism

Morphological differentiation between the genders of same species is referred as sexual dimorphism. Male and females look alike (similar) in most of the nematode species.

» Males being smaller than females.

» Sexual dimorphism has been observed in number of genera of order Tylenchida.

» The females on maturity assume pear shape in *Meloidogyne*, globular shape in *Globodera*, reniform/kidney shape in *Rotylenchulus reniformis* or saccate in *Tylenchulus semipenetrans*.

» The swelling of female nematodes increases the reproductive potential of the organism.

Body is tubular and divided into three regions 1. Outer body tube or body wall. 2. Body cavity or Pseudocoelom in which excretory, nervous and reproductive system 3. Inner body tube or alimentary canal.

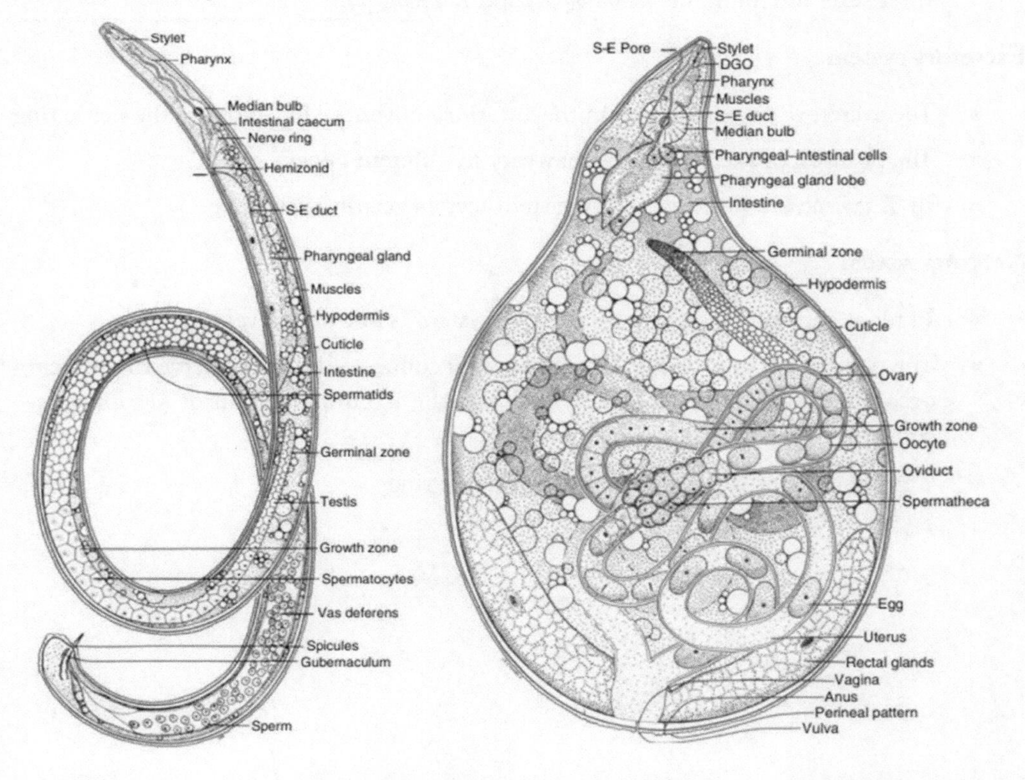

***Meloidogyne*:** (A) male and (B) female body (Source: A: Eisenback and Hunt (2009); B: Karssen and Moens 2006).

Reproductive system

- Nematodes possess either one ovary (monodelphic) or two ovaries (didelphic)
- Monodelphic ovary is always, anteriorly directed. Prodelphic E.g., *Pratylenchus*, Ditylenchus; Posteriorly directed. E.g., *Xiphenema*
- In didelphic ovaries, if both the ovaries are anteriorly directed and vulva is terminal
- in position – didelphic prodelphic. E.g., *Meloidogyne*, *Heterodera* and *Globodera*
- Whereas two ovaries are opposite to one another, such that one is anteriorly directed and the condition, as found in the case of *Tylenchorhynchus*, *Hoplolaimus* and *Helicotylenchus*
- Sexual reproduction – amphimetic reproduction
- Parthenogenetic reproduction common in *Meloidogyne* and *Tylenculus semipenetrans*
- Inter sexes are found in *Meloidogyne* and *Ditylenchulus*

Excretory system

- The excretory pore is located in the anterior midventral line close to the nerve ring
- The position of excretory pore may vary in different genera
- In *T. semepentrans* the excretory system secrets gelatinous matrix

Nervous system

- In plant parasitic nematodes – nervous system is not well developed
- The parts of nervous system located are well connected with the nerve ring (circum oesophageal commisure) which encircles the isthumus region of oesophagus – considered to be as brain of nematode
- Central nervous system comprises the nerve ring

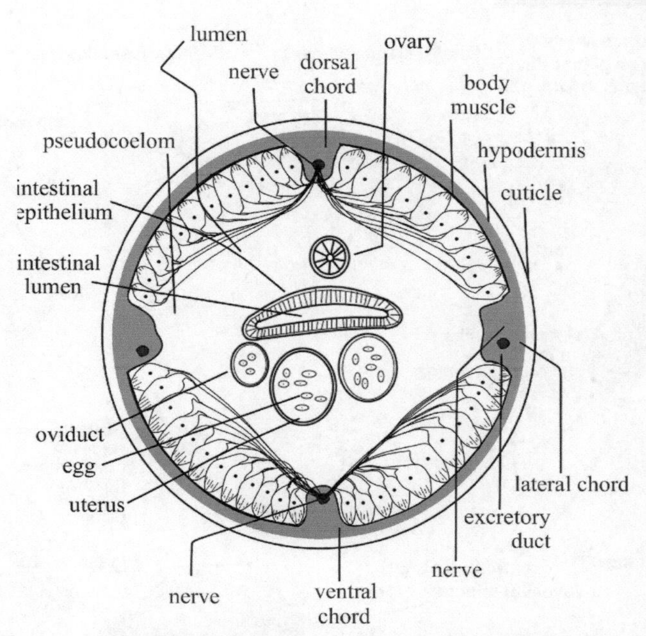

Cross section of plant parasitic nematode body

Biology of plant parasitic nematodes

Typically, there are six stages within a nematode's life cycle: an egg, four juvenile stages and the adult stage. Each juvenile stage and the adult are separated by a moulting phase. The first stage occurs in the egg. For most plant parasitic nematodes, the first moult also occurs in the egg so that it is the second-stage juvenile which hatches.

- The life cycle of nematode has six stages (J-Juvenile or larva), Egg- J1 – J2 – J3 – J4 – Adult. The first stage juvenile undergo molting with in the egg after hatching and second stage juveniles will be released into the soil, which is the infective stage.
- The larva undergoes series of four moults. The period of growth between moults are called juvenile stages.
- Larva or juvenile undergo moulting thrice and then develops into adult. The first moult occurs within the egg shell and the second stage juvenile comes out by rupturing the egg shell as J2.
- Multinucleate, dense ctoplasmic, enlarged nuclei with several mitochondria and golgi bodies – Syncytium (Giant cell) – Metabolically more active.

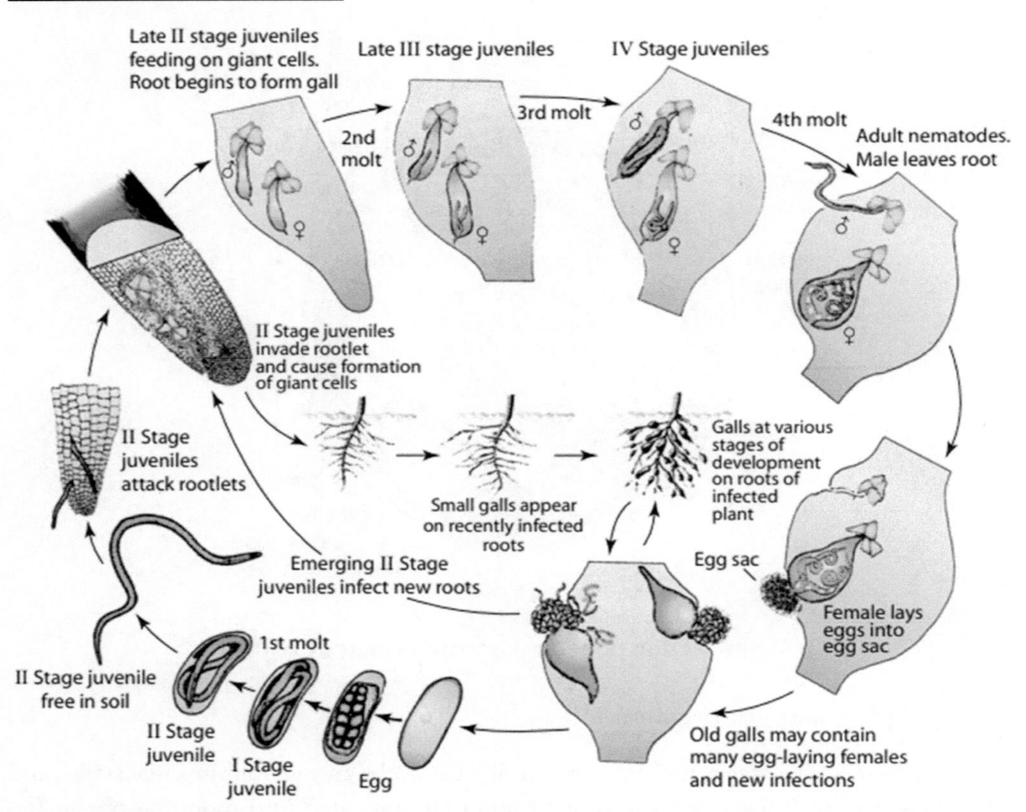

Biological and disease cycle of root-knot nematode (Source: G.N. Agrios, 2005).

Important life cycle parameters of major nematode pests

Nematodes	Life cycle	Giant cell (Syncytium)
Root knot nematode	30 days	Present
Cyst nematode	1 year	Present
Citrus nematode	6 to 8 week	Not present
Renifrom nematode	25 days	Present
Burrowing nematode	25 days	-
Lesion nematode	45 to 60 days	-

Practice Questions

1. ………..nematode produce white tip in roots
2. ………..is known as seed gall nematode
3. ………..nematode facilitates yellow slime disease in wheat
4. ………..is scientific name of root-knot nematode
5. ………nematode produces cysts in potato
6. Name some important plant pathogenic nematodes.
7. Explain the symptoms of root-knot nematode and rhizobia galls
8. Write about the sexual dimorphism in nematodes
9. Describe the organs of nematodes
10. Write the impact and significance of plant parasitic nematodes
11. Write the historical develpments of nematology in India

Suggested Readings

Perry, R. N., & Moens, M. (Eds.). (2013). *Plant nematology*. CABI.

Dropkin, V. H. (1989). *Introduction to plant nematology* (No. Ed 2nd). John Wiley and Sons Inc.

Jenkins, W. R., & Taylor, D. P. (1968). Plant nematology. *Soil Science*, *106*(4), 326.

Ravichandra, N. G. (2013). *Plant nematology*. IK International Pvt Ltd.

Coyne, D. L. (2007). *Practical plant nematology: a field and laboratory guide*. IITA.

Turner, S. J., & Subbotin, S. A. (2013). Cyst nematodes. *Plant nematology*, (Ed. 2), 109-143.

Karssen, G., Wesemael, W., & Moens, M. (2013). Root-knot nematodes. *Plant nematology*, (Ed. 2), 73-108.

Khan, M. R. (2008). *Plant nematodes: methodology, morphology, systematics, biology and ecology*. CRC Press.

Chapter - 20

Ecological Classification of Plant Parasitic Nematodes

Nematodes feed on all parts of the plant, including roots, stems, leaves, flowers and seeds. Many plant-parasitic nematodes feed on the roots of plants. The feeding process damages the plant's root system and reduces the plant's ability to absorb water and nutrients. Typical nematode damage symptoms are a reduction of root mass, a distortion of root structure and/or enlargement of the roots. Nematode damage of the plant's root system also provides an opportunity for other plant pathogens to invade the root and thus further weakens the plant. Direct damage to plant tissues by shoot-feeding nematodes includes reduced vigor, distortion of plant parts, and death of infected tissues depending upon the nematode species.

According to feeding habits, the nematodes can be divide into

1. Above ground feeders

a. Feeding on flower buds, leaves and bulbs

» Seed gall nematode: *Anguina tritici*

» Leaf and bud nematode: *Aphelenchoides*

» Stem and bulb nematode: *Ditylenchus*

b. Feeding on tree trunk

» Red ring nematode: *Rhadinaphelenchus cocphilus*

» Pine wilt nematode: *Bursaphelenchus xylophilus*

2. Below ground feeders

According to feeding habits, the nematodes can be divide into

a. Ectoparasitic nematodes,
b. Semi endoparasitic nematodes
c. Endoparasitic nematodes

A. Endoparasite: The entire nematode is found inside the root and the major portion of nematode body found inside the plant tissue

i. Migratory endoparasite: They move in the cortial parenchyma of host root. While migrating they feed on cells, multiply and cause necrotic lesions

ii. Sedentary endoparasite: penetrate the root lets up to cortex and become sedentary throughout the life.

Nematode-fungus interaction was first observed by Atkinson (1892) in cotton with *Fusarium* wilt + *Meloidogyne* spp.

Sedentary	**Migratory**
» Cyst nematode: *Heterodera* spp., *Globodera* spp.	» Lesion nematode - *Pratylenchyus* spp.
» Root–knot nematode: *Meloidogyne* spp.	» Burrowing nematode - *Radopholus simili,*
	» Root lesion nematode *Hischmanniela* spp.

B. Semi-endoparasite: Anterior part of the nematode, head and neck being permanently fixed in the cortex and the posterior part extents free into the soil.

i. Citrus nematode: *Tylenchulus semipenetrans*

ii. Reniform nematode: *Rotylenchulus reniformis*

C. Ectoparasite: Live freely in the soil and move closely or on the root surface feed intermittently on the epidermis and root hairs near the root tip.

i. Migratory ectoparasites: These nematodes spend their entire life cycle free in the soil. When the roots are disturbed they detach themselves.

ii. Sedentary ectoparasites: the attachment of nematode to the root system is permanent but for this, when the roots are disturbed they detach themselves.

Sedentary	**Migratory**
» Sheath nematodes: *Hemicriconemoides* spp. *Hemicycliphora* spp. *Cacopaurus* spp.	» Needle nematode: *Longidorus* spp. » Dagger nematode: *Xiphinema* spp. » Stubby nematode: *Trichodorus* spp. » Pin nematode: *Paratylenchus* spp.

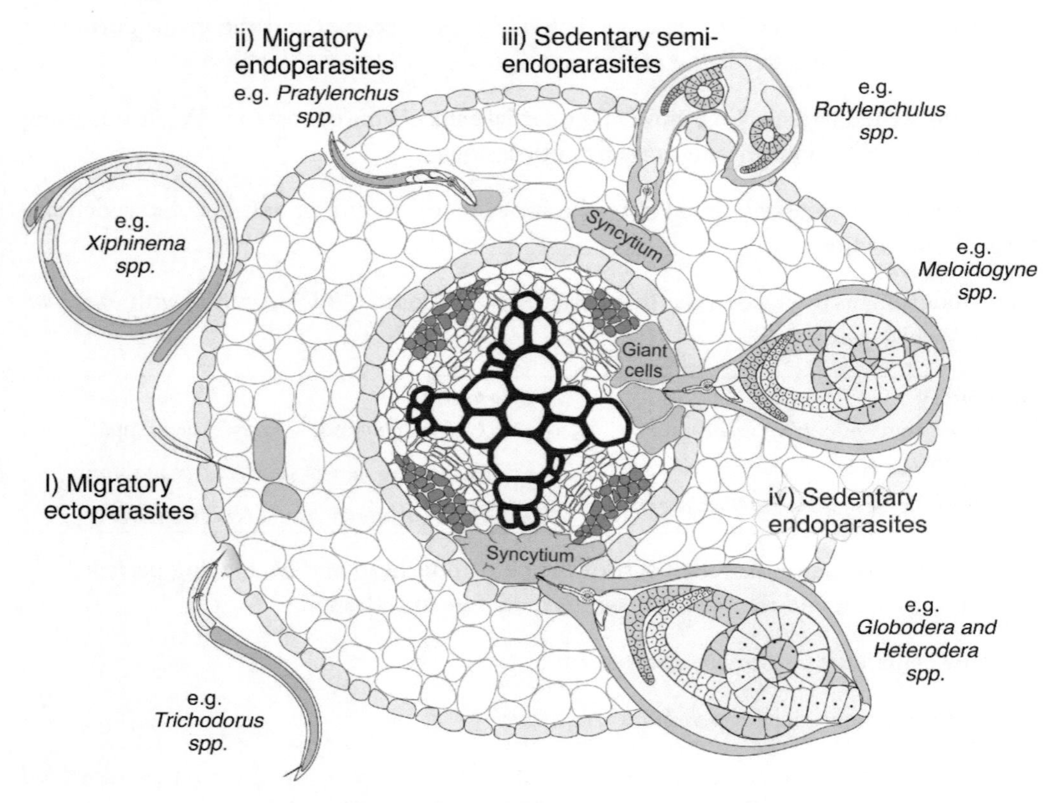

Lifestyles of major plant-parasitic nematode

Important diseases of Nematodes on crops

Crop	Disease	Nematode
Rice	Ufra disease	*Ditylenchus angustus*
	Mentek disease	*Hirshmaniella oryzae*
Wheat	Earcockle	*Anguina tritici*
	Tundu (or) yellow slime	*A. tritici + Clavibacter tritici*
	Molya disease	*Heterodera avenae*
Citrus	Spreading decline	*Radopholuss similis*
	Slow decline, Die back	*Tylenchulus semipenerans*
Banana	Black head, Tip over disease	*Radopholus similis*
Pepper	Yellowing, slow eilt	*Radopholuss similis*
Strawberry	Spring crimp, Summer crimp	*Aphelenchoides fragrae*
Sugarbeet	Bearding	*Meloidogyne* sp.
Onion	Bloat	*Ditylenchus dipsaci*
Other symptoms	Curly tip, Fish hook symptom	*Xiphinema* spp
	Stubby root	*Paratrichodorus* sp.

Practice Questions

1. Write the differences between sedendary and migratory parasitic nematodes
2. …………is migratory endoparasite in pulses
3. Explain the ecological classification of nematodes

Suggested Readings

Williamson, V. M., & Gleason, C. A. (2003). Plant–nematode interactions. *Current opinion in plant biology*, *6*(4), 327-333.

Barker, K. R., Pederson, G. A., & Windham, G. L. (1998). *Plant and nematode interactions*. American Society of Agronomy.

Khan, M. W. (Ed.). (2012). *Nematode interactions*. Springer Science & Business Media.

Siddique, S., Coomer, A., Baum, T., & Williamson, V. M. (2022). Recognition and Response in Plant–Nematode Interactions. *Annual Review of Phytopathology*, *60*.

Ali, M. A., Abbas, A., Azeem, F., Javed, N., & Bohlmann, H. (2015). Plant-nematode interactions: from genomics to metabolomics. *International Journal of Agriculture and Biology*, *17*(6).

Chapter - 21

Nematode Disease Symptoms on Crop Plants

Most of the plant parasitic nematodes affect the root portion of plants except *Anguina* spp., *Aphelenchus* spp., *Aphelenchoides* spp., *Ditylenchus* spp., *Rhadinaphelenchus cocophilus* and *Bursaphelenchus xylophilus.* Nematodes suck the sap of the plants with the help of stylet and causes leaf discolouration, stunted growth, reduced leaf size and fruits and lesions on roots, galls, reduced root system and finally wilting.

Symptoms of nematode diseases can be classified as

I. Symptoms produced by above ground feeding nematodes

II. Symptoms produced by below ground feeding nematodes

I. Symptoms produced by above ground feeding nematodes

1. **Leaf discolouration:** The leaf tip become white in rice due to rice white tip nematode *Aphelenchoides besseyi*, yellowing of leaves in Chrysanthemum due to Chrysanthemum foliar nematodes, *A. ritzembasoi*.
2. **Dead or devitalized buds:** In case of straw berry plants infected with *A. fragariae*, the nematodes affect the growing point and kill the plants and result in blind plant.
3. **Seed galls:** In wheat, *Anguina tritici* larva enters into the flower primordium and develops into a gall. The nematodes can survive for longer period (even upto 28 years) inside the cockled wheat grain.
4. **Twisting of leaves and stem:** In onion, the basal leaves become twisted when infested with *D. angustus*.
5. **Crinkled or distorted stem and foliage:** The wheat seed gall nematode. *A. tritici* infests

the growing point as a result distortion in stem and leaves take place.

6. **Necrosis and discolouration:** The red ring disease on coconut caused by *Rahadinaphelenchus cocophilus*. Due to the infestation, red coloured circular area appears in the trunk of the infested palm.
7. **Lesions on leaves and stem**: Small yellowish spots are produced on onion stem and leaves due *D. dipsaci*, and the leaf lesion caused by *A. ritzemabosi* on Chrysanthemum.

II. Symptoms produced by below ground feeding nematodes

The nematodes infest and feed on the root portion and exhibit symptoms on below ground plant parts as well as on the above ground plants parts and they are classified as

a. Above ground symptoms

b. Below ground symptoms

a. Above ground symptoms

1. **Stunting**: Reduced plant growth, and the plants can not able to withstand adverse conditions. Patches of stunted plants appears in the field. (E.g.,) in potato due to *Globodera rostochiensis*, in gingelly, due to *Heterodera cajani* and in wheat by *Heterodera avenae.*
2. **Discolouration of foliage**: Patchy yellow appearance in coffee due to *Pratylenchus coffeae, G. rostochiensis* infested potato plants show light green foliage. *Tylenchulus semipenetrans* induce fine mottling on the leaves of orange and lemon trees.
3. **Wilting:** Day wilting due to *Meloidogyne* spp. i.e. In hot weather the root – knot infested plants tend to droop or wilt even in the presence of enough moisture in the soil. Severe damage to the root system due to nematode infestation leads to day wilting of plants.

b. Below ground symptoms

1. **Root galls or knots:** The characteristic root galls are produced by root – knot nematode, *Meloidogyne* spp. False root galls are produced by *Nacobbus batatiformis* on sugar beet and tomato. Small galls are produced by *Hemicycliophora arenaria* on lemon roots. *Ditylenchus radicicola* cause root galls on wheat and oats. *Xiphinema diversicaudatum* cause galls on rose roots.
2. **Root lesion:** The penetration and movement of nematodes in the root causes typical root lesions E.g., Necrotic lesions induced by *Pratylenchus* spp on crossandra; the burrowing nematode, *Radopholus similes* in banana. Similarly, *Pratylenchus coffeae* and *Helicotylenchus multicinctus* cause reddish brown lesion on banana root and corm. The rice root nematode also causes brown lesions on rice root.

3. **Reduced root system:** Due to nematode feeding the root tip growth is arrested and the root produce branches. This may be of various kinds such as coarse root, stubby root and curly tip.
   a. **Coarse root:** *Paratrichodorus* spp. infestation arrest the growth of lateral roots, and leads to an open root system with only main roots without lateral roots.
   b. b. **Stubby roots:** The lateral roots produce excessive rootless (E.g., *P. christei*)
   c. c. **Curly tip**: In the injury caused by *Xiphinema* spp. the nematode retard the elongation of roots and cause curling of roots known as "Fish book' symptom.
4. **Root proliferation**: Increase in the root growth or excessive branching due to nematode infestation. The infested plant root produced excessive root hair at the point of nematode infestation. (E.g.,) *Trichodorus christei, Nacobbus* spp., *Heterodera* spp. *Meloidogyne hapla* and *Pratylenchus* spp. etc.
5. **Root rot:** The nematodes feed on the fleshly structure and resulting in rotting of tissues (E.g.,) Yam nematode *Scutellonema bradys* and in potato *Ditylenchus destructor* cause root rot.
6. **Root surface necrosis:** The severe injury caused by *T. semipenetrans* on citrus leads to complete decortications of roots and results in root necrosis.
7. **Cluster of sprouts on tubers:** On the tubers, clusters of short and swollen sprouts are formed due to *D. dipsaci* infestation in many tuber plants.

Practice Questions

1. .... nematode causes stunting in potato
2. False root galls are produced by............
3. The rice root nematode also causes .... lesions on rice root
4. The red ring disease on coconut caused by ....
5. Name the causal organism and main symptoms of ear cockle disease and 'tundu' disease of wheat.

Suggested Readings

Mashela, P. W., De Waele, D., Dube, Z., Khosa, M. C., Pofu, K. M., Tefu, G., & Fourie, H. (2017). Alternative nematode management strategies. In *Nematology in South Africa: a view from the 21st century* (pp. 151-181). Springer, Cham.

Khan, M. R. (2015). Nematode diseases of crops in India. In *Recent advances in the diagnosis and management of plant diseases* (pp. 183-224). Springer, New Delhi.

Stirling, G., & Stanton, J. (1997). Nematode diseases and their control. *Plant pathogens and plant diseases*, 505-517.

Reddy, P. P. (2021). Nematode diseases of crop plants: an overview. *Nematode Diseases of Crops and their Management*, 3-32.

Hillocks, R. J., & Wydra, K. (2001). Bacterial, fungal and nematode diseases. In *Cassava: biology, production and utilization* (pp. 261-280). Wallingford UK: CABI.

Chapter - 22

Liberation and Dispersal of Plant Pathogens

Liberation of pathogen: Spores formed in the infected host, have to be liberated (take off), travelled (the flight) and deposited (landing) on the suitable host to continue its infectious life cycle. Spore liberation is a physical process by which the spores produced are released from the fruiting bodies. Energy is required to liberate the spores and this energy can be within the fungus itself (active liberation) or it can be aided by the external source, such as raindrop wind etc (passive liberation).

Two types of spore liberation

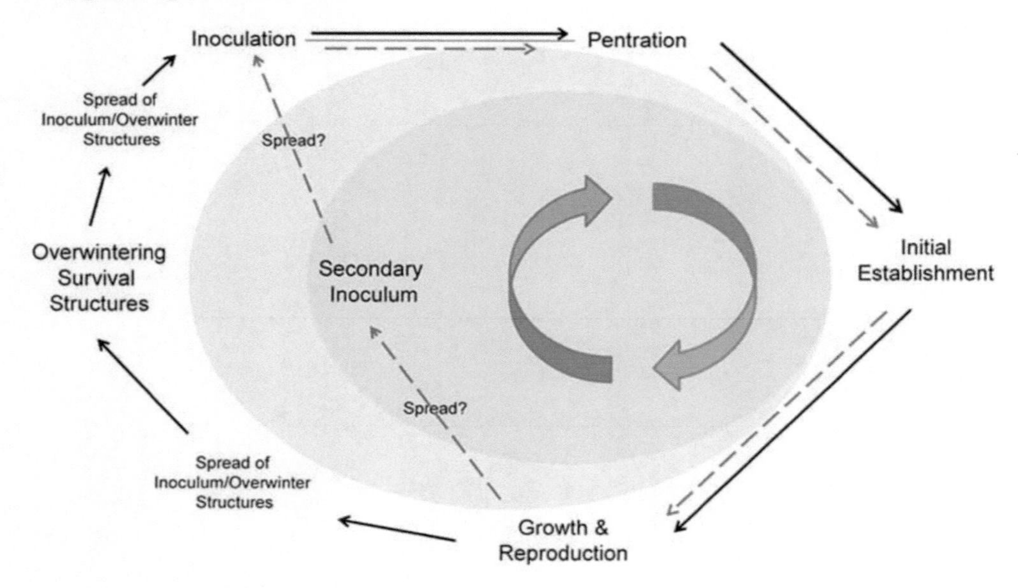

1. Active liberation

Water dependent mechanisms

a. Squirt gun mechanism: Squirt gun mechanisms are responsible for launching spores at the highest speeds, pressurized by osmosis and are most common in the Ascomycota, including lichenized species, but have also evolved among the Zygomycota. The ascomycetes squirt the ascospores through the ostiole. Ascospores may be discharged singly (Sordariomycetes) or simultaneously (Pezizomycetes). E.g., *Podospora fumicola* discharges all eight of its asccospores in a single projectile.

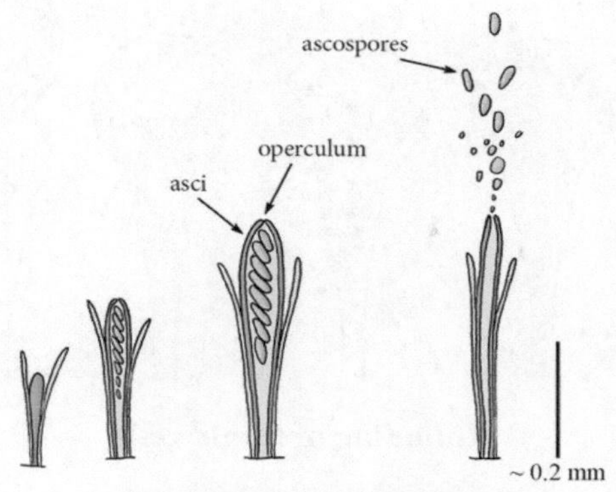

Squirt gun mechanism

b. Squirting mechanism (Fluid pressure catapult mechanism): Sporangiophore development showing the sterigma (stalk), balloon-like vesicle, and the sporangium at the tip. When a critical pressure (of about 0.55 MPa relative to ambient) is reached, the sporangium breaks free from the sporangiophore and is propelled forward by a cell sap jet that is powered by the contracting vesicle wall. Collapse of the sporangiophore after discharge of the sporangium. It is a water dependent squirting mechanism occurs in dung-inhabiting fungus *Pilobolus kleinii*, and in the weak plant parasite like *Nigrospora sphaerica*.

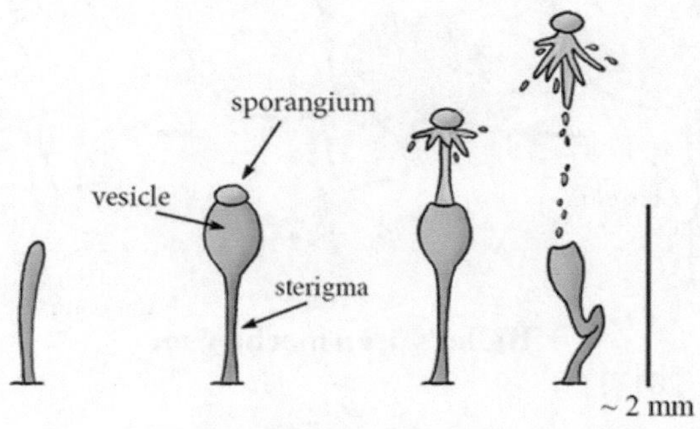

Squirting mechanism

c. Rounding of turgid cells: Mostly, in entomophthorous fungal conidia consists of their being forcibly discharged from the conidiophores. The mechanism of this discharge involves violent rounding off of turgid elements (liberation due to rounding off of turgid cells). The rounding off of elements comprises special parts of the conidial wall (papilla) and of the conidiophore wall (columella). This type of mechanism occurs in the chromista (e.g, Ballospores of *Entomophthora coronata* and *Sclerospora phillipiensis*, in the ascomycetes conidia of *Xylosphaeria furcata*, and in the basidiomycetous aecidiospores, and basidiospores of *Gymnosporangium nidus-avis*.

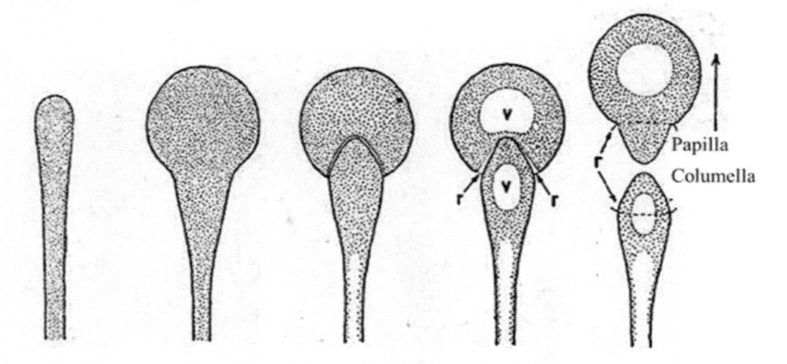

Rounding of turgid cells

d. Buller's drop mechanism (Momentum catapult mechanism): Buller's drop is generated by condensation of water from the humid air surrounding the spore. Condensation is driven by the presence of free sugars on the spore surface that lower its water potential. This is a ballistospore discharge discharge mechanism in basidiomycete fungi, it was first studied by A. H. R. Buller. E.g., Ballistospore of *Tilletia caries* and *Auricularia tabascens*.

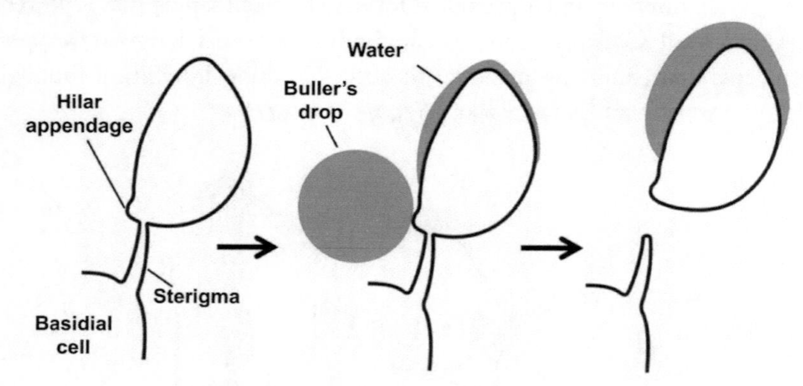

Buller's drop mechanism

Source: Piepenbring (2015).

Dry mechanisms

a. **Hygroscopic mechanism**: The sporangiophore of *Phytophthora infestans* twists violently with changing relative humidity to discharge the sporangia into the air.

b. **Water ruptures mechanism**: The spores of Deightoniella torulosa, the incitant of black leaf and fruit spot of banana are discharged by this mechanism. In dry condition water evaporates from the conidiophore and the thin cell wall in the apical region is sucked inwards as the cell dries but returns to its original position when a gas bubble forms. This sudden movement flicks of the conidium from the conidiophore and gets dispersed. E.g., *Deightoniella torulosa, Corynespora cassicola* and *Alternaria* sp.

c. **Electrostatic charges**: The spores of *Peronospora parasitica*, the causal organism of downy mildew of crucifers are dispersed by this mechanism. As the host plant dries, it becomes charged and repels the similarly charged sporangia.

2. Passive liberation

Water dependent mechanisms

The spores of some fungi are dispersed in water or on the surface of water or rain drops. The chemical composition of the cell wall of these spores makes them non wettable and hence they stick on to the raindrops.

The spores of some fungi are dispersed in water or on the surface of water or rain drops.

a. Slimy matrix: The spores adhere to the raindrops and forms many secondary droplets containing the inoculums and when the raindrop falls on the slimy matrix, the spores adhere to the raindrops and forms many secondary droplets containing the inoculums. E.g., *Fusarium* and *Colletotrichum*

b. Splash cup mechanism: The spores are produced in a splash cup. When raindrops hit the cup, spores are splashed out and dispersed away from it, for example: birds nest fungus

c. Mist pickup: Cladosporium

d. Rain tap and puff: Puccinia recondita

e. Bellows mechanism: *Geastrum*

f. Rain splash: *Colletotrichum lindemuthianum*

g. Drip splash, bubble scavenging and droplet launching in aquatic fungi

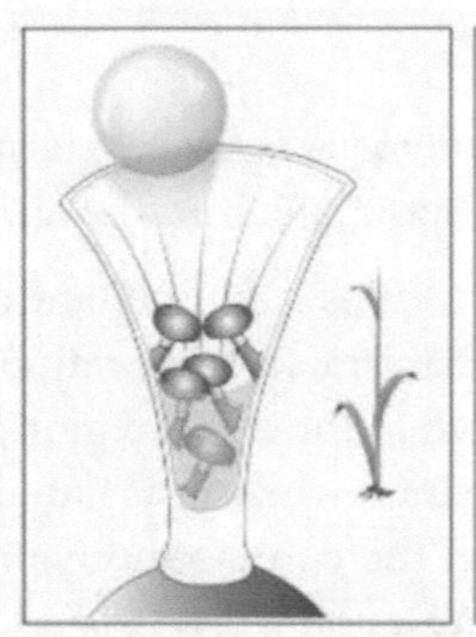

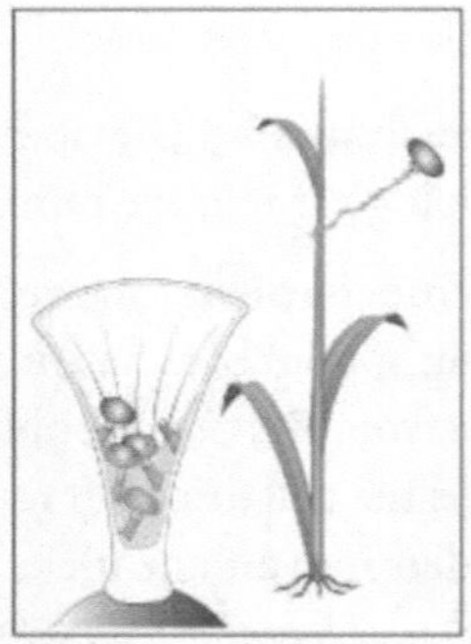

Splash cup mechanism

Dry mechanisms

The spore shedding by gravity, upward movement of spores by convention current or dislodging of conidia by mechanical disturbance. The liberated spores fall from air under the action of gravity (sedimentation) on a susceptible host and germinate under favorable condition. E.g., *Erysiphe graminis.*

Dispersal: Transport of spores or infectious bodies, acting as inoculum, from one host to another host at various distances resulting in the spread of disease, is called dissemination, dispersal or transmission of plant pathogens.

In fungi, productions of asexual and sexual spores follow the active vegetative growth of the fungus in or on the host tissues and are dispersed mechanically in time and space by various means. In bacterial diseases, the bacterial cells come out on the host surface as ooze or the tissues may be disintegrated so that the bacterial mass is exposed and then dispersed by various physical and biological agencies. Viral diseases which have no such organs are transmitted by insects, mites, phanerogamic parasites, nematodes and human beings.

The two links in the infection chain of an animate pathogen, *viz.*, survival through dormant structures and the dispersal of the pathogen are very closely bound with each other. Actually, the dormant structures provide means of *dispersal in time*, i.e., the pathogen is retained viable over a period of time enabling it to be transported through physical agencies without being harmed.

Dispersal of plant pathogens

The dispersal of infectious plant pathogens occurs through 2 ways

1. Direct or Active or Autonomous dispersal
2. Indirect or passive dispersal

1. Active (Soil, seed or planting material)

Soil as means of active dispersal: Facultative parasites and facultative saprophytes

a. Dispersal in soil (movement of pathogen in soil or by growth)

i. **Contamination of soil:** Contamination of the soil takes place by gradual spread of the pathogen from an infested area to a new area.

ii. **Growth and spread of a pathogen in soil:** Once the pathogen has reached the soil it can grow and spread based on its ability to multiply and spread. Among characters of the pathogen its adaptability to soil environment including its saprophytic survival ability are most important. The survival ability of the pathogen is governed by high growth rate, rapid spore germination, better enzymatic activity, capability to produce antibiotics and tolerance to antibiotics produced by other soil-microorganisms. On the basis of this competitive saprophytic ability the pathogens in soil can be of three types. Specialized facultative parasites (Saprophytes) can pass their life in soil in the absence of host plants, but they depend more on the residues of the host plant (E.g., *Armillariella mellea, Ophiobolous graminis* etc.). Unspecialized facultative parasites can pass their entire life in the soil (*Pythium* sp., and *Phytophthora* sp.). The soil borne obligate parasites such as *Plasmodiophora brassicae*, *Synchytrium endobioticum* require the presence of active host.

iii. **Persistence of the pathogen in soil**: The pathogens persist in the soil as dormant structures like oospores (*Pythium, Phytophthora, Sclerospora* etc.), Chlamydospores (*Fusarium*), smut spores (*Ustilago*) and sclerotia (*Rhizoctonia, Sclerotium*).

» Contamination of soil (from affected area to new area)

» Growth and spread of pathogen in soil: E.g., *Pythium, Phytophthora, Plasmodiophora, Synchytrium*

» Germination of resting structures: E.g., Oospores, chlamydospores, smutspores, sclerotia, etc.,

» Colonization through rhizomorph: E.g., *Armillaria, Fomes, Poria, Rosellinia.*

b. Dispersal by soil (Movement of soil containing pathogen)

The pathogen is dispersed by the soil during cultural operations through the agricultural implements, irrigation water, worker's feet etc. Propagules of fungi and the plant debris containing the fungal and bacterial pathogens thus spread through out the field. The transfer of soil from one place to another along with propagating materials is the most important method of dispersal of pathogen. For Example, transfer of papaya seedlings from a nursery infested with *Pythium aphanidermatum* (causal agent of stem or foot rot of papaya) can introduce the pathogen in new pits for transplanting the seedlings. Similarly grafts of fruit trees transported with soil around their roots can transmit pathogens present in the nursery to the orchards.

- During the cultural operations by implements, irrigation, workers, etc.
- Nursery to main field – stem and foot rot of papaya (*Pythium aphanidermatum*)
- Soil with graft of fruit trees – main field.

Seed and seed materials as means of active dispersal

Since most of the cultivated crops are raised from seed the transmission of diseases and transport of pathogens has much importance. The dormant structures of the pathogen (E.g., seeds of *Cuscuta*, sclerotia of ergot fungus, smut sori, etc.) are found mixed with seed lots and they are dispersed as seed contaminants. The bacterial cells or spores of fungi present on the seed coat (such as in smuts of barley, sorghum, etc.) are transported to long distances. Dormant mycelium of many fungi present in the seed is transmitted to long distances. There are three types of dispersal by seed, *viz.*, contamination of the seed, externally seed borne and internally seed borne.

a. Contamination of the seed (During harvest)

Seed borne pathogens move in seed lot as separate contaminants without being in intimate contact with the viable crop seeds. The seeds of the pathogen or parasite and the host are mixed during harvest of the crop. In many cases, the identity of the seeds of the two entities (host and the pathogens) is difficult to separate.

- Smut of pearlmillet (*Tolyposporium penicillariae*)
- Ergot of pearlmillet (*Claviceps fusiformis*)
- Seeds of *Cuscuta cernua*

b. Externally seed-borne

Close contact between structure of the pathogen and seeds is established where the pathogen gets lodged in the form of dormant spores or bacteria on the seed coat during growth of the crop or at the time of harvest and threshing.

- Stinking smut (bunt) of wheat – *Tilletia caries* (viable for 18 years)
- Smut of oat (*Ustilago avenae*) – 13 years
- Bacterial blight of cotton

c. Internally seed-borne

The pathogen may penetrate into the ovary and cause infection of the embryo while it is developing.

- Loose smut of wheat (*Ustilago tritici*)

» Bacterial blight of rice (*Xanthomonas oryzae* pv. *oryzae*)

» *Urdbean leaf crinkle virus*, TMV on tomato.

» **Seed materials:** Potato late blight and black scurf, Sugarcane red rot, Citrus canker, *Citrus tristeza virus*, *Banana bunchy top virus*.

Differentiate seed infection and infestation

Seed infection: The seed in infected only when the pathogen has grown in or on it for sometime and established its relationship with the seed tissues. E.g., Loose smut of wheat, where the fungus grows in the embryonic tissues and becomes dormant when the seed enters dormancy.

Seed infestation: When the fungus or the pathogen is present on the seed coat and in the seed lot, it is only transport of the pathogen and the seed is infested.

2. Passive dispersal

a. Animate agents

Insects: They can disseminate bacteria, fungi, viruses, mycoplasmas, spiroplasmas, etc. It is further divided into two types.

i. **Epizoic:** External transmission of pathogens by insects by hairs and honey dew like excretions of insects attract pathogen. E.g., Sugary disease of sorghum and cumbu, fire blight (*Erwinia amylovora*) by Bees and ants, Citrus canker (*Xanthomonas axonopodis* pv. *citri*) by leaf miner, *Puccinia graminis* by honey bees (Pycniospores), Sheath blight of paddy by rice mealy bug.

ii. **Endozoic:** Internal nature of transmission by insect mouth. E.g., Dutch elm disease (*Ceratostomella ulmi*) by Elm boat beetles, Cucumber wilt (*Erwinia tracheiphila*) by cucumber beetle (*Doliops duodecimpunctata*) insect gut (survival and transmission).

Major insect vectors

a. Aphids (stylet-borne viruses)

» *Toxoptera citricidus* - *Citrus tristeza virus*

» *Myzus persicae* - *Potato virus Y*

» *Acyrthosiphon pisum* - *Soybean mosaic virus*

b. Leaf hoppers (circulatory / propagative)

» GLH: *Nephotetix virescens* - *Rice tungro virus*

» Leaf hopper: *Hishimonus phycitis* - Brinjal little leaf

» Leaf hopper: *Orosius albicintus* - Gingelly phyllody

c. White flies (*Bemisia tabaci*) - *Bhendi yellow vein mosaic virus*, *Cassava mosaic virus*, chillies leaf curl, *Tobacco leaf curl virus*

d. Thrips - *Frankliniella schultzei, Scirtothrips dorsalis, Thrips tabaci* - *Tomato spotted wilt virus*

e. Mealy bugs - *Pseudococcus saccharifolii* (*Ca.* Phytoplasma) - Sugarcane spike, *Planococcoides* sp. - *Cocoa swollen shoot virus*

f. Grasshoppers - *Potato virus X, Tobacco mosaic virus*

Lady bird beetle **(***Epilachna* sp.), Earwig, Bugs, leafminer, weevils - *Cowpea mosaic virus*

Mites

Aceria cajani – Pigeonpea sterility mosaic virus

Fungi transmitted by insects

» Dutch elm disease: Elm bark beetle

Bacteria transmitted by insects

» Corn wilt: *Xanthomonas campestris* pv. *stewartii*

» Corn flea beetle: *Cheatomia pulicaria*

» Cucumber wilt: *Xanthomonas campestris* pv. *tracheiphila*

» Cucumber beetle: *Diaphorina vitara*

Chitrids fungi

» *Olpidium brassicae* – Tobacco necrosis

» *Spongospora subterranea* – Potato mobtop

» *Synchytrium endobioticum* – potato virus

Nematodes (Fungi, bacteria, viruses)

» Fungi – *Phytophthora, Fusarium, Rhizoctonia* spp.

» Bacteria – Wheat – yellow ear rot – ear cockle nematode/ seed gall (*Cornebacterium tritici*) (*Anguina tritici*)

» Viruses – Both polyhedral & tubular viruses are transmitted by

- Nematode transmitted polyhedral viruses (NEPO) - *Xiphinema*, *Longidorus* and *Paralongidorus*
- Nematode transmitted tubular viruses (NETU) - *Trichodorus* and *Paratrichodorus*

Human beings (Anthropochory)

a. Transportation of seed / planting materials (Autonomous and Passive)

Citrus canker (Asia to USA)

Fire blight of apple (USA to Europe)

Grapevine Powdery mildew (USA to France)

Late blight of Potato (Europe to India)

Rice blast (South EastAsia to Tamil Nadu)

Panama wilt of banana (Panama to Bombay)

Bunchy top of banana (Sri Lanka to Kerala)

b. By adopting farming practices

c. Through clothings (*Tobacco mosaic virus*)

d. Use of contaminated implements (*Banana bunchy top virus*)

e. Diseased grafting budding materials

Farm and wild animals

- Downy mildew of pearlmillet and sorghum (Oospores)
- Grain smut, Loose smut and head smut of sorghum, *Tobacco mosaic virus*

Birds

Moreover, spores of chestnut blight fungus, *Endothea parasitica* are disseminated by more than 18 species of birds. Cleistothecia of many powdery mildew fungi are carried by feathers of birds.

- Gaint mistletoe (*Dendrophthoe* spp.)
- Dodder (*Cuscuta* spp.)
- Chestnut blight (*Endothia parasitica*)

Phanerogamic parasites

Phanerogamic parasites transmit the viruses by acting as a bridge between the diseased and healthy plants. E.g., Dodder (*Cuscuta California, C. campesris, C. subinclusa* etc.)

- *Cuscuta subinclusa* – Cucumber mosaic virus
- *Cuscuta california* – Tobacco mosaic virus, Tobacco rattle virus, Tomato spotted wilt virus
- *Cuscuta campestris* - Tomato bushy stunt virus

b. Inanimate agents

i. **Wind (Anemochory)**: Damping off, Potato wart (*Synchytrium endobioticum*), *Striga*, powdery mildews, downy mildews, rusts, smut, sooty mould, leaf spot, blast, Fire blight of apple (*Erwinia amylovora*), etc.

ii. **Water (Hydrochory):** *Fusarium, Ganoderma, Macrophomina, Phytophthora, Plasmodiophora, Pythium, Rhizoctonia, Sclerotium, Verticillium, Sclerotinia.*

Practice Questions

1. Ballistospores are dispersed by.........mechanism
2. ....is water loving parasite attacking cereals
3. Slimy matrix depended dispersal is common in...........fungi
4. Squirting gun mechanism is common in ....fungi
5. Pilobolus is ejected in the atmosphere by..........mechanism
6. Write the differences between internal and external seed borne pathogens
7. Explain about the method of transmission of soil borne fungi
8. Write about the spore liberation process in the diseased leaves.
9. What is the essential difference between distribution and dissemination?
10. Give examples of plant diseases in which dissemination of the plant pathogens is dependent upon special types of insects.

Suggested Readings

Golan, J. J., & Pringle, A. (2017). Long-distance dispersal of fungi. *Microbiology spectrum*, *5*(4), 5-4.

Sakes, A., van der Wiel, M., Henselmans, P. W., van Leeuwen, J. L., Dodou, D., & Breedveld, P. (2016). Shooting mechanisms in nature: a systematic review. *Plos one*, *11*(7), e0158277.

Malloch, D., & Blackwell, M. (1992). Dispersal of fungal diaspores. *The fungal community: its organization and role in the ecosystem*, 147-171.

Madden, L. V. (1997). Effects of rain on splash dispersal of fungal pathogens. *Canadian Journal of Plant Pathology*, *19*(2), 225-230.

Ingold, C. T., & Hudson, H. J. (1993). Dispersal in fungi. In *The biology of fungi* (pp. 119-131). Springer, Dordrecht.

Chapter - 23

Survival of Plant Pathogens

The means of survival are the first action in infection chain or disease cycle. The initial infection that occurs from the sources of pathogen survival (Infected host as a reservoir of inoculum, saprophytic survival outside the host or dormant spores and other structures in or on the host or outside the host) in the crop is primary infection and the propagules that cause this infection are called primary inoculum. After initiation of the disease in the crop, the spores or other structures of the pathogen are sources of secondary inoculum and cause secondary infection, thereby spreading the disease in the field. E.g., The oomycete (*Phytophthora infestans*) causing late blight of potato survives in seed tubers or in soil. Infected tubers bring the primary infection in the field while primary inoculum present in soil causes primary infection of the crop from healthy seed. The primary inoculum may also be brought by wind from neighboring fields or long distances. Then the fungus produces spores on leaves. These spores are dispersed by wind and water and reach healthy plant surfaces to cause new infections. This is secondary infection. The primary infection initiates the disease and secondary infection spreads the disease.

1. Survival in soil

Soil is the major substrate for the survival of pathogens in its active or inactive stages over a period of time. The dormant structure of fungi includes resting propagules like sclerotia, rhizomorph, chlamydospores; sexual spores like oospores, zygospores and fruiting body like acervuli, pycnidia, ascoma etc.

Based on their survival, they are grouped into soil inhabitant (soil transient) or root inhabitant (soil invaders).

a. Soil inhabitant: Those organisms which survive indefinitely in the soil as saprophytes in the absence of the host plant.

Chromista: *Pythium*

Fungi: *Macrophomina, Rhizoctonia*

Bacteria: *Agrobacterium, Pseudomonas, Streptomyces, Ralstonia*

b. Root inhabitant: These are more specialized parasites that survive in soils in close association with their hosts. The active saprophytic phase remains as long as the host tissue in which they are living as parasites is not completely decomposed.

Fungi: *Fusarium, Verticillium, Gaumannomyces, Phymatotrichum omnivorum*

c. Rhizosphere colonizers: Those organisms which colonize the dead substrates in the root region and continue to live like that for a longer period which are more tolerant to soil antagonism. E.g., Leaf mold in tomato: *Cladosporium fulvum*

2. Survival in association with living plant.

a. The infected host serving as reservoir of active inoculum is grouped into

a. Seed: Seed may be externally or internally infected by plant pathogens during the course of development and maturation in fruit or pod. Most seed borne pathogens survive as long as seed remains viable.

E.g., The pathogen of loose smut of wheat, *Ustilago nuda tritici*, enters the stigma and style and infects the young seed, in which it survives as mycelium. *Pseudomonas syringae* pv. *tomato* has been shown to survive in dried tomato seed for 20 years.

b. Collateral hosts / Alternative hosts (wild hosts of same families): Collateral hosts are those which are susceptible to the plant pathogens of crop plants and provide adequate facilities for their growth and reproduction of these pathogens during offseason. Weeds which survive and live during non-cropping season provide for the continuous growth and multiplication of the pathogen. Thus the weed hosts help to bridge the gap between two crop seasons. E.g., The fungal pathogen for blast disease of rice, *Pyricularia grisea* (Teleomorph: *Magnaporthe grisea*) can infect the grass weeds like *Brachiaria mutica*, *Dinebra retroflexa*, *Leersia hexandra*, *Panicum repens* etc., and survive during off-season of rice crop. As soon as a fresh rice crop is raised, the conidia (inoculum) liberated from the weed host disseminated by wind infects the fresh rice crop.

c. Alternate hosts (Wild hosts of other families): The role of alternate hosts is not as important as of collateral hosts. However, when a pathogen has very wide host range (as

Sclerotium rolfsii, Rhizoctonia solani, Fusarium moniliforme etc.) and is tolerant to wide range of weather conditions the alternate hosts become very important source for survival of the pathogen. These alternate hosts are very important for the completion of the life cycle of heteroecious rust pathogens. In temperate regions the alternate host of *Puccinia graminis* var. *tritici* (black or stem rust pathogen of wheat), the barberry bush (*Berberis vulgaris*) grows side by side with the cultivated host. In such areas this wild host belonging to a different family is important for survival of the fungus.

Cultivated hosts	Rust fungus	Alternate host
Barley	*Puccinia hordei*	*Ornithogalum* spp.
Cherry	*Puccinia cerasi*	*Eranthis* spp.
Cotton	*Puccinia stakmanii*	*Bouteloua* spp.
Gooseberry	*P. caricis* var. *grossulariata*	*Carex* spp.
Oat	*Puccinia coranata*	*Rhamnus* spp.
Pear	*Gymnosporangium* spp.	*Juniperus* spp.
Plum	*Tranzschelia discolor*	*Anemone coronaria*
Red currant	*Cronartium ribicola*	*Pinus* spp.
Sorghum	*Puccinia sorghi*	*Oxalis stricta, O. corniculata, O. bowiei*
Wheat	*Puccinia graminis*	*Berberis* spp.
	Puccinia recondita	*Thalictrum* spp., *Isopyrum* spp., *Anchusa* spp.

d. Self sown crops: Self sown crops, voluntary crops and early sown crops are reservoirs of many plant pathogens. E.g., Self sown rice plants harbour the pathogen (*Rice tungro virus*) as well as vector (*Nephottetix virescens*).

e. **Ratoon crops:** Sometimes ratoon crops also harbour the plant pathogens. E.g., Sugarcane mosaic.

f. Survival by latent infection: Latent infection refers to the conditions in which the plant pathogens may survive for a long time in plant tissue without development of visual symptoms. E.g., *Xylella fastidiosa*, the causal agent of pierce's disease of grapevine infect different weeds without developing visible symptoms.

3. Survival as saprophytes: Many soil and foliar pathogens survives in infected plant debris in soil as saprophytes for long time.

4. Survival as specialized resting structures and dormant spores

a. Phytopathogenic bacteria: The plant bacteria also do not produce resting spores or similar structures. They continuously live in their active parasitic stage in the living host or as active saprophytes on dead plant debris. E.g., Endospores.

b. **Nematodes**: They survive in the form of active parasitic phase on a living host and also survive through dormant structures, i.e., eggs, cysts, knots, galls, formed in host tissues. These structures may be present in soil or in seed lots.

c. Phanerogamic parasites: They survive in dormant state for many years through seeds. Eg; Seeds of Orobanchae survive in soil for more than 7 years.

d. **Soil borne fungi:** Dormant spores - Conidia (Peach leaf curl pathogen, *Taphrina deformans*), Chlamydospores (Wilt pathogen, *Fusarium* sp.), oospores (Downy mildew fungi), perithecia (Apple scab pathogen, *Venturia inaequalis*), thickened hypha, sclerotia (Cottony rot fungus, *Sclerotinia sclerotiorum*), microsclerotia (*Verticillium*), Rhizomorphs (*Armillaria mellea*) etc.

d. Seed borne fungi

- » **Externally seed borne**: Dormant spores on seed coat E.g., Covered smut of barley, grain smut of jowar, bunt of wheat, etc.
- » **Internally seed borne**: Dormant mycelium under the seed coat or in the embryo E.g., Loose smut of wheat (*Ustilago nuda tritici*)

5. Survival in association with insects, nematodes and fungi

The corn flea beetle, *Cheatocnema pulicaria* carries inside its body, the corn wilt pathogen, *Xanthomonas stewartii* and thus helps in over wintering. Plant viruses like wheat mosaic, tobacco necrosis, tobacco rattle and tobacco ringspot viruses survive with nematodes or fungi found in the soil between crop seasons. Tobacco ringspot is associated with the nematode *Xiphinema americana*. The fungi, *Polymyxa graminis* (Wheat soil borne mosaic and Barley yellow mosaic) and *Spongospora subterranea* (Potato mop top virus) carry the viruses internally and transmit them through the resting spore.

- » Nematode transmitted fungi: *Fusarium oxysporum* f.sp. *cubense*
- » Nematode transmitted bacteria: *Anguina tritici*
- » Nematode transmitted virus: Tobacco necrosis virus,
- » Fungal transmitted virus: *Tobacco ringspot virus* (NEPO), *Tobacco rattle virus* (NETO).

Practice Questions

1. ....is hard resting structure produced by fungi
2. ....alternate host for sorghum rust

3. .... infection refers to the conditions in which the plant pathogens may survive for a long time in plant tissue without development of visual symptoms
4. Self sown rice plants harbour the pathogen .... as well as vector ....
5. *Phytophthora infestans* causing late blight of potato survives in .... and........
6. Explain the different survivability nature of plant pathogens

Suggested Readings

Willetts, H. J. (1971). The survival of fungal sclerotia under adverse environmental conditions. *Biological Reviews*, *46*(3), 387-407.

Engeli, H., Edelmann, W., Fuchs, J., & Rottermann, K. (1993). Survival of plant pathogens and weed seeds during anaerobic digestion. *Water Science and Technology*, *27*(2), 69-76.

Mayer, A. M., Staples, R. C., & Gil-ad, N. L. (2001). Mechanisms of survival of necrotrophic fungal plant pathogens in hosts expressing the hypersensitive response. *Phytochemistry*, *58*(1), 33-41.

Schuster, M. L., & Coyne, D. P. (1974). Survival mechanisms of phytopathogenic bacteria. *Annual Review of Phytopathology*, *12*(1), 199-221.

Shlevin, E., Mahrer, Y., Kritzman, G., & Katan, J. (2004). Survival of plant pathogens under structural solarization. *Phytoparasitica*, *32*(5), 470-478.

Chapter - 24

Types of Parasitism

Parasitism: Organisms which derive the susbtrate materials needed for growth and reproduction from living plants (host or suscept) are called parasites. It is further classified into several types

Ectoparasite:

» Most of the parts live on the external surface of the host

» Uptake nutrients from epidermal and mesopyll cells

E.g., *Erysiphe polygoni*

Endoparasite

» Most of the organal parts live on the internal tissues of the host. It grows in the subcuticular, parenchyma tissues or vascular bundles.

» Intercellular – grow between the host cells and draw nutrients. E.g., *Erysiphe polygoni*

» Intracellular – grow inside the host cell and draw nutrients. E.g., *Leveillula taurica*

Endobiotic parasite

» Completely live only in the internal tissues of the host.

E.g., *Synchytrium endobioticum*

Destructive parasite

» Draw nutrition from the host and destroy / kill the host

E.g., *Rhizoctonia, Macrophomina, Sclerotium, Fusarium, Verticillium*

Balanced parasite

» Draw nutrition without killing the host. E.g., *Colletotrichum*

Facultative parasite

» Usually saprophyte

» Mode of life and under certain condition – became a parasite

E.g., *Pythium, Fusarium, Rhizopus*

Facultative saprophyte

» Usually a parasite, under certain condition became a saprophyte

E.g., *Phytophthora, Mucor, Venturia*

Mode of nutrition in fungi

Complicating our understanding of life cycles is the fact that microorganisms interact with plants in a variety of drastically different ways, ranging from commensal to mutualistic to pathogenic to saprophytic.

There are three extreme lifestyles for pathogens: biotrophic, hemibiotrophic and necrotrophic. Purely biotrophic pathogens are typically obligate parasites, although necrotrophic pathogens can also exist as saprophytes.

Parasites: Organisms which live within or outside another organism for their nutrition either completely or for a part of their life.

» **Pathogen**: If a parasite damages the host then they are called as pathogens.

» All pathogens are not parasites and all parasites need not be pathogens.

1. Biotrophs: Biotrophic pathogens are live and reproduce only in living tissue. If one of their spore's lands on soil or dead plant material, it will die.

a. **Obligate parasite/True parasite**: An organism that can live, multiply and complete their lifecycle only on another living organism. These kinds of parasites which cannot be grown on dead or artificial food. E.g., Powdery mildew fungi (*Blumeria* (*Erysiphe*) *graminis*), downy mildew parasites (*Plasmopara viticola*), *Cladosporium fulvum*, Candidatus *Phytoplasma* and viruses.

b. **Facultative saprophytes:** Organisms which are usually parasites but have ability to become saprophytes. E.g., *Ustilago maydis*

c. **Obligate inapproporiate**: The biotophic parasites which can be cultured away from their host on artificial media. E.g., Rust (*Puccinia triticina, Uromyces fabae*) and smut fungi (*Ustilago maydis*).

2. Necrotroph (Perthotrophs or perthophytes): A parasite is a necrotroph when it kills the host tissues in advance of penetration and then lives saprophytically. E.g., *Phytophthora infestans*, *Pythium ultimum*, *Agrobacterium tumefaciens*. Also, possess enzyme and toxin secretion during infection process. E.g., *Sclerotium rolfsii, Botrytis cinerea*, *Cochliobolus heterostrophus*, *Fusarium oxysporum* and *Claviceps* spp.

Necrotrophs	Biotrophs
Opportunistic, unspecialized ('non-obligate') pathogens	Specialized ('obligate') pathogens
Host cells killed rapidly	Cause little damage to the host plant; host cells not killed rapidly, but can induce hypersensitive cell death in incompatible interactions
Entry unspecialized via wounds or natural openings	Entry specialized, E.g., direct entry (powdery mildews) or through natural openings (rusts)
Secrete copious cell-wall-degrading (lytic) enzymes and toxins	Few if any lytic enzymes or toxins are produced
Appressoria/haustoria not normally produced	Possess appressoria or haustoria
Seldom systemic	Often systemic
Usually attack weak, young or damaged plants	Plants of all ages and vigour attacked
Wide host range	Narrow host range
Easy to culture axenically	Not easily cultured axenically
Survive as competitive saprotrophs	Frequently survive on host or as dormant propagules
Controlled by quantitative resistance genes (example is *Septoria nodorum* blotch caused by *Stagonospora nodorum*)	Controlled by specific (gene-for-gene) resistance genes (for example, tomato leaf mould, the rusts, powdery and downy mildews
Growth in host intercellular and intracellular through dead cells	Growth intercellular
Controlled by jasmonate- and ethylene-dependent host defence pathways	Controlled by salicylate-dependent host defence pathways

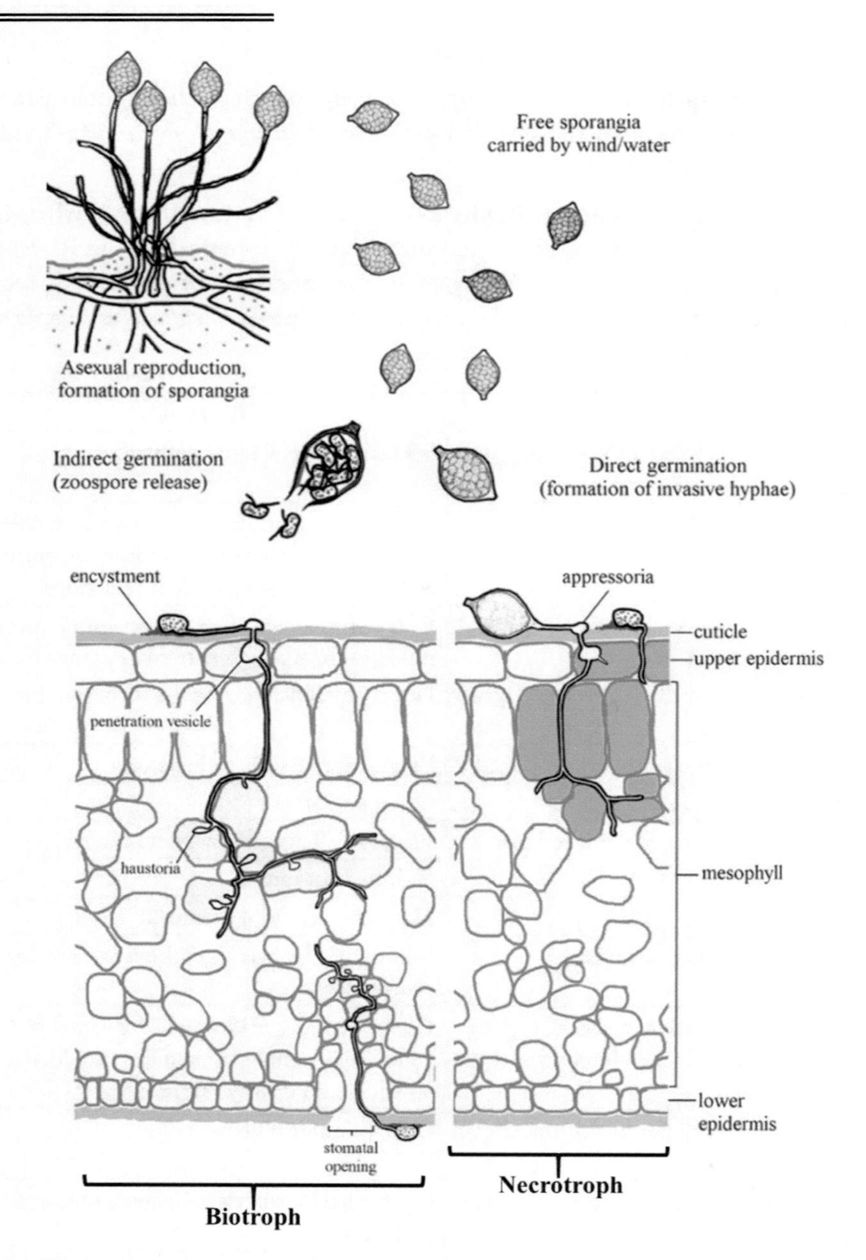

3. Hemibiotroph (Facultative Saprophyte): The parasites which attack living tissues in the same way as biotrophs but will continue to grow and reproduce after the tissue is dead as necrotroph called as facultative saprophytes. E.g., Leaf spoting fungi such as *Venturia ineaqulis*, *Mycosphaerella graminicola*, *Alternaria solani*, *Helminthosporium*, *Magnaporthe grisea*, *Cladosporium fulvum*, *Colletotrichum lindemuthianum* etc.,

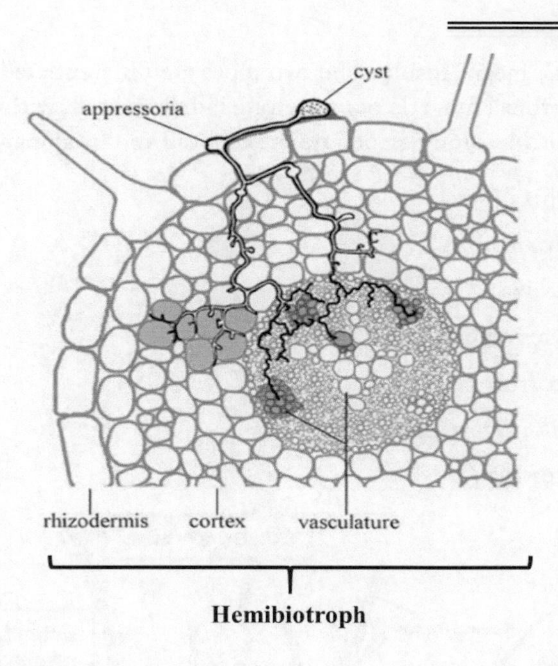

4. Saprophytes: (sapros=rotten, phytos=plant) Organisms which obtain nutrition on from dead organic matter either completely or for a part of their life. A large number of fungi fall under this category. E.g., *Saprolegnia, Rhizopus, Mucor, Alternaria.*

a. **Obligate saprophytes:** (obligate= to bind itself) Organisms which can never grow on living organisms or can never obtain their food from living source. They get their food only from dead organic matter. E.g., *Mucor, Agaricus*.

b. **Facultative parasite:** (facultas= ability) Organisms which are usually saprophytic but have ability to become as parasites. E.g., *Pythium aphanidermatum, Fusarium solani, Rhizoctonia solani.*

Symbiosis

Mutual beneficial relationship between two dissimilar organisms which living together in close association. E.g., Lichens, AM fungi, *Rhizobium.*

The roots of most plants are colonized by symbiotic fungi to form mycorrhiza, which play a critical role in the capture of nutrients from the soil and therefore in plant nutrition. Frank (1885) was probably the first to recognize the widespread nature of associations between plant roots and mycorrhizal fungi. Major mycorrhizal types have been described based on their structure and function.

Mycorrhizas are commonly divided into ectomycorrhizas and endomycorrhizas *ecto* means outside the root, *endo* means inside. The two types are differentiated by the fact that the hyphae of ectomycorrhizal fungi do not penetrate individual cells within the root, while the hyphae of endomycorrhizal fungi penetrate the cell wall and invaginate the cell membrane.

a. Ectomycorrhiza
» Arbutoid mycorrhiza

b. Endomycorrhiza
» Arbuscular mycorrhiza
» Ericoid mycorrhiza
» Monotropoid mycorrhiza
» Orchid mycorrhiza

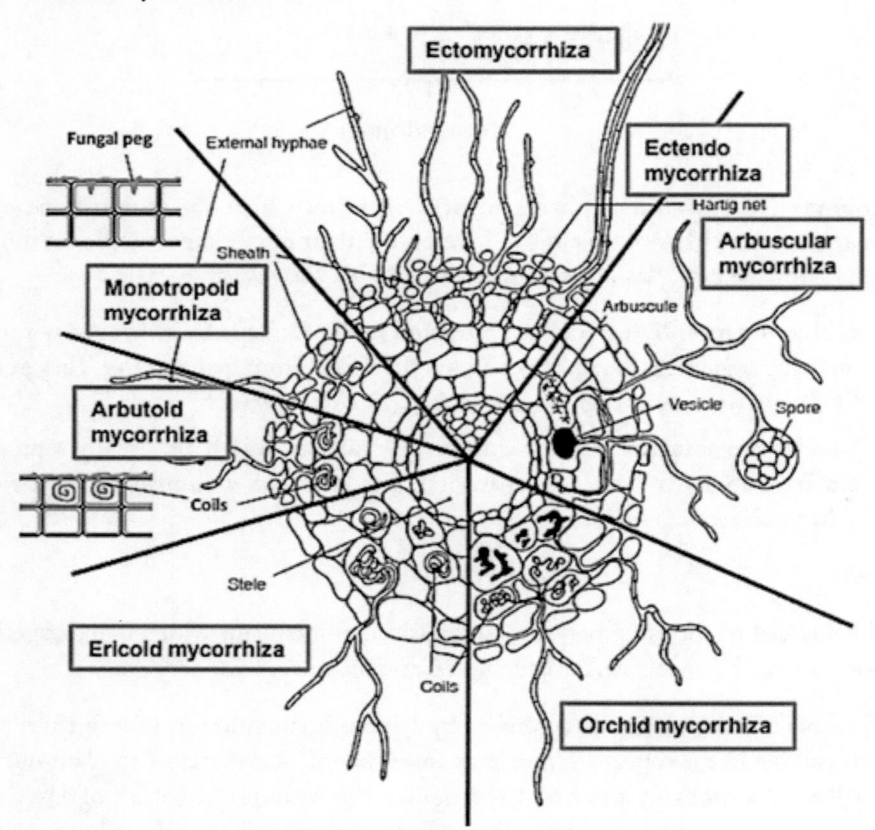

Mycorrhizal interactions in roots

Source: Selosse & Le Tacon (1998)

Practice Questions

1. Write the mechanisms of mycorrhizal associations in plants
2. Write difference between parasite and saprophyte
3. Write difference between biotroph and necrotroph

Suggested Readings

Kabbage, M., Yarden, O., & Dickman, M. B. (2015). Pathogenic attributes of Sclerotinia sclerotiorum: switching from a biotrophic to necrotrophic lifestyle. Plant science, 233, 53-60.

van Kan, J. A. (2006). Licensed to kill: the lifestyle of a necrotrophic plant pathogen. *Trends in plant science*, *11*(5), 247-253.

Pandaranayaka, E. P., Frenkel, O., Elad, Y., Prusky, D., & Harel, A. (2019). Network analysis exposes core functions in major lifestyles of fungal and oomycete plant pathogens. *BMC genomics*, *20*(1), 1-15.

Luttrell, E. S. (1974). Parasitism of fungi on vascular plants. *Mycologia*, *66*(1), 1-15.

Chapter - 25

Variability and Resistance in Plant Pathogens

Successful management of plant disease is mainly dependent on the accurate and efficient detection of plant pathogens, amount of genetic and pathogenic variability present in pathogen population, development of resistant cultivars and deploying of effective resistance gene in different epidemiological region. Genetic analysis of pathogen populations is fundamental to understanding the mechanisms generating genetic variation, host-pathogen co-evolution, and in the management of resistance.

Variability: It is the property of an organism to change its characters from one generation to the other. Characteristics of individuals within a species are not fixed and variation is introduced- segregation and recombination.

Physiological specialization: it happens with in the species of a pathogen there exist certain individuals that are morphologically similar but differs with respect to their physiology, biochemical characters and pathogenicity and are differentiated on the basis of their reaction on certain host genera or cultivars.

Physiologic race: individuals with in the species of a pathogen that morphologically similar but difffer with respect to their pathogenicity on particular set of host varieties

Forma specialis (f. sp.): individuals with in the species of a pathogen that morphologically similar but difffer with respect to their pathogenecity on particular host genera.

E.g., *Puccinia graminis* f.sp. *tritici*

Variability: it is the property of an organism to change its characters from one generation to the other.

Variation: When progeny of an individual show variation in characters from parents such a progeny is called a variant

Pathotype: A pathotype is a population of a parasite species in which all individuals have a stated pathosystem character (pathogenicity or parasitic ability) in common.

Biotype: Progeny developed by variant having similar heredity is called a biotype or a subgroup of individuals with in the species, usually characterized by the possession of single or few characters in common.

Variability in Plant Pathogens

» Fungi produced from sexual spores (oospores, ascospores, basidiospores) show much variability.

» The frequency and degree of variability among the progeny are reduced greatly when fungi produced asexually.

Stages of variation in fungi, bacteria and viruses

Characters	Fungi	Bacteria	Virus
Morphology and biochemistry	**Genus**	**Genus**	**Genus**
	↓	↓	↓
	Species	Species	Species
	↓	↓	↓
Host	Variety or special form	Variety or pathovar	Type
	↓	↓	↓
Differential varieties or symptoms	Race	Race	Strain
	↓	↓	↓
Localized field population	Isolate	Isolate	Isolate
	↓	↓	↓
Clonal population	Single spore-derived biotype	Single spore-derived strain	Single local isolate

1. Process of variability in fungi: Mechanism of variability in case of fungi includes mutation, recombination, heterokaryosis, parasexulaism, and heteroploidy

a. Mutation

Mutation is a process in which there is a change in the genetic material of an organism occur either through naturally or through induced factors, which is then transmitted in a hereditary fashion to the progeny. It is a more or less abrupt change in the genetic material of an organism, i.e. DNA and the change is heritable to the progeny.

» Mutations represent changes in the sequence of bases in the DNA either through substitution of one base for another or through addition or deletion of one or many base pairs.
» May be by amplification of particular segment of DNA to multiple copies by insertion or excision of a transposable element into coding or regulatory sequences of the gene.
» Mutations are spontaneous.
» It is fast and expressed soon in single celled organism mostly recessive
» Also reported in the extra nuclear DNA (cytoplasmic DNA).

Gassner and Straib (1993) were the first to suggest mutation as a mechanism for formation of new races in *Puccinia striiformis*.

b. Recombination

Recombination in plant pathogens is the similar process to that of sexual reproduction. It occurs either through a process of somatic hybridization, in which nuclear and cytoplasmic material get exchanged. In most of cases nuclear exchange may be followed by nuclear fusion and recombination also called as parasexual cycle. The exchange of cytoplasmic as well as nuclear fusion leads to increased genotypic diversity in a pathogen population, but their importance varies both within and among species.

When two haploid nuclei (1N) containing different genetic material unite to form diploid (2N) nucleus called a Zygote, when undergo meiotic division produce new haploid. Recombination of genetic factor occurs during meiotic division of zygote as a result of cross over in which part of chromatid of one chromosome of a pair are expressed with that of the other.

» Recombination can also occur during mitotic division of cell in the course of growth of the individual and is important in fungi *Puccinia graminis*.
» The majority of fungi being haploid, have only a very brief diploid phase, very often undergoing meiosis very soon after karyogamy.
» Some Oomycetes are predominantly diploid.
» Many Basidiomycetes are functionally diploid by virtue of their extended dikaryon phase.
» In case of homothallic fungi which are essentially self-compatible and in a single

thallus. In these fungi, the extent of out-crossing will be close to zero and there will be little opportunity for recombination between genetically different individuals.

» On the other hand, there are homothallic fungi with physiologic distant mycelia upon which are produced compatible male and female gametes.

c. Gene and genotype flow

Certain genes move from one population to another geographically separated population is known as Gene flow. Migration of one pathogen population from one place to another leads to development of new species of which are either absent or not on many occasions (E.g., the introduction of *Cryphonectria parasitica* to North America, *Phytophthora infestans* to Europe, and *Puccinia striiformis* to Australia.

d. Heterokaryosis

Heterokaryosisis a condition in which cells of fungal hyphae or parts of hyphae contain two or more nuclei those are genetically different. For example, in Basidiomycetes, the dikaryotic state is found to be completely different from the haploid mycelium and spores of the fungus. In *Puccinia graminis* var. *tritici*, the dikaryotic mycelium can grow in both barberry and wheat but the haploid mycelium can grow only in barberry not on wheat. Similarly, the haploid basidiospores can infect barberry but not wheat. However, the dikaryotic aeciospores and uredospores can infect wheat but not barberry. Heterokaryotic condition arises by mutation, anastomosis and inclusion of dissimilar nuclei in spores after meiosis in heterothallic fungi. Cellular events preceding successful anastomosis opened the way to genetic tests on compatibility/incompatibility, which showed genetic exchange between different genotypes.

Rust fungi had dikaryotic phase differ from haploid mycelium and spores

» Haploid phase infects barberry

» Dikaryotic phase infects wheat

e. Heteroploidy

Heteroploidy is the existance of cell tissues or whole organisms with numbers of chromosomes per nucleus. Heteroploids may be haploids, diploid, triploid or tetraploids i.e. have one or more. Extra chromosomes from normal euploid number E.g., N+1. Like, in *Verticillium alboatrum* – wilt disease of cotton – heteroploid lose their ability to infect.

f. Parasexuality

A process in which plasmogamy, karyogamy and haploidization takes place in sequence but not at specified points in the life cycle of an individual. First discovered in 1952 by

Pontecorvo and Roper in the University of Glasgow in *Aspergillus nidulans*, the imperfect stage of *Emericella nidulans*. The parasexual cycle (genetic recombination without meiosis). Stages of the parasexual cycle are numbered as follows (1) Hyphal conjugation (plasmogamy). (2) Heterokaryosis. (3) Nuclear fusion (karyogamy). (4) Mitotic recombination and nondisjunction. (5) Haploidization and nuclear segregation leading to homokaryosis.

» In many rusts, including *Puccinia graminis* var. *tirtici, Puccinia coronata* and in some smuts *Ustilago hordei* and *Ustilago maydis.*

» In rust fungi as *Puccinia graminis* var. *tritici,* mitotic recombination may represent a most important method of generating new races where sexual stage of the fungus is rare.

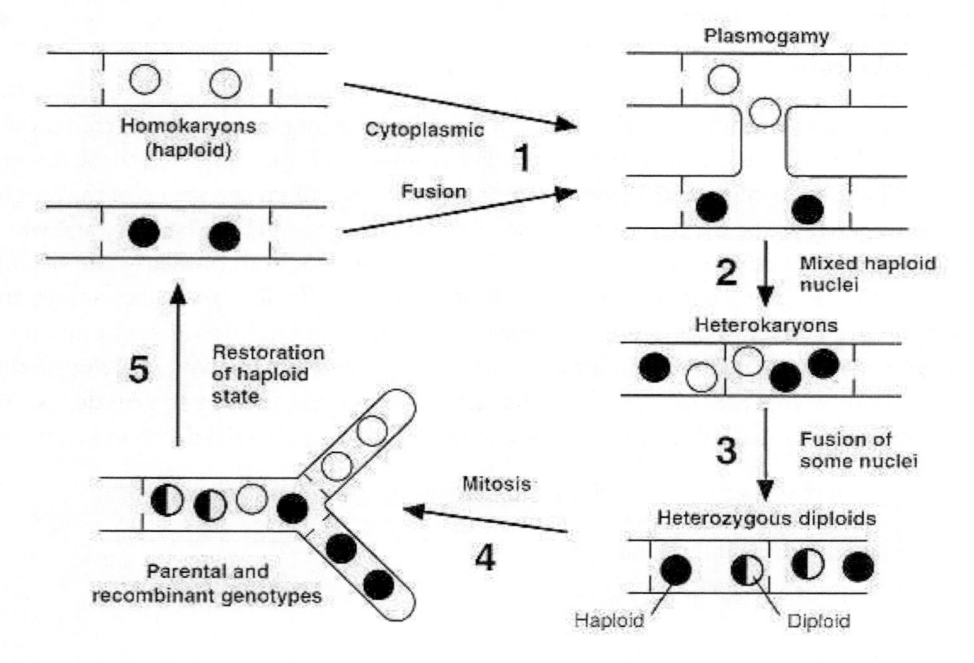

The sequences of events in a complete parasexual cycle are as follow:

» Formation of heterokaryotic mycelium.

» Fusion between two nuclei.

» Fusion between like nuclei.

» Fusion between unlike nuclei.

» Multiplication of diploid nuclei

» Occasional mitotic crossing over during the multiplication of the diploid nuclei.

» Sorting out of diploid nuclei Occasional haploidization of diploid nuclei.

» Sorting out of new haploid strains.

2. Process of variability in bacteria: mutation, conjugation, transformation and transduction.

a. Mutation: A gene will mutate spontaneously, about once in a hundred million cell divisions. Such bacteria are called mutants. Most of these mutants die, but a when a mutant can adapt itself to the environment more readily; it may emerge as a new variant. Chromosomal mutations may lead to emergence of drug resistance in bacteria.

b. Transformation: Some bacteria have ability to uptake naked DNA fragment from the surrounding environment. When such a DNA confers new property to the bacterium, it is termed transformation. Change from R form to S form as demonstrated by Griffith is due to transformation.

c. Conjugation: Transfer of genetic material (usually plasmids) from one bacterium to another through the mediation of sex pili is known as conjugation. Any property that is coded on a transmissible plasmid can be transferred to a recipient bacterium. Properties such drug resistance mediated by beta-lactamases, bacteriocin production etc can be transferred by conjugation.

» Transfer of DNA from one bacterial cell to another

» Donor cell (F+ or Hfr) transfers DNA to recipient cell (F-)

» In these two compatible bacteria come in contact and exchange the portion o plasmid or chromosome through Conjugation Bridge or pilus.

d. Transduction: Transfer of genetic material through mediation of bacteriophage is known as transduction.

e. Transposition: Variations in the flagellar antigens are due to transposons.

3. Process of variability in virus: recombination, reassortment and mutation

a. Mutation: The rate of spontaneous mutation is a key parameter to understanding the genetic structure of populations over time. Mutation represents the primary source of genetic variation on which natural selection and genetic drift operate. RNA viruses show mutation rates that are orders of magnitude higher than those of their DNA-based hosts and in the range of 0.03-2 per genome and replication round.

b. Recombination and reassortment: Recombination has been associated with the expansion of viral host range, increases in virulence, the evasion of host immunity and the evolution of resistance to antiviral. When two strains of the same virus are inoculated into the same host plant, one or more new virus strains are recovered with properties (virulence, symptomatology, etc) different from those of either of the original strains introduced into the host.

Plant susceptibility

Plant susceptibility to pathogens is usually considered from the perspective of the loss of resistance. However, susceptibility cannot be equated with plant passivity since active host cooperation may be required for the pathogen to propagate and cause disease.

Plant susceptibility to pathogens is usually considered from the perspective of the loss of resistance. However, susceptibility cannot be equated with plant passivity since active host cooperation may be required for the pathogen to propagate and cause disease. This cooperation consists of the induction of reactions called susceptible responses that transform a plant from an autonomous biological unit into a component of a pathosystem.

Plant susceptibility can be divided into two main categories depending on their consequences.

- The first category includes susceptibility factors are necessary for the multiplication of a pathogen inside the host; without these factors, the proliferation of a microorganism in plant is impossible, and disease symptoms do not occur.
- The second category of susceptibility factors includes that host are not required for pathogen multiplication but are necessary for the manifestation of disease symptoms. Absence of susceptibility factors of this category does not hamper pathogen reproduction in the host plant but prevents the pathological process.

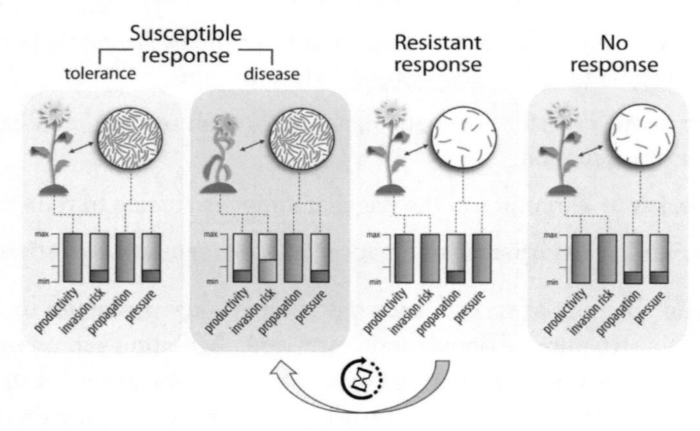

Crop responses to the pathogen

The consequences of different types of susceptible responses compared to the defence response and pathogen insensitivity. The columns show the relative levels of plant productivity, risk of invasion of the alternative pathogens, pathogen propagation in plants, and selective pressure exerted on a pathogen.

Susceptibility (S) gene is any plant gene that facilitates the infection process or supports compatibility with a pathogen. Plant S-gene products provide signals attracting pathogens and increasing their aggressiveness and contribute to the accommodation of pathogens. E.g., SWEET gene: encodes a sucrose-efflux transporter that is hijacked by Xanthomonas oryzae, causing bacterial blight in rice.

Plant S-genes were divided into three main categories:

- Genes that facilitate host recognition and penetration;
- Genes that encode negative regulators of immune signalling; and
- Genes that fulfill metabolic or structural requirements of a pathogen that allow its proliferation.

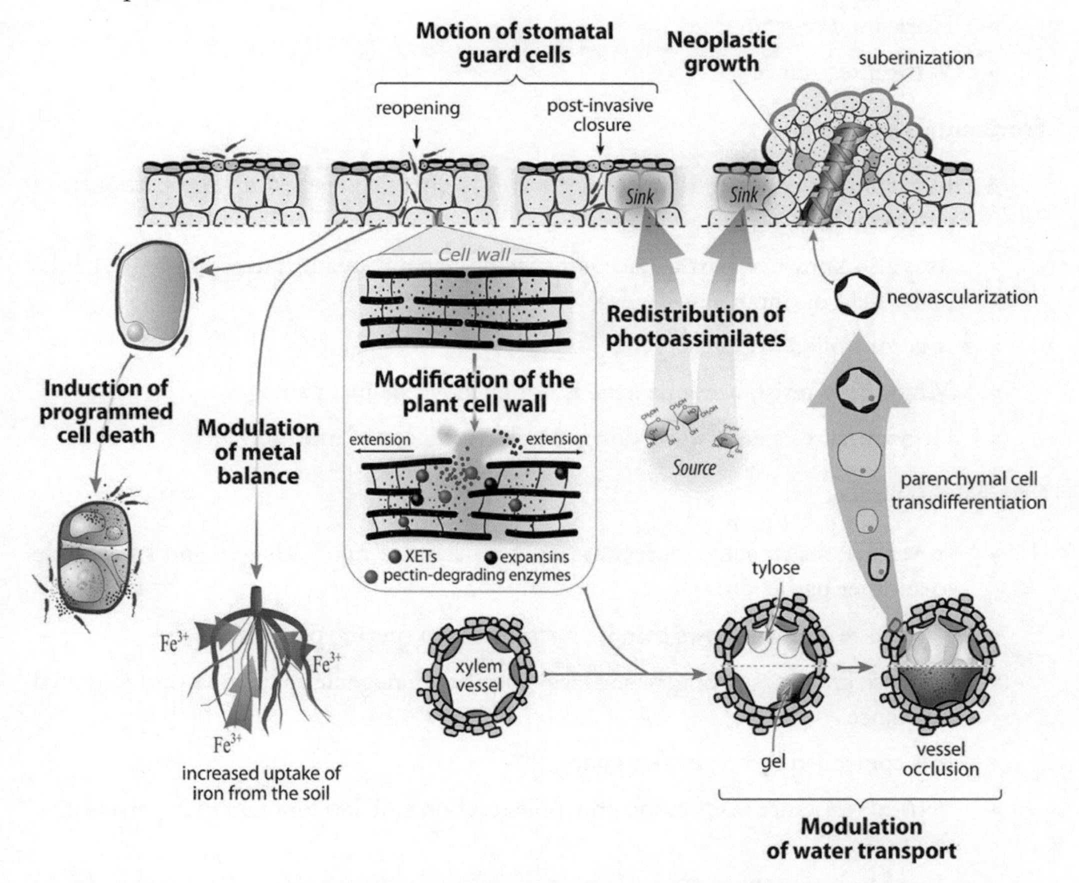

Schematic representation of pathogen-induced plant susceptible responses. XET, xyloglucan endotransglucosylases/hydrolase. (Source: Gorshkov and Tsers, 2021)

Host resistance

Generally, plants are resistant to pathogen due to various reasons that they may be non host for pathogen or they possess R genes against avr genes or the plant escape the disease due to unfavorable environment and tolerance. There are two types of resistant, True resistance and apparent resistance.

True resistance

Disease resistance that is controlled genetically by the presence of one, a few, or many genes for resistance in the plant is known as true resistance.

- » Horizantal resistance
- » Vertical resistance

Horizantal resistance

- » Horizontal resistance is nonspecific and is effective against all the pathogen of particular host
- » It is also known as partial, race nonspecific, general, quantitative, polygenic, adult-plant, field or durable resistance
- » It is controlled by several gene
- » Minor gene resistance- one gene is not effective against pathogen
- » Horizontal resistance slows down the development of infection loci

Vertical resistance

- » In vertical resistance, a variety strongly resistant to one pathogen and susceptible to another pathogen
- » Vertical resistance shows complete resistance to specific pathogen
- » It is also known as strong, major, race- specific, monogenic, qualitative or differential resistance
- » It is controlled by one or few genes
- » Vertical resistance leads to incompatible reaction and development of hypersensitive reaction
- » E.g., wheat has 20 to 40 genes for *P. recondita*.

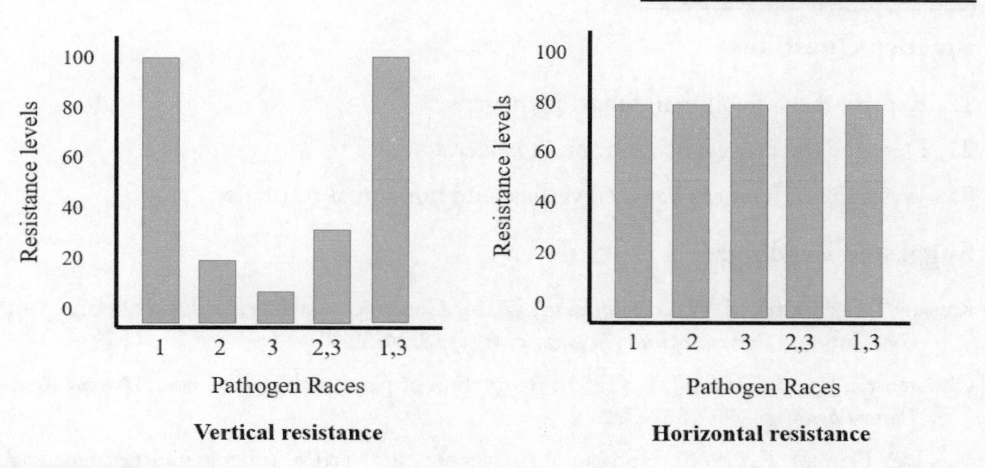

Apparent resistance

The apparent resistance to disease of plants known to be susceptible is generally a result of disease escape or tolerance to disease.

Disease escape

Disease escape occurs whenever genetically susceptible plants do not become infected because the three factors necessary for disease (susceptible host, virulent pathogen, and favorable environment) do not coincide and interact at the proper time or for sufficient duration.

Example:

» Plant escapes *Fusarium* and *Rhizoctonia* infection at low temperature likewise from *Pythium* and *Phytophthora* infection in dry soil condition

» Plants escape disease because they are interspersed with other types of plants that are insusceptible to the pathogen and because the amount of inoculum that reaches them is much less than if they were in monocultural plantations.

Disease tolerance: Tolerance results from specific, heritable characteristics of the host plant that allow the pathogen to develop and multiply in the host while the host, either by lacking receptor sites for or by inactivating or compensating for the irritant excretions of the pathogen, still manages to produce a good crop. Tolerant plants are, obviously, susceptible to the pathogen, but they are not killed by it and generally show little damage.

The repression of crop susceptibility leads not to plant resistance but to plant tolerance - a state in which a plant does not show a significant reduction in fitness despite high pathogen numbers in plants. Plant tolerance to pathogens is associated with the development of latent infections.

Practice Questions

1. Explain the variability in fungal pathogens
2. Describe the process of variation in bacteria
3. Write the differences between vertical and horizontal resistance

Suggested Readings

Karasov, T. L., Horton, M. W., & Bergelson, J. (2014). Genomic variability as a driver of plant–pathogen coevolution?. *Current opinion in plant biology*, *18*, 24-30.

Christensen, J. J., & DeVay, J. E. (1955). Adaptation of plant pathogen to host. *Annual Review of Plant Physiology*, *6*(1), 367-392.

Van Der Plank, J. E. (1966). Horizontal (polygenic) and vertical (oligogenic) resistance against blight. *American Potato Journal*, *43*(2), 43-52.

Parlevliet, J. E. (1977). Plant pathosystems: An attempt to elucidate horizontal resistance. *Euphytica*, *26*(3), 553-556.

Keane, P. J. (2012). Horizontal or generalized resistance to pathogens in plants. *Plant Pathology. Rijeka, Croatia: InTech*, 327-362.

Chapter - 26

Pathogenesis

Pathogenicity is the ability of the pathogen to cause disease.

Pathogenesis is the chain of events that lead to development of disease in the host (or) sequence of progress in disease development from the initial contact between the pathogen and its host to the completion of the syndrome.

It is the third link in the infection chain after survival and dispersal of inoculum. Infection process means establishment of pathogen in the host plant. Entry and colonization of pathogen in the host tissues is known as establishment and the infective propagules coming in contact with the host are known as inoculum.

Disease Cycle

After the initial interaction between host and pathogen, a sequence of particular events leads to the development of a disease and, eventually, an epidemic. This sequence of occurrences is referred to as the disease cycle. Frequently, a disease cycle is practically identical to a pathogen's life cycle, with the important difference being that a disease cycle largely depicts the disease as it develops due to the interaction between the host plant and the pathogen.

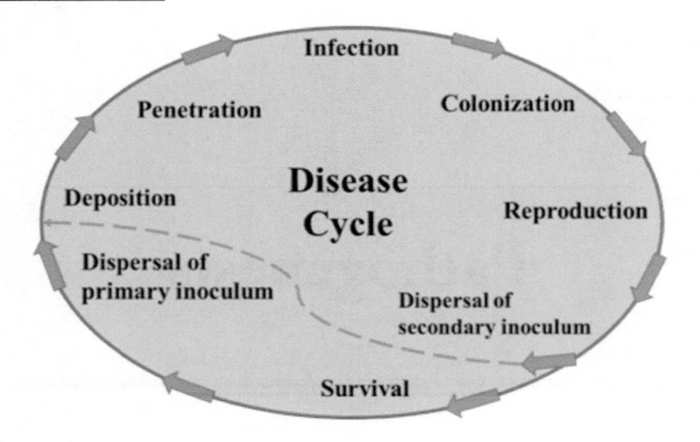

Factors governing disease cycle

Inoculum: It is the part of the pathogen that on contact with a susceptible host plant causes infection (or) the infective propagules which, on coming in contact with the host plant causes an infection are known as inoculum.

Inoculum may for example be spores, mycelium fragments, resting structures, virus particles, nematode cysts, and bacterial cells.

- » In case of specialized pathogens as rusts and powdery mildews, very few or even one spore is capable of causing infection successfully.
- » In case of non-specialized pathogens such as *Pythium, Phytophthora, Rhizoctonia* and *Sclerotium* require high density of inoculum on the surface of susceptible host for successful infection.

Inoculum potential: It is the inoculum needed for successful infection. It is a function of inoculum density and their capacity. The energy of growth of a parasite available for infection of a host at the surface of the host organ to be infected (or) The resultant of the action of environment, the vigour of the pathogen to establish an infection, the susceptibility of the host and the amount of inoculum present.

- » It is defined as the resultant of the action of environment, the vigour of pathogen to establish an infection, susceptibility of the host and amount of inoculum present (Dimond and Horsfall, 1960).
- » It is defined as the energy of growth of a parasite available for infection of a host at the surface of the host organ to be infected (Garret, 1960).

Incubation period: The period of time (or time lapse) between penetration of a host by a pathogen and the first appearance of symptoms on the host. It varies with pathogens, hosts and environmental conditions.

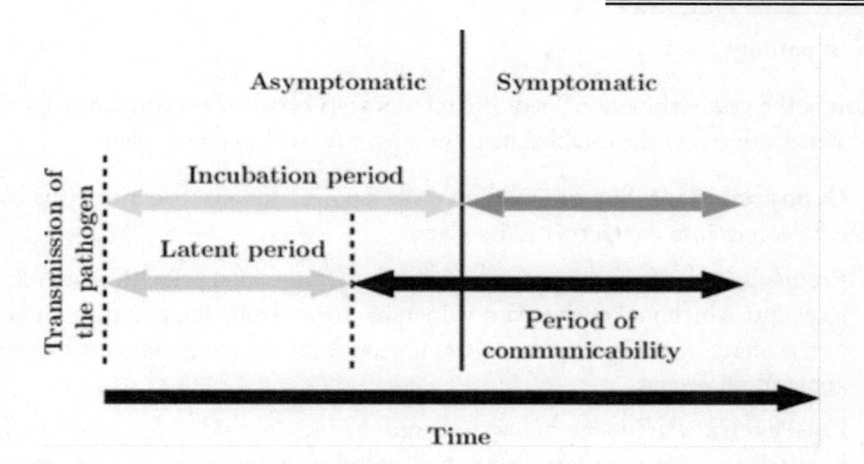

Hypersensitivity: The hypersensitive response is considered a biochemical defense mechanism as distinguished from a structural mechanism. It is an excessive sensitivity of plant tissues to certain pathogens. In many instances, the nucleus disintegrates as soon as the pathogen makes contact with the cell and moves toward it. Dark, resin-like granules eventually form in the infested cytoplasm, initially near the pathogen entry point and then across the cytoplasm. The invading hypha starts to deteriorate as the cell cytoplasm continues to brown, and death begins to set in. Most of the time, the invasion is prevented because the hypha does not develop from such cells. When a cell comes into contact with bacteria, the hypersensitive response causes all of the cell membranes to be destroyed. This is followed by the desiccation and necrosis of the bacterially infected leaf tissues.

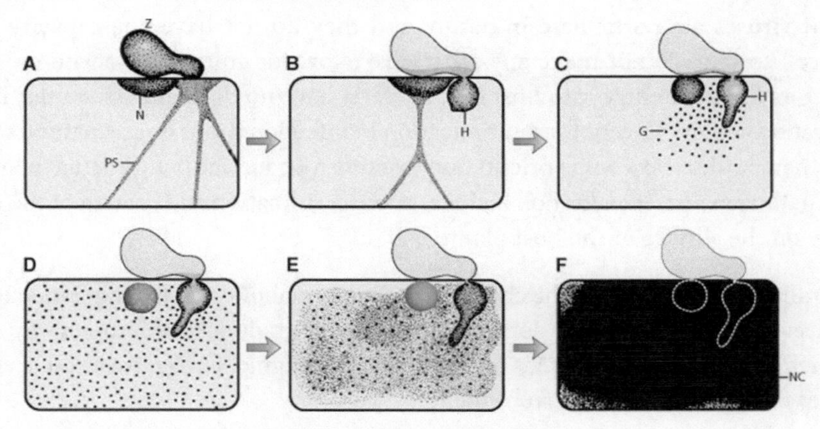

Stages in developing the necrotic defense reaction in a host cell.

N, nucleus; PS, protoplasmic strands; Z, zoospore; H, hypha; G, granular material; NC, necrotic cell. (Source: Tomiyama, 1956).

Process of pathogenesis

Infection is the establishment of parasitic relationship between two organisms, following entry or penetration (or) the establishment of a parasite within a host plant.

a. **Deposition**: Inoculum or infection units (spores, mycelia, bacteria, virus particles, etc.) comes into contact with the plant.

b. **Predisposition:** It is the action of set of environments, prior to penetration and infection, which makes the plant vulnerable to attack by the pathogen. It is related to the effect of environments on the host, not on the pathogen, just before actual penetration occurs.

c. **Pre-Penetration**: Inoculum may undergo changes that aid in the penetration of the host, for example, spore germination. Depending upon the plant pathogen activity, the plant pathogens are classified into two categories as active invaders and passive invaders

Active invaders	Passive invaders
Pathogens which make an aggressive effort to gain entry into intact host cells. They do not require help of any External agency to gain entry into host cells. E.g., Phyto-pathogenic fungi, Phanerogamic parasites	No aggressive effort. Require help of External agencies like insect vectors or wounds caused by agricultural implements. E.g., Plant viruses, Phyto-pathogenic bacteria

Plant viruses are particulate in nature and they do not have any capacity to enter the host cell so they do not make any aggressive effort for entry, but depend on different insect vectors for their entry into host cell. Bacteria have no dormant structures; hence no prepenetration activity Except for multiplication in infection drops on the natural openings. However, nematodes show some orientation towards root surface before actual penetration. In fungal pathogens, pre-penetration includes spore germination and growth of the resulting germ tube on the surface of the host plant.

Germination is essentially the change from low metabolic rate to a high metabolic rate and involves a change from near dormancy to intense activity; for this an energy source is needed such as a carbohydrate or fat reserve in the propagule. Fungal invasion is chiefly by germ tubes or structures derived from them.

In some fungi like *Rhizoctonia solani* and *Armillariella mellea*, the hypha act in a concerted way to achieve the penetration. In *Rhizoctonia solani*, the fungus on coming in contact with root surface, first forms infection cushions and appressoria and from these multiple infections takes place by means of infection pegs. In *Armillariella mellea*, the fungus hyphae form the rhizomorphs (aggregation of hyphae into rope like strands) and only these can cause infection.

d. **Penetration:** Pathogens penetrate plant surfaces by direct penetration or indirectly through wounds or natural openings. Bacteria enter plants mostly through wounds and less frequently through natural openings. Viruses, viroids, mollicutes, fastidious bacteria enter through wounds made by vectors. Fungi, nematodes and parasitic higher plants enter through direct penetration and less frequently through natural openings and wounds.

A) Indirect penetration

1. Wounds: Wounds caused by farm operations, hail storms, or insect punctures, etc., will help in the entry of different plant pathogens into the host cells. Organisms which cause storage diseases and ripe rots will enter through the wounds caused by farm operations. E.g., *Rhizopus, Gloeosporium, Aspergillus, Penicilium, Colletotrichum, Diplodia*, etc.

Weak parasites enter through the wounds caused by hail storms and freezing E.g., *Macrophomina phaseolina.*

Pathogen causing brown rot of fruits (*Sclerotinia fructicola*) enters through the wounds caused by insect punctures. Similarly, causal organism of Dutch elm disease (*Ceratostomella ulmi*) enters through the wounds caused by elm bark beetle.

2. Natural openings

a) Stomata: There is variation in the behaviour of germ tube at the time of penetration through the stomata. In *Puccinia graminis* var. *tritici*, the uredospore germinates and forms a germ tube which on approaching stoma swells at the tip to form an appressorium in the stomatal aperture. From the appressorium a blade like wedge grows through the stomatal slits and swells inside to form a sub-stomatal vesicle from which the haustoria penetrating the cells are produced. In *Peronospora destructor* infecting onion leaves, the germ tube continues to grow after the formation of first appressorium. In *Pseudoperonospora cubensis*, the hyphae penetrate the stomatal aperture and swell to form a sub-stomatal vesicle from which in turn other hyphae grow to form haustoria in the adjacent cells of the leaves. *Mycosphaerella musicola* forms a small structure called stomatopodium over the pore of the stoma after growing for few days on the surface of the leaf. A hypha then arises from it which grows into the sub-stomatal chamber and swells to form a vesicle, which in turn gives rise to hyphae which invade palaside tissues.

Other Examples: *Xanthomonas campestris* pv. *malvacearum* (Black arm of cotton), *Xanthomonas phaseoli* (Bacterial leaf spot of green gram), *Phytophthora infestans* (Late blight of potato), *Albugo candida* (White rust of crucifers) and uredospores of *Puccinia graminis* var. *tritici* (Black stem rust of wheat).

b) Lenticels: *Sclerotinia fructicola* (Brown rot of fruits), *Streptomyces scabies* (Scab of potato), *Phytophthora arecae* (Mahali disease of arecanut)

c) Hydathodes: *Xanthomonas campestris* pv. *campestris* (Black rot of crucifers)

B) Direct penetration: Most fungi, nematodes and parasitic higher plants are capable of penetrating the host surface directly. However, the plants are provided with different mechanisms of defense which include structural features of the host, presence of chemical coverings on the cell walls, and anti-infection biochemical nature of the protoplasm. Hence, the pathogen should have mechanisms to overcome these barriers for direct penetration.

Examples of direct penetration of fungi

Pathogen	Disease	Mode of entry
Plasmodiophora brassicae	Club root of crucifers	Root tissue
Pythium spp.	Damping off of vegetables	Seed coat and emerging seedlings
Ustilago nuda var. *tritici*	Loose smut of wheat	Stigma and young ovary walls
Claviceps spp.	Sugary disease	Stigma
Erysiphe spp.	Powdery mildew	Epidermis
Sclerotinia spp.	Brown rot of stone fruits	Floral parts
Tilletia carries	Wheat bunt	Young seedlings
Phytophthora infestans	Late blight of potato	Cuticle and cell wall
Botrytis cinerea	Petal fire in flowers	Cuticle

a) Breakdown of physical barriers: Viruses have no physical force or enzyme system of their own to overcome structural or chemical barriers of the host and therefore come in contact with the host protoplasm only through wounds. Bacteria are mostly weak parasites and cannot employ force to effect penetration. Fungi and nematodes are the only group of plant pathogens that employ force for direct penetration of the host. Fungi penetrate host plants directly through a fine hypha produced directly by the spore or mycelium or through a penetration peg produced by an appressorium. These structures Egert pressure on the surface which results in stretching of the epidermis which becomes thin. Then the infection peg punctures it and affects its entry.

b) Breakdown of chemical barriers: the host is provided with defense mechanisms against invasion which include i) presence of cuticular layer on the epidermis, ii) lack of suitable nutrients for the pathogen in the host cells, iii) presence of inhibitory or toxic substances in the host cells, iv) exudation of substances toxic to pathogen or stimulatory to antagonists of the pathogen. E.g., The glands in leaf hairs of bengalgram contain maleic acid which is antifungal and provide resistance to infection by the rust fungus (*Uromyces ciceris arietini*). Similarly, protocatecheuic acid and catechol in the red scales of onion provide resistance to onion smudge pathogen, *Colletotrichum circinans*. To overcome these physical and chemical barriers, the fungi produce various enzymes, toxins organic acids and growth regulators.

Through non-cutinized surfaces:

a) Seedlings: Grain smut of jowar (*Sphacelotheca sorghi*), Loose smut of jowar (*Sphacelotheca cruenta*), Downy mildew of jowar and bajra (*Sclerospora graminicola*), Wheat bunt disease (*Tilletia caries*, *Tilletia foetida*)

b) Root hairs: Wilt causing fungi (*Fusarium* sp.), Club root of cabbage (*Plasmodiophora brassicae*), Root rot of cotton (*Phymatotrichum omnivorum*)

c) Buds: Pea rust fungi (*Uromyces pisi*), Witches broom of cherries (*Taphrina cerasi*)

d) Flowers: Loose smut of wheat (*Ustilago nuda tritici*), Long smut of jowar (*Tolyposporium ehrenbergi*), Bunt of rice (*Neovossia horrida*), Ergot of rye (*Claviceps purpurea*)

e) Leaves: Basidiospores of white pine blister rust fungus (*Cronartium ribicola*) germinate and grow down into branches and leaves, where aecia are produced.

d) Nectaries: Fire blight of apple (*Erwinia amylovora*)

e) Stalk ends: *Penicillium italicum*, *Theilaviopsis paradoxa* (Post harvest disease fungi)

Through cutinized surfaces:

a) Cuticle: Leaf spot of spinach (*Cercospora beticola*), early blight of solanaceous plants (*Alternaria solani*), Tikka disease of groundnut (*Cercospora personata*)

e. Post-penetration infection

The establishment of a pathogen inside a host. The pathogen suppresses host defences.

Invasion: Infection is the process by which pathogens establish contact with the susceptible cells or tissues of the host and derive nutrients from them. A parasitic relationship is formed between host cytoplasm and parasite cytoplasm. During infection, pathogens grow and multiply within the plant tissues.

f. Colonization: The pathogen invades the plant and acquires nutrients from the host. The pathogen spreads in the host tissue away from the initial site of infection. Colonization of plant tissues by the pathogen, and growth and reproduction of the pathogen are two concurrent stages of disease development.

Fungi spread into all parts of host organs, either by growing directly through the cells as an intracellular mycelium or by growing between the cells as an intercellular mycelium. During establishment, pathogen produces different substances which include enzymes, toxins, growth hormones and polysaccharides which will help in colonization of the host.

- In ectoparasites the main body of the pathogen lies on the surface of the host with only feeding organs (haustoria) penetrating the tissues, for example most of the powdery mildew fungi.
- Some fungal parasites develop both external and internal mycelium E.g., *Rhizoctonia solani.*
- The endophytic parasites or endoparasites grow subcuticularly (*Diplocarpon rosae*, black spot of rose), in parenchyma tissues (most fungal and bacterial pathogens as well as many nematodes) or in vascular tissues (vascular wilt parasites).
- Some pathogens are endobiotic, *i.e.,* mycelium is not produced and the thallus is entirely present within a host cell E.g., *Synchytrium endobioticum.*
- Bacteria invade tissues intercellularly, but also grow intracellularly when parts of the cell walls dissolve.
- Viruses, viroids, mollicutes and fastidious bacteria invade tissues by moving from cell to cell intracellularly. Infection caused by microbes may be local (involve single cells or few cells or small area) or systemic (pathogen spreads and invades most or all susceptible cells and tissues throughout the plant E.g., *Sclerospora graminicola*). The time interval between inoculation and appearance of disease symptoms is called the incubation period.

Reproduction: The pathogen produces new infectious units, for example, fungal spores or virus particles.

Survival: Resting of the pathogen in the absence of host plants (overwintering, summer survival, etc.).

Dispersal: After invasion and colonization of the host, the reproductive propagules come out of the host to maintain the continuity of the infection chain or disease cycle and escape death due to overcrowding. Once the pathogens Exit from the host, they survive and are disseminated to other hosts and continues the infection cycle.

- Viruses can exist only with the living protoplasm and hence disseminated through their animate vectors like insects, fungi, nematodes, etc.
- The bacteria ooze out in the form of slime on the host surface from where they can be disseminated through water and insects.
- Most plant pathogenic fungi grow out on the host surface and produce repeating

spores (secondary inoculum), usually asexually, under favourable conditions. The spores thus formed are disseminated through wind, water, soil, seed, vegetative propagating material, agricultural implements, etc.

Practice Questions

1. Write the inoculum potential of pathogens
2. Describe the hypersensitivity reaction due to pathogen invasion
3. Write the direct penetration mechnisms in fungi

Suggested Readings

Wheeler, H. (2012). *Plant pathogenesis* (Vol. 2). Springer Science & Business Media.

Oku, H. (2020). *Plant pathogenesis and disease control.* CRC Press.

Vanderplank, J. E. (2012). *Genetic and molecular basis of plant pathogenesis* (Vol. 6). Springer Science & Business Media.

Chapter - 27

Enzymes in Disease Development

Enzymes are large protein molecules which catalyze all inter-related reactions in the living cell. Most pathogens derive energy principally from enzymatic break down of food materials from host tissue.

Composition of the cell wall: A plant cell wall is arranged in layers, functionally cell wall is divided into three regions, *viz.*, middle lamella (made of pectins), primary cell wall (cellulose, pectic substances) and secondary cell wall (entirely cellulose). and contains cellulose microfibrils, hemicellulose, pectin, lignin, and soluble protein. The cell wall surrounds the plasma membrane and provides the cell tensile strength and protection.

Middle lamella acts as intercellular cement which binds the cells together in tissue system. Pectin or pectic substances are major chemical constituents of wall layers and entire middle lamella, where as in other layers, cellulose is found in good amounts. Besides these two major components, other components such as hemicelluloses, lignin and some amount of protein are also present. Main components of cell wall are pectic substances, cellulose, hemicelluloses, lignin and small quantity of protein.

The epidermis of plants is covered by cuticle; whose major chemical substance is cutin in addition to cuticular wax.

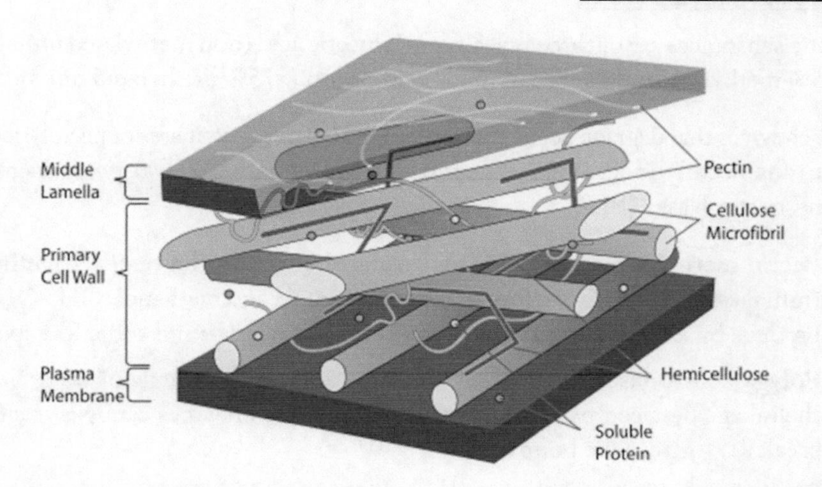

Plant plasma membrane and cell-wall structure

(Source: Sticklen, M.B., 2008)

Cuticular wax: Plant waxes are found as granular or rod like projections or as a continuous layer outside / within the cuticle. Wax formation is a continuous process and it is not a terminal phase in the development of leaf. Cuticular waxes are made up of long chain molecules of paraffin, hydrocarbons, alcohols, ketones and acids. Most of the fungi and parasitic higher plants penetrate wax layers by means of mechanical force alone. Some pathogen has been found to produce enzymes causing degradation of cuticular wax. E.g., *Puccinia hordei*, *Pestalotia malicola.*

Cutin: It is insoluble polyester of unbranched derivatives of C16 and C18 hydroxyl fatty acids. Cutin is admixed with waxes on upper side and with pectin and cellulose on the lower side. Cutinases break cutin molecules and release monomers as well as oligomers from insoluble cutin polymer. Cutinases reaches its highest concentration at penetrating point of the germ tube and at infection peg of appressorium forming fungi. E.g., *Colletotrichum gloeosporioides, Sphaerotheca pannosa, Venturia inaequalis, Helminthosporium victoriae, Botrytis cinerea, Erysiphe graminis,* and *Streptomyces scabies.*

Pectic substances: These are major components of middle lamella (intercellular cement that holds in place the cells of plant tissues). They also make up a large portion of primary cell wall in which they form an amorphous gel filling the spaces between cellulose microfibrils. Pectic substances are polysaccharides consisting mostly of d-galactouronic acid units with a-1,4-glycosidic bonds.

These chains are esterified with methyl groups or linked with other carboxyl groups in calcium and magnesium salt bridges.

Pectic substances are of three types, namely, pectic acid (non methylated units), pectinic acid (<75% methylated galacturonan units) and pectin (>75% methylated units).

The enzymes that degrade pectic substances are known as pectinases or pectolytic enzymes. Pectinases and pectolytic enzymes are pectin methyl esterases (PME's), polygalactouronases (PG's) and pectin lyases (PL's).

a. **Pectin methyl esterases (PMEs)**: Breaks ester bonds and removes methyl groups from pectin leading to the formation of pectic acid and methanol (CH_3-OH). PMEs produce by *Erwinia amylovora*, *Xanthomonas campestris* and some leaf spot fungi.
b. **Polygalacturonases**: Split pectin chain by adding a molecule of water and breaks thelinkage between two galacturonan units. These enzymes catalyze reactions that break a1,4-glycosidic bonds.
c. **Pectin lyases**: Split pectin chain by removing a molecule of water from the linkage, thereby breaking it and releasing products with unsaturated double bonds. These pectin enzymes can be Exopectinases (break only terminal linkage) or endopectinases (break pectin chain to random sites). Pectin degradation results in liquefaction of the pectic substances and weakening of cell walls, leading to tissue maceration. E.g., Soft rot bacterium, *Erwinia caratovora* subsp. *caratovora* and other fungi like *Botrytis cinerea, Sclerotium rolfsii*, etc.

Cellulose: Cellulose is a polysaccharide, made of chains of **ß-D-glucopyranose** units (where C1 is linked to C). Glucose chains are held by hydrogen bonds. Cellulose occurs in all higher plants as the skeletal substance of cell walls in the form of microfibrils.

Primary and secondary wall consists of a matrix in which a large number of microfibrils are embedded. These microfibrils are like bundles of iron bars in a reinforced concretebuilding. In some parts of microfibrils the chains are arranged in an orderly fashion attaining crystalline form, when arranged in less orderly fashion, it attains amorphous form. If the proportion of crystalline portion is more, the resistance of the host to pathogen is more. The space between microfibrils and between micelles or cellulose chains is filled with pectins, hemicelluloses and also lignin at maturity.

» Cellulose is insoluble in crystalline form (native form), and soluble in amorphous form (modified cellulose). The enzymatic breakdown of cellulose results in final production of glucose molecules.

» Cellulose is degraded by cellulases. Cellulase one (C4) attacks native cellulose by cleaving cross-linkages between chains. A second cellulase (C) also attacks native cellulose and breaks into shorter chains. These shorter chains are then attacked by C2 enzyme, which degrade them into disaccharide, cellobiose. Finally, cellobiose is degraded by the enzyme, **ß-glucosidase** into glucose.

- Cellulase degrading enzymes play a role in softening and degradation of cell wall material and facilitate easy penetration and spread of pathogen in the host. E.g., Basidiomycetes fungi

Hemicellulose: These are the major constituents of primary cell wall and also seen in middle lamella and secondary cell wall. The hemicellulose polymers include primarily xyloglucan but also glucomannans, galactomannans, arabinogalactans, etc.

- Hemicelluloses link the ends of pectic polysaccharides and various points of the cellulose microfibrils.
- Hemicellulases degrade hemicelluloses and depending on the monomer released from polymer on which they act, they are termed as xylanase, galactanase, glucanase, arabinase, mannose, and so on. E.g., *Sclerotinia sclerotiorum, Sclerotinia fructigena.*

Lignin: Lignin is found in the middle lamella, as well as in the secondary cell wall of xylem vessels and the fibres that strengthen plants. It is an amorphous, three-dimensional polymer made up of basic structural unit, phenylpropanoid. Lignin forms by oxidative condensation (C-C and C-O bond formation) between phenylpropanoid units or substituted cinnamyl alcohols (p-coumaryl alcohol, coniferyl alcohol and sinapyl alcohol). White rot fungi (Basidiomecetes) secrete one or more ligninases which enable them to utilize lignin. E.g., *Xylaria, Chaetomium, Alternaria, Cephalosporium*, etc.

Cell wall proteins: Cell wall proteins are similar to other proteins, except that they are rich in aminoacid, hydroxy proline. Five classes of structural proteins are found in cell walls: extensins, proline-rich proteins (PRP's), glycine-rich proteins (GRP's), Solanaceous lectins and arabinogalactan proteins (AGP's). Proteins are degraded by means of enzymes, proteases or proteinases or peptidases.

Lipids: Various types of lipids occur in all plant cells. The most important ones are phospholipids and glycolipids. These lipids contain fatty acids, which may be saturated or unsaturated. Lipolytic enzymes, called lipases (phospholipases, glycolipases) hydrolyze lipids and release fatty acids.

Starch: Starch is the main reserve polysaccharide found in plant cells. It is a glucose polymer and Egists in two forms: amylose, a linear molecule, and amylopectin, a highly branched molecule. Starch is degraded by enzyme, amylases.

Examples

Substrate	Enzymes	Pathogen
Cutin	Cutinases	*Colletotrichum gloeosporioides*, *Fusarium solani* f. sp. *pisi*.
Cellulose	Cellulases (Cl, C2, Cx, and ß-glucosidase)	*Ralstonia solanacearum*, *Erwinia carotovora*, *E. chrysanthemi*, *C. lindemuthianum*, *F. oxysporum* f. sp. *lycopersici*, *Sclerotinia sclerotiorum*, *Botrytis cinerea*, *Ascochyta pisi*, *Rhizoctonia solani*.
Pectinases	Hydrolases/ Pectin methyl esterase (PME), Polygalacturonases (PG) (endo- or exo-), Pectin methyl galacturonases (PMG) (endo- or exo-) Lyases: Polygalacturonic acid trans-eliminases (exo- or endo-), (Pectin methyl trans-eliminases (exo- or endo-)	*E. carotovora*, *E. chrysanthemi*, *Pythium* spp.: *R. solani*, *Verticillium albo-atrum*, *Sclerotium rolfsii*, *S. sclerotiorum*, *B. cinerea*.
Hemicellulose	Hemicellulases	*S. rolfsii*, *R. solani*, *S. sclerotium*, *F. roseum*, *E. carotovora*, *E. chrysanthemi*.
Protein	Proteinases	*R. solani*, *Penicillium expansum*, *Pseudomomas lachrymans*, *E. carotovora*, *E. chrysanthemi*.
Phospholipids	Phospholipase B, Phospholipase C	*S. rolfsii*, *Thielaviopsis basicola*, *B. cinerea*, *E. carotovora*.
Lignin	Ligninases	*Heterobasidion annosus*, *B. cinerea*.
Suberin	Esterases (Cutinases) to degrade aliphatic component	*F. solani* f. sp. *pisi*.

Practice Questions

1. Pectinases are majorly produced by.........bacteria
2. Cellulose in the plant cell wall is breached by...........enzyme
3. Explain the role of enzymes in plant cell wall degradation and infection
4. What do you understand by cell wall degrading enzymes? How are these involved in the production of disease syndrome? Explain their role with reference to soft rots in vegetables or fruits.
5. How will you classify cell wall degrading enzymes on the basis of their substrate utilization and mode of action?

6. What are "macerating principles"? Explain their role in disease development.\
7. Explain the following:
   i. Role of proteolytic enzymes in pathogenesis
   ii. Role of cutinolytic enzymes in pathogenesis
   iii. Role of lignolytic enzymes in pathogenesis.
8. How cuticle acts as a physical or chemical barrier to infection.

Suggested Readings

Kubicek, C. P., Starr, T. L., & Glass, N. L. (2014). Plant cell wall-degrading enzymes and their secretion in plant-pathogenic fungi. *Annu Rev Phytopathol*, *52*(1), 427-451.

Van den Ende, G., & Linskens, H. F. (1974). Cutinolytic enzymes in relation to pathogenesis. *Annual Review of Phytopathology*, *12*(1), 247-258.

Goto, M. (2012). *Fundamentals of bacterial plant pathology*. Academic Press.

Alghisi, P., & Favaron, F. (1995). Pectin-degrading enzymes and plant-parasite interactions. *European Journal of Plant Pathology*, *101*(4), 365-375.

Collmer, A., & Keen, N. T. (1986). The role of pectic enzymes in plant pathogenesis. *Annual review of phytopathology*, *24*(1), 383-409.

Chapter - 28

Toxins in Disease Development

Definition: Toxin can be defined as a microbial metabolite excreted (exotoxin) or released by lysed cells (endotoxin) which in very low concentration is directly toxic to the cells of the suscept (host).

The term toxin is used for a product of the pathogen, its host, or pathogen host interaction which even at very low concentration directly acts on living host protoplasm to influence disease development or symptom expression.

Toxins are different from enzymes in that they do not attack structural integrity of host tissues but affect the metabolism of the host because the toxins will act on protoplast of the cell.

Difference between enzymes and toxins

Enzymes	Toxins
All enzymes are proteinaceous in nature	All are secondary metabolites, polysaccharide proteins and organic acids
Single/ few genes invovled	Large number of genes are invovled in pathway process
High molecular weight	Low molecular weight
Sensitive to high temperature	Tolerant / resistant to high temperature

Toxin hypothesis (Luke and Wheeler, 1955):

1. A toxin should produce all symptoms characteristic of the disease

2. Sensitivity to toxin will be correlated with susceptibility to pathogen
3. Toxin production by the pathogen will be directly related to its ability to cause disease.

Except, victorin, the toxic metabolite of *Cochliobolus victoriae*, the vast majority of toxins associated with plant diseases fail to Eghibit all the above characters.

Classification of toxins (Wheeler and Luke, 1963) According to the source of origin, toxins are divided into 3 broad classes namely, pathotoxins, vivotoxins and phytotoxins.

1. **Pathotoxins**: These are the toxins which play a major role in disease production and produce all or most of the symptoms characteristic of the disease in susceptible plants. Most of these toxins are produced by pathogens during pathogenesis.

E.g., Victorin: *Cochliobolus victoriae* (*Helminthosporium victoriae*), the causal agent of Victoria blight of oats. This is a host specific toxin.

Other Examples are:

a) Selective

» T- toxin: *Helminthosporium maydis* race T

» HC-toxin: *Helminthosporium carbonum*

» HS- toxin: *Helminthosporium sacchari*

» Phyto-alternarin: *Alternaria kikuchiana*

» PC- toxin: *Periconia circinata*

b) Non-selective

» Tentoxin: *Alternaria tenuis*

» Tabtoxin or wild fire toxin: *Pseudomonas tabaci*

» Phaseolotoxin: *Pseudomonas syringae* pv. *phaseolicola*

c) Produced by plant or plant × pathogen interaction

» Amylovorin: *Erwinia amylovora* (Fire blight of apple and pears)

2) Phytotoxins: These are the substances produced in the host plant due to host-pathogen interactions for which a causal role in disease is merely suspected rather than established. These are the products of parasites which induce few or none of the symptoms caused by the living pathogen. They are non-specific and there is no relationship between toxin production and pathogenicity of disease-causing agent. E.g., Alternaric acid – *Alternaria solani*

3) Vivotoxins: These are the substances produced in the infected host by the pathogen and

/ or its host which functions in the production of the disease, but is not itself the initial inciting agent of the disease.

- Fusaric acid: Wilt causing *Fusarium* sp.
- Lycomarasmin: *Fusarium oxysporum* f.sp. *lycopersici*
- Piricularin: *Pyricularia oryzae*

Classification based on specificity of toxins

1. **Host specific / Host selective toxins (Pathogenecity factor)**: These are the metabolic products of the pathogens which are selectively toxic only to the susceptible host of the pathogen. E.g., Victorin, T-toxin, Phyto-alternarin, Amylovorin.
2. **Non-specific / Non-selective toxins (Virulance factor):** These are the metabolic products of the pathogen, but do not have host specificity and affect the protoplasm of many unrelated plant species that are normally not infected by the pathogen. E.g., Ten-toxin, Tab-toxin, Fusaric acid, Piricularin, Lycomarasmin and Alternaric acid.

Differentiate host – specific and non-host specific toxins

Host specific	Non-host specific
1. Selectively toxic only to susceptible the physiology of those plants that are normally not infected by the pathogen	1. No host specificity and can also affect host of the pathogen
2. Primary determinants of disease	2. Secondary determinants of disease
3. Produce all the essential symptoms the disease	3. Produce few or none of the symptoms of the disease
4. E.g., Victorin, T- toxin	4. E.g., Tentoxin, Tabtoxin

Host-specific or host-selective toxins

- A host-specific or host-selective toxin is a substance produced by a pathogenic microorganism that, at physiological concentrations, is toxic only to the hosts of that pathogen and shows little or no toxicity against nonsusceptible plants.
- Most host-specific toxins must be present for the producing microorganism to be able to cause disease.
- Host-specific toxins have been shown to be produced only by certain fungi (*Cochliobolus, Alternaria, Periconia, Phyllosticta, Corynespora*, and *Hypoxylon*), although certain bacterial polysaccharides from *Pseudomonas* and *Xanthomonas* have been reported to be host specific.

a. Victorin, or HV Toxin

» Victorin, or HV-toxin is produced by the fungus *Cochliobolus* (*Helminthosporium*) victoriae, which causes Victoria blight of oats.

» This fungus appeared in 1945 after the introduction and widespread use of the oat variety Victoria and its derivatives, all of which contained the Vb gene for resistance to crown rust disease.

» *C. victoriae* infects the basal portions of susceptible oat plants and produces a toxin that is carried to the leaves, causes a leaf blight, and destroys the entire plant.

» Victorin is a complex chlorinated, partially cyclic pentapeptide.

» The primary target of the toxin seems to be the cell plasma membrane, where victorin seems to bind to several proteins.

» Victorin also functions as an elicitor that induces components of a resistance response that include many of the features of hypersensitive response and lead to programmed cell death.

b. T Toxin [HMT Toxin]

» T toxin is produced by race T of *Cochliobolus heterostrophus* (anamorph: *Bipolaris maydis*, earlier called *Helminthosporium maydis*), the cause of southern corn leaf blight.

» First appeared in the United States in 1968, it spread throughout the corn belt by 1970, attacking only corn that had the Texas male-sterile (Tms) cytoplasm.

» The ability of *C. heterostrophus* race T to produce T toxin and its virulence to corn with Tms cytoplasm are controlled by one and the same gene.

» T toxin does not seem to be necessary for the pathogenicity of *C. heterostrophus* race T, but it increases the virulence of the pathogen.

» The T toxin apparently acts specifically on mitochondria of susceptible cells, which are rendered nonfunctional, and inhibits ATP synthesis.

c. HC Toxin

» Race 1 of *Cochliobolus carbonum* (anamorph: *Bipolaris* (*Helminthosporium*) *zeicola*) causes northern leaf spot and ear rot disease in maize.

» It also produces the host-specific HC toxin, which is toxic only on specific maize lines.

» The mechanism of action of HC toxin is not known, but this is the only toxin, so far, for which the biochemical and molecular genetic basis of resistance against the toxin is understood.

» Resistant corn lines have a gene (Hm1) coding for an enzyme called HC toxin reductase that reduces and thereby detoxifies the toxin.

» Susceptible corn lines lack this gene and, therefore, cannot defend themselves against the toxin.

Alternaria toxins

» Several pathotypes of *Alternaria alternata* attack different host plants and on each they produce one of several multiple forms of related compounds that are toxic only on the particular host plant of each pathotype.

» Some of the toxins and the hosts on which they are produced and affect

» AK toxin causing black spot on Japanese peat fruit,

» AAL toxin causing stem canker on tomato,

» AF toxin on strawberry,

» AM toxin on apple,

» ACT toxin on tangerine,

» ACL toxin on rough lemon, the HS toxin on sugar cane.

» As an example of *A. alternata* toxins, the AM toxin is produced by the apple pathotype of *A. alternata*, known previously as *A. mali*, the cause of *Alternaria* leaf blotch of apple.

Non-host selective toxins

a. Tabtoxin

» Tabtoxin is produced by the bacterium *Pseudomonas syringae* pv. *tabaci*, which causes the wildfire disease of tobacco.

» Tabtoxin is a dipeptide composed of the common amino acid threonine and the previously unknown amino acid tabtoxinine.

» Tabtoxin as such is not toxic, but in the cell it becomes hydrolyzed and releases tabtoxinine, which is the active toxin.

» Toxin-producing strains cause necrotic spots on leaves, with each spot surrounded by a yellow halo.

» Sterile culture filtrates of the organism, as well as purified toxin, produce symptoms identical to those characteristic of wildfire of tobacco not only on tobacco, but in a large number of plant species belonging to many different families.

» Strains of *P. syringae* pv. *tabaci* sometimes produce mutants that have lost the ability to produce the toxin.

b. Phaseolotoxin

- Phaseolotoxin is produced by the bacterium *Pseudomonas syringae* pv. *phaseolicola*, the cause of halo blight of bean and some other legumes.
- Phaseolotoxin is a modified ornithine–alanine–arginine tripeptide carrying a phosphosulfinyl group.
- Soon after the tripeptide is secreted by the bacterium into the plant, plant enzymes cleave the peptide bonds and release alanine, arginine, and phosphosulfinylornithine, which is the biologically functional moiety of phaseolotoxin.
- The localized and systemic chlorotic symptoms produced in infected plants are identical to those produced on plants treated with the toxin alone so they are apparently the results of the toxin produced by the bacteria.
- Infected plants and plants treated with purified toxin also show reduced growth of newly expanding leaves, disruption of apical dominance, and accumulation of the amino acid ornithine.
- Phaseolotoxin plays a major role in the virulence of the pathogen by interfering with or breaking the disease resistance of the host toward not only the halo blight bacterium, but also several other fungal, bacterial, and viral pathogens.

c. Tentoxin

- Tentoxin is produced by the fungus *Alternaria alternata* (previously called *A. tenuis*), which causes spots and chlorosis in plants of many species.
- Tentoxin is a cyclic tetrapeptide that binds to and inactivates a protein (chloroplast-coupling factor) involved in energy transfer into chloroplasts.
- The toxin also inhibits the light-dependent phosphorylation of ADP to ATP. In sensitive species, tentoxin interferes with normal chloroplast development and results in chlorosis by disrupting chlorophyll synthesis.
- An additional but apparently unrelated effect of tentoxin on sensitive plants is that it inhibits the activity of polyphenol oxidases, enzymes involved in several resistance mechanisms of plants.
- Both effects of the toxin, namely stressing the host plant with events that lead to chlorosis and suppressing host resistance mechanisms, tend to enhance the virulence of the pathogen.

d. Cercosporin

- Cercosporin is produced by the fungus *Cercospora* and by several other fungi.

- It causes damaging leaf spot and blight diseases of many crop plants, such as Cercospora leaf spot of zinnia and grey leaf spot of corn.
- Cercosporin is unique among fungal toxins in that it is activated by light and becomes toxic to plants by generating activated species of oxygen, particularly single oxygen.

Other non-host-specific toxins

- Fumaric acid - produced by *Rhizopus* spp. in almond hull rot disease
- Oxalic acid - *Sclerotium* and *Sclerotinia* spp. in various plants they infect and by *Cryphonectria parasitica*, the cause of chestnut blight
- Alternaric acid, alternariol, and zinniol - *Alternaria* spp. in leaf spot diseases of various plants
- Ceratoulmin - *Ophiostoma ulmi* in Dutch elm disease
- Fusicoccin - *Fusicoccum amygdali* in the twig blight disease of almond and peach trees
- Ophiobolin - several *Cochliobolus* spp. in diseases of grain crops
- Pyricularin- *Pyricularia grisea* in rice blast disease
- Fusaric acid and lycomarasmin - *Fusarium oxysporum* in tomato wilt

Effect of toxins on host tissues

a. Changes in cell permeability: Toxins kill plant cells by altering the permeability of plasma membrane, thus permitting loss of water and electrolytes and also unrestricted entry of substances including toxins. Cellular transport system, especially, H exchange at the cell membrane is affected.

b. Disruption of normal metabolic processes: Increase in respiration due to disturbed salt balance

- Malfunctioning of enzyme system E.g., Piricularin inhibits polyphenol oxidase
- Uncoupling of oxidative phosphorylation

c. Interfere with the growth regulatory system of host plant E.g., Restricted development of roots induced by *Fusarium moniliforme*

Practice Questions

1. Write the differences between host-selective and non-host selective toxins.
2. What are toxins? How are these involved in the development of vascular wilt syndrome?
3. What is role of toxins in host-specificity?

4. The discovery of host-specific toxin raised hopes that biochemical basis of pathogenicity, and conversely host resistance, might turn out to be simple, determined by a right compound of major effect. Overall these hopes have not been fully realized. Why?
5. What can be the evidences for involvement of a host specific toxin in plant disease?

Suggested Readings

Wheeler, H., & Luke, H. H. (1963). Microbial toxins in plant disease. *Annual Reviews in Microbiology*, *17*(1), 223-242.

Yamane, H., Konno, K., Sabelis, M., Takabayashi, J., Sassa, T., & Oikawa, H. (2010). Chemical defence and toxins of plants.

Logrieco, A., Moretti, A., & Solfrizzo, M. (2009). Alternaria toxins and plant diseases: an overview of origin, occurrence and risks. *World Mycotoxin Journal*, *2*(2), 129-140.

Kimura, M., Anzai, H., & Yamaguchi, I. (2001). Microbial toxins in plant-pathogen interactions: Biosynthesis, resistance mechanisms, and significance. *The Journal of general and applied microbiology*, *47*(4), 149-160.

Durbin, R. (Ed.). (2012). *Toxins in plant disease*. Elsevier.

Scheffer, R. P., & Livingston, R. S. (1984). Host-selective toxins and their role in plant diseases. *Science*, *223*(4631), 17-21.

Chapter - 29

Plant Growth Regulators in Disease Development

Plant growth hormones, are naturally occurring small organic molecules or substances which influence physiological processes in plants at very low concentrations.

Growth regulators

Growth regulators are of two types

1. Growth promoting substances and 2. Growth inhibiting substances

Auxins, gibberellins and cytokinins are growth promoting substances, whereas, dormin, ethylene and abscissic acid are growth inhibiting substances. The imbalance in growth promoting and growth inhibiting substances causes hypertrophy (Excessive increase in cell size) and atrophy (decrease in cell size). Symptoms may appear as tumors, galls, knots, witches broom, stunting, Excessive root branching, defoliation and suppression of bud growth.

1. Growth promoting substances:

a) Auxins: Indole-3-acetic acid (IAA) is the naturally occurring auxin. It is continuously produced in young meristematic tissue and moves rapidly to older tissues. If auxin concentration is more, its concentration is reduced by the enzyme, IAA oxidase.

Functions: IAA regulates cell elongation and differentiation, also affects permeability of the membrane, increases respiration, and promotes synthesis of mRNA.

How disease is induced?

Increased IAA results in hypertrophy and decreased IAA results in atrophy. Increased IAA may be due to inhibition of IAA oxidase.

E.g., *Ralstonia solanacearum* (*Pseudomonas solanacearum*), the causal agent of wilt of Solanaceous plants, induces a 100-fold increase in IAA level in diseased plants. Increased plasticity of cell walls as a result of high IAA levels renders the pectin, cellulose and protein components of the cell wall more accessible to pathogen degradation. Increase in IAA levels may also inhibit lignifications of tissues.

Increased IAA levels have been reported in plants infected with the following pathogens *Phytophthora infestans* (Late blight of potato), *Ustilago maydis* (Maize smut), *Plasmodiophora brassicae* (Club root of crucifers), *Sclerospora graminicola* (Downy mildew of sorghum), *Agrobacterium tumefaciens* (Crown gall of apple), and *Meloidogyne incognita* (Root knot nematode).

b) Gibberellins: First isolated from *Gibberella fujikuroi* (Conidial stage: *Fusarium moniliforme*), the causal agent of bakanae or foolish seedling disease of rice. Infected seedlings show abnormal elongation due to excessive elongation of internodes. Best known gibberellin is Gibberellic acid.

Functions: Cell elongation, stem and root elongation, promote flowering and growth of fruits. It also induces IAA synthesis. IAA and GA act synergistically. E.g., *Sclerospora sacchari*, the causal agent of downy mildew of sugarcane induces GA production.

c) Cytokinins: Kinetin was the first compound isolated from herring sperm DNA and does not occur naturally in plants. Cytokinins, such as zeatin and isopentenyl adenosine (IPA) have been isolated from plants.

Functions: Cytokinins are necessary for cell growth and differentiation. It inhibits breakdown of proteins and aminoacids and thereby inhibit senescence and they have the capacity to direct the flow of aminoacids and other nutrients towards high cytokinin concentration. Cytokinin activity increases in club root, in crown galls and in rust infected bean leaves. E.g., Green islands are formed around infection in bean (*Phaseolus vulgaris*) leaves infected by *Uromyces phaseoli*.

2. Growth inhibiting substances

a) Ethylene (CH2=CH): Ethylene exerts a variety of effects on plants, viz., chlorosis, leaf abscission, epinasty, stimulation of adventitious roots, fruit ripening and increased permeability of cell membranes.

E.g., Ethylene is involved in premature ripening of fingers in banana infected by *Pseudomonas solanacearum,* the causal agent of moko disease of banana. Ethylene was also detected in leaf epinasty symptom of the vascular wilt syndrome. E.g., *Fusarium oxysporum* f.sp. *lycopersici* (Wilt in tomato).

b) Abscissic acid: It Exerts dormancy in seeds, closure of stomata, inhibition of seed germination and growth and stimulated germination of fungal spores. It is one of the factors involved in stunting of plants.

c) Dormin / Abscissin II: Dormin induces dormancy by converting developing leaf primordia of a bud into bud scales. It acts as an antagonist of gibberellins and masks the effect of IAA. However, the exact role of dormin is not known.

The plant hormones ethylene, jasmonic acid, and salicylic acid (SA) play a central role in the regulation of plant immune responses. In addition, other plant hormones, such as auxins, abscisic acid (ABA), cytokinins, gibberellins, and brassinosteroids, that have been thoroughly described to regulate plant development and growth, have recently emerged as key regulators of plant immunity. Also, the phytohormone systems mediated by auxin, cytokinins, gibberellins, and abscisic acid were involved physiologically in plant susceptibility towards pathogen invasion.

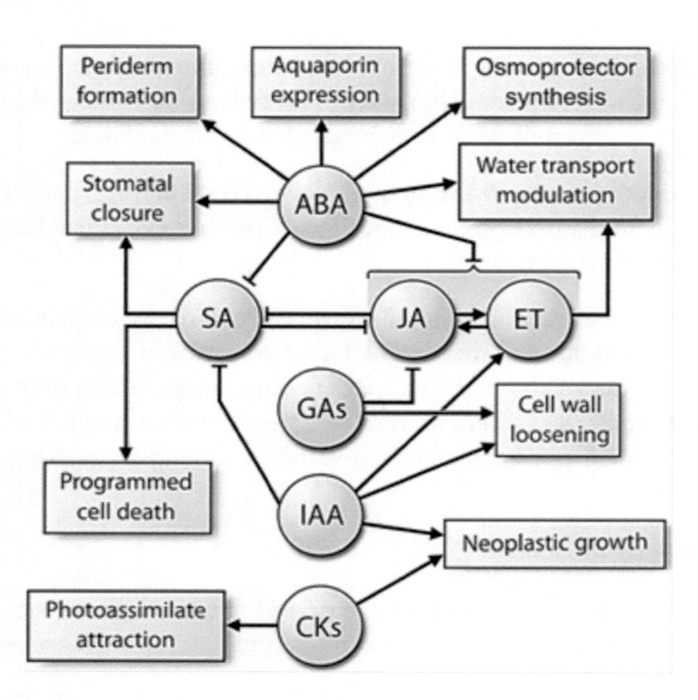

Phytohormone crosstalk and phytohormone-mediated susceptible responses. ABA, abscisic acid; CKs, cytokinins; ET, ethylene; GAs, gibberellins; IAA, indole-3-acetic acid (auxin); JA, jasmonic acid; SA—salicylic acid.

Examples

PGRs	Pathogens
Cytokinins	Increased concentration: crown gall (*Agrobacterium tumefaciens*), fasciation disease (*Corynebacterium fascians*), white rust of crucifers (*Albugo candida*), bean rust (*Uromyces phaseoli*), club roots of crucifers (*Plasmodiophora brassicae*), western pine blister rust (*Cronartium fusiforme*), white pine blister rust (*C. ribicola*), root knot of tobacco (*Meloidogyne incognita*). Reduced concentration: Verticillium wilt of tomato and cotton, root knot of tomato (*M. incognita*)
Auxins	Increased concentration: crown gall (*A. tumefaciens*), peach leaf curl (*Taphrina deformans*), wheat stem rust (*Puccinia graminis* f. sp. *tritici*), wheat powdery mildew (*Erysiphe graminis*), safflower rust (*Puccinia carthami*) white rust of crucifers (*A. candida*), downy mildew of crucifers (*Peronospora parasitica*), tomato wilt (*Verticillium albo-atrum*), banana wilt (*F. oxysporum* f. sp. *cubense*) bacterial wilt of solanaceous crops (*Pseudomonas solanacearum*), Oleander knot (*P. syringae* pv. *savastanoi*). Reduced concentration: Mango malformation (*F. oxysporum*), Tobacco mosaic (TMV), Potato leaf roll virus (PLRV), Curly top of sugarbeet (SBCTV).
Ethylene	Increased concentration: fruit rot of citrus (*Pencillium digitatum*), tomato wilt (*F. oxysporum* f. sp. *lycopersici*), wilt of tulip (*F. oxysporum* f. sp. *tulipae*). Verticillium wilt of cotton (*V. albo-atrum*).
Gibberellins	Increased concentration: "Bakanae" disease of rice (*Gibberella fujikuroi*), creeping thistle rust (*Puccinia punctiformis*) Reduced concentration: Anther smut of sea campion (*Ustilago violacea*)

Practice Questions

1. ....hormone reduces the virus infectivity in growin tip
2. ....hormone isolated from the pathogen of foot rot of rice
3. Increased concentration auxin result in ....disease in apple
4. ....hormone are necessary for cell growth and differentiation
5. Green islands are formed around infection in bean due the........hormone

Suggested Readings

Sequeira, L. (1963). Growth regulators in plant disease. *Annual Review of Phytopathology*, *1*(1), 5-30.

Yamada, T. (1993). The role of auxin in plant-disease development. *Annual review of phytopathology*, *31*(1), 253-273.

Gaspar, T., Kevers, C., Penel, C., Greppin, H., Reid, D. M., & Thorpe, T. A. (1996). Plant hormones and plant growth regulators in plant tissue culture. *In vitro Cellular & Developmental Biology-Plant*, *32*(4), 272-289.

González-Lamothe, R., El Oirdi, M., Brisson, N., & Bouarab, K. (2012). The conjugated auxin indole-3-acetic acid–aspartic acid promotes plant disease development. *The Plant Cell*, *24*(2), 762-777.

Chapter - 30

Effect of Pathogens on Physiological Functions of the Plants

The symptoms observed in a diseased plant depend on the effect of the pathogen on the physiology of the plant. Phytopathogen infection leads to changes in secondary metabolism based on the induction of defence programmes as well as to changes in primary metabolism which affect growth and development of the plant. Therefore, pathogen attack causes crop yield losses even in interactions which do not end up with disease or death of the plant. Pathogen attack first initiates a series of rapid changes resulting in a decline in photosynthesis and an increase in respiration, photorespiration, and invertase enzyme activity.

Plant–pathogen interactions

Pathogens can also be divided according to the environment in which they occur and the tissues which they infect. A common classification is to distinguish between above- and below-ground tissues as the primary target of the pathogen. Related to this, above-ground tissue might be green, assimilate-producing tissue, typically source leaves, or assimilate-importing tissue such as flowers. Pathogens infecting source tissue will encounter different conditions related to primary metabolism as well as to defence responses compared with those pathogens infecting sink or assimilate-producing tissue such as roots, flowers, and sink leaves.

Effect on photosynthesis

Photosynthesis is the essential function of plants and any pathogen that interferes with it will cause disease that may appear as chlorosis (yellowing) and necrosis (browning and death) of the leaves and stems. Even mild impairment of photosynthesis weakens the plant and increases susceptibility to other pests and pathogens.

In leaf spot, blight, and other kinds of diseases in which there is destruction of leaf tissue, photosynthesis is obviously reduced because of the reduction, through death, of the photosynthetic surface of the plant. Even in other diseases, however, plant pathogens reduce photosynthesis, especially in the late stages of diseases, by affecting the chloroplasts and causing their degeneration. The overall chlorophyll content of leaves in many fungal and bacterial diseases is reduced, but the photosynthetic activity of the remaining chlorophyll seems to remain unaffected. In plants infected by vascular pathogens, chlorophyll is reduced and photosynthesis stops even before the eventual wilting of the plant. Most virus, mycoplasma, and nematode diseases induce varying degrees of chlorosis. In the majority of such diseases photosynthesis of infected plants is reduced qreatly, in advanced stages of the disease the rate of photosynthesis being no more than one-fourth the normal rate.

- » Reduction of chloroplast numbers. E.g., Mosaic
- » Reduction in chlorophyll content. E.g., Mosaic, Powdery mildew, Downy mildew.
- » Chloroplast abnormalities. E.g., Mosaic, Rust
- » Reduction in photochemical activity
- » Stimulating CO_2 incorporation at early stage of infection, but declined after virus infection for several days.
- » Reduction in sucrose content and starch accumulation. E.g., PRSV, AVSvd.

Effect on transpiration

In plant diseases in which the pathogen infects the leaves, transpiration is usually increased. This is the result of destruction of at least part of the protection afforded the leaf by the cuticle, increase in permeability of leaf cells, and dysfunction of stomata. Diseases like the rusts, mildews, and apple scab destroy a considerable portion of the cuticle and epidermis and this results in uncontrolled loss of water from the affected areas. If water absorption and translocation cannot keep up with the excessive loss of water, loss of turgor and wilting of leaves follows. The suction force of excessively transpiring leaves is abnormally increased and may lead to collapse and/or dysfunction of underlying vessels through production of tyloses and gums. Decrease the water flow by polysaccharides accumulation in the tyloses produced by *Fusarium* - obstruction of normal water translocation.

Effect on respiration

Respiration is the process by which cells, through enzymatically controlled oxidation (burning) of the energy-rich carbohydrates and fatty acids, liberate energy in a form that can be utilized for the performance of various cellular processes.

Most, if not all, infectious diseases **increase respiration**, this being a general reaction of the plant to most types of stress. The majority of the increase in respiration occurs in the infected host tissue and appears to be a basic response to injury. Consequences of increased oxygen uptake and enhanced activity of respiratory enzymes include: a slight increase in temperature, accumulation of metabolites around points of infection, and even an increase in dry weight of the host tissue.

- Respiration increases during pathogen attack.
- Rate continues to rise during multiplication and sporulation of the pathogen.
- Declines to normal or below-normal levels rate increases more rapidly in resistant varieties.
- Increased respiration depletes plant's reserves.
- Changes metabolism.

E.g., Rust, Smut, Powdery Mildew and Downy Mildew

The increased respiration in diseased plants is apparently brought about, at least in part, by uncoupling of the oxidative phosphorylation.

The increased respiration of diseased plants can also be explained as the result of increased metabolism in the plant. In many plant diseases, growth is first stimulated, protoplasmic streaming increases, and materials are synthesized, translocated, and accumulated in the diseased area.

Effect on translocation of water and nutrients

All living plant cells require an abundance of water and an adequate amount of organic and inorganic nutrients in order to live and to carry out their respective physiological functions. Plants absorb water and inorganic (mineral) nutrients from the soil through their root system. These are generally translocated upward through the xylem vessels of the stem and into the vascular bundles of the petioles and leaf veins, from which they enter the leaf cells. The minerals and part of the water are utilized by the leaf and other cells for synthesis of the various plant substances, but most of the water evaporates out of the leaf cells into the intercellular spaces and from there diffuses into the atmosphere through the stomata. On the other hand, nearly all organic nutrients of plants are produced in the leaf cells, following photosynthesis, and are translocated downward and distributed to all the living plant cells by passing for the most part through the phloem tissues.

Pathogens can affect **translocation of water and nutrients** through the vascular system of the host plant. This might be an effect on transpiration through the aerial parts of the plant or poor uptake of nutrients and water through diseased roots. Consequentially sluggish translocation through the vascular system will itself lead to wilting and chlorosis, and possibly necrosis, 'upstream' of the disease focus.

a. Effect on absorption of water by roots

Many pathogens, such as the damping-off fungi, the root-rotting fungi and bacteria, most nematodes, and some viruses cause an extensive destruction of the roots before any symptoms appear on the aboveground parts of the plant. Root injury affects directly the amount of functioning roots and decreases proportionately the amount of water absorbed by the roots. Some vascular parasites, along with their other effects, seem to inhibit root hair production, which reduces water absorption. These and other pathogens also alter the permeability of root cells, an effect that further interferes with the normal absorption of water by roots.

- Damping-off causing fungi: Cause injury to roots
- Root-rot causing fungi, and bacteria: Inhibit root hair production

Xylem gets destroyed: It is mainly by rot or canker pathogens and gall Formation

Xylem gets clogged: Growth of vascular wilt pathogens in xylem.

b. Effect on translocation of water through the xylem

Fungal and bacterial pathogens that cause damping-off, stem rots, and cankers may reach the xylem vessels in the area of the infection and, if the affected plants are young, may cause their destruction and collapse. Affected vessels may also be filled with the bodies of the pathogen and with substances secreted by the pathogen or by the host in response to the pathogen, and may become clogged. Whether destroyed or clogged the affected vessels cease to function properly and allow little or no water to pass through them. Certain pathogens, such as the crown gall bacterium (*Agrobacterium tumefaciens*), the clubroot fungus (*Plasmodiophora brassicae*), and the root-knot nematode (*Meloidogyne* sp.) induce gall formation in the stem and/or the roots. The enlarged and proliferating cells near or around the xylem exert pressure on the xylem vessels, which may be crushed and dislocated and, thereby, become less efficient in transporting water.

Effect on cell mebrane

Changes in plasma membrane and organelle membrane permeability are often the first detectable responses of plant cells to infection by pathogens, and this often leads to loss of electrolytes (calcium and potassium ions in particular). The permeability change is a response to toxins produced by the invading fungus.

- The victorin, a cyclic peptide, is produced by *Cochliobolus* (which used to be called *Helminthosporium*) *victoriae* in oat leaves. Victorin binds to a membrane protein and changes membrane permeability, eventually causing chlorosis and necrotic stripes ('leaf spots') on the leaves; so, this is a host-specific (or host-selective) toxin.

- The HMT-toxin of *Cochliobolus heterostrophus* Race T, which causes leaf blight in maize. This toxin increases membrane permeability of mitochondria to protons by reacting with unique site(s) on the inner mitochondrial membrane of Texas (T) cytoplasmic male-sterile maize.

Effect on transcription and translation in host cells

This is clearly aimed at changing the host metabolism in ways that benefit the pathogen. It may lead to a simple adaptation of plant secondary metabolism to enhance the production of chemicals that favour the fungus in some way; for example, by attracting vectors to transport the fungus to other hosts. Equally interesting are cases of excess production of plant hormones that cause, for example, plant tissue proliferation. Changes in plant hormones can result from the influence of the pathogen on metabolism of the host or from production of a plant growth hormone or its analogue by the pathogen.

- *Albugo candida* causes increased indole acetic acid (IAA) production by infected plants of the genus *Brassica.* IAA, the original plant auxin, controls plant cell elongation, apical dominance (prevents lateral bud formation), prevents abscission, and promotes continued growth of fruit tissues, cell division in vascular and cork cambium, and formation of lateral and adventitious roots.
- The gibberellin plant hormones were actually isolated in the 1930s from rice suffering the bakanae disease disease (Japanese for 'foolish seedling'), which is caused by *Gibberella fujikuroi* (*Fusarium moniliforme*). The fungus itself produces a surplus of gibberellic acid, which acts as a growth hormone for the plant. It causes hypertrophy leading to etiolation and chlorosis, and the plants finally collapse and die. In the normal plant gibberellin growth regulators affect cell elongation in stems and leaves, seed germination, dormancy, flowering, enzyme induction, and leaf and fruit senescence.

Practice Questions

1. What is transcription and translation? How does the disturbance in any of these lead to drastic changes in structure and function of the affected cell?
2. Define photosynthesis. How is this process affected as a result of infection by a pathogen?
3. Define respiration. How is this process affected due to infection by a pathogen?
4. What are the various mechanisms responsible for augmented respiration in diseased plant?
5. Explain the differences in respiration between resistant and susceptible varieties.

Suggested Readings

Lumsden, R. D. (1979). Histology and physiology of pathogenesis in plant diseases caused by Sclerotinia species. *Phytopathology*, *69*(8), 890-895.

Berger, S., Sinha, A. K., & Roitsch, T. (2007). Plant physiology meets phytopathology: plant primary metabolism and plant–pathogen interactions. *Journal of experimental botany*, *58*(15-16), 4019-4026.

Goodman, R. N., Király, Z., & Wood, K. R. (1986). *The biochemistry and physiology of plant disease*. University of Missouri Press.

Edgerton, C. W. (1908). The physiology and development of some anthracnoses. *Botanical Gazette*, *45*(6), 367-408.

Huang, J. S. (2013). *Plant pathogenesis and resistance: biochemistry and physiology of plant-microbe interactions*. Springer Science & Business Media.

Chapter - 31

Plant Disease Epidemiology

A similar definition of an epidemic is the dynamics of change in plant disease in time and space. The study of epidemics and of the factors that influence them is called epidemiology. Epidemiology is concerned simultaneously with populations of pathogens and host plants as they occur in an evolving environment, i.e., the classic disease triangle. As a result, epidemiology is also concerned with population genetics of host resistance and with the evolutionary potential of pathogen populations to produce pathogen races that may be more virulent to host varieties or more resistant to pesticides. Epidemiology, however, must also take into account other biotic and abiotic factors, such as an environment strongly influenced by human activity, particularly as it relates to disease management.

Epiphytology or Epidemiology of plant diseases is essentially a study of the rate of multiplication of a pathogen and spread of the disease caused by it in a plant population. Epidemiology deals with outbreaks and spread of diseases in a population.

Some plant disease epidemics, e.g., wheat rusts, southern corn leaf blight, and grape downy mildew, have caused tremendous losses of produce over rather large areas. Others, e.g., chestnut blight, Dutch elm disease, and coffee rust, have threatened to eliminate certain plant species from entire continents. Still others have caused untold suffering to humans. The Irish potato famine of 1845-1846 was caused by the Phytophthora late blight epidemic of potato, and the Bengal famine of 1943 was caused by the *Cochliobolus* (*Helminthosporium*) brown spot epidemic of rice.

Importance of epidemiology: Knowledge of epidemiology is useful in forecasting of a disease and also for the management of a disease

The terms like compound interest and simple interest diseases were given by Vanderplank (1963) in his book “Plant Disease Epidemics and control”.

No.	Simple interest disease/ Monocyclic	Compound interest disease/ Polycyclic
1.	Rate of increase of disease is mathematically analogous to simple interest in money (Interest is added only at the end; interest does not get interest).	Rate of increase of disease is mathematically analogous to compound interest in money (Interest is added periodically to the capital; interest gets interest).
2.	Pathogen produces spores at very slow rate	Pathogen produces spores at rapid rate
3.	Propagules disseminate by soil or seed	Propagules disseminate by air
4.	Incubation period and sporulation period is long	Incubation period and sporulation period is short
5.	There is only one generation of the pathogen in the life of a crop	There are several generations of the pathogen in the life of a crop
6.	E.g., Smuts of wheat, barley and sorghum	E.g., Rusts of cereals

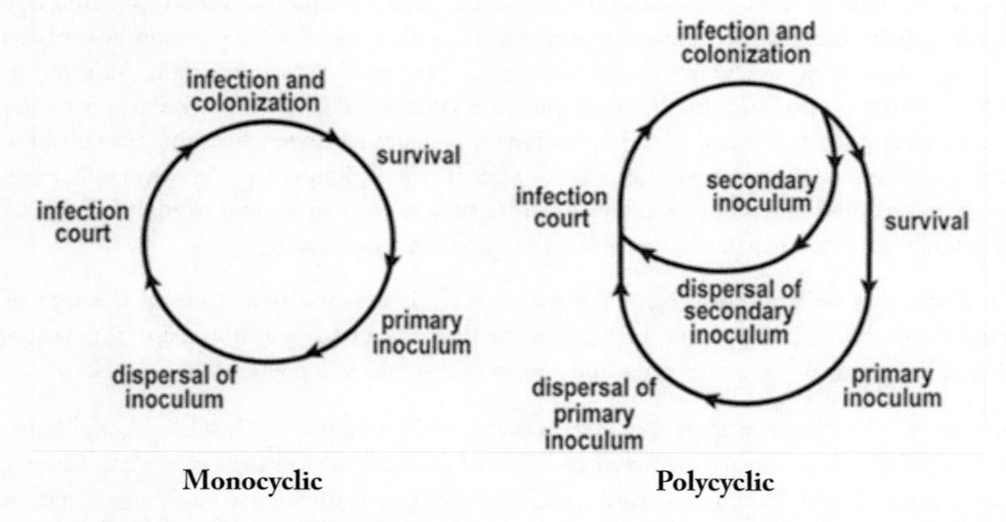

Monocyclic **Polycyclic**

Plant diseases are caused by infectious agents: plant pathogens. What we observe as a disease is the outcome of the interaction between the host plant, the pathogen and the environment. The concept is described as the disease triangle.

Disease triangle: The disease triangle is a conceptual model that shows the interactions between the environment, the host and an infectious agent. This model can be used to predict epidemiological outcomes in plant health, both in local and global levels. The interaction of susceptible host plant, virulent pathogen, and favourable environmental conditions leads to the development of the disease.

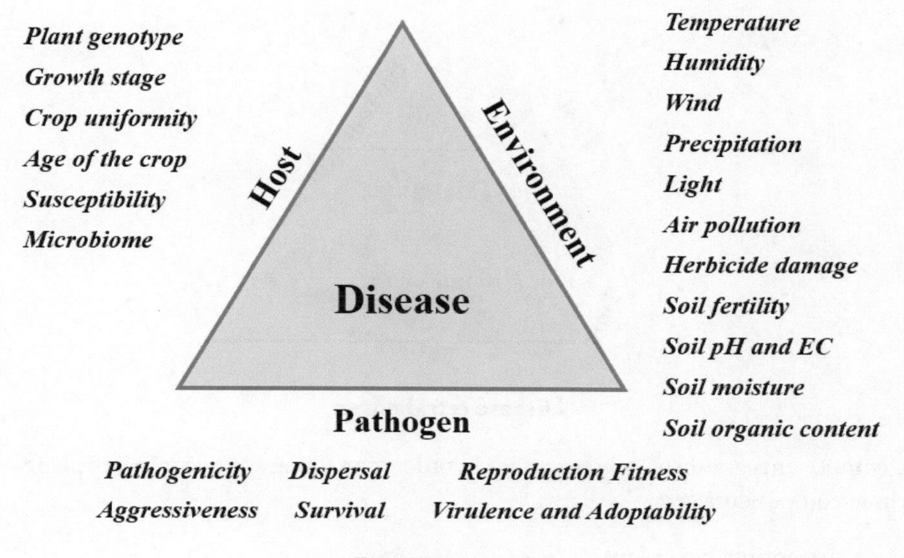

Disease triangle

Disease pyramid: The disease triangle can be expanded to include two more components, time and humans. The amount of each of the three components of disease and their interaction in the development of the disease are influenced by the fourth component, time. The effect of time on disease development becomes apparent when we consider the importance of time of year, the duration and frequency of favourable temperature and rain, the time of appearance of the vector, the duration of the cycle of a particular disease. If the four components of disease pyramid could be quantified, its volume would be proportional to the amount of disease in a plant or in a plant population.

Thus addition of time component to the disease triangle results into a tetrahedron or disease pyramid. Humans affect disease development in various ways. Furthermore, humans can affect the plant–pathogen interaction through crop rotation, choice of plant species and cultivar, their level of resistance, time of planting, and density of planting. In particular, the use of genetically homogenous plant genotypes over large areas may create very favourable conditions for a pathogen to spread and multiply, thereby leading to severe epidemics.

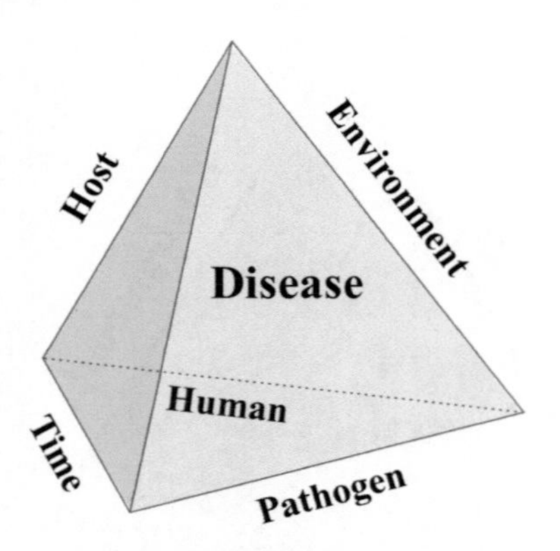

Disease tetrahedran

Three components are absolutely necessary in order for a disease to occur in any plant system. The three components are:

1. a susceptible host plant
2. a virulent pathogen
3. a favorable environment

A disease will develop if a susceptible host plant is intimately associated with a virulent plant pathogen under favourable environmental conditions when all three of these factors are present. There will be a moderate to high quantity of disease when there is a significant degree of overlap.

If the host is resistant to a pathogen, even when the pathogen is present under favorable environmental conditions, a disease will not occur.

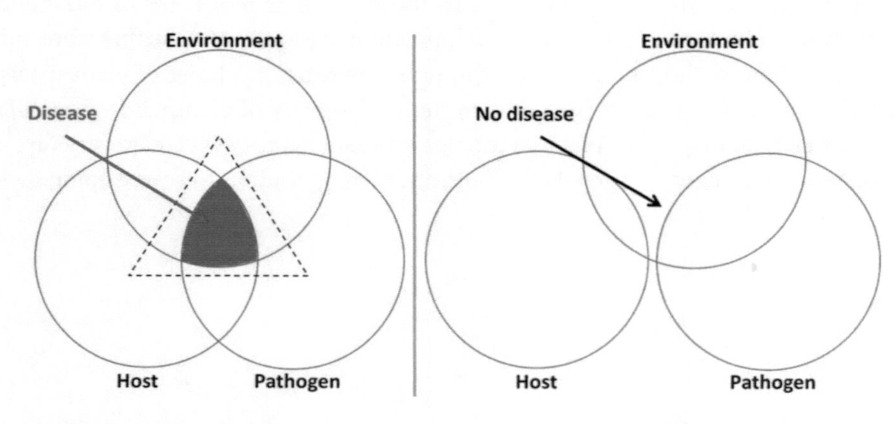

Favourable Interactions **Unfavourble Interactions**

It is important to remember that within each of the three components such as host, pathogen, and environment, there are numerous variables that may affect both the incidence and severity of the disease. The interactions of three components of disease, i.e., the host, pathogen, and environment, can be visualized as a disease triangle. The length of each side is proportional to the total of the characteristics of each component that favour disease.

Essential components/conditions for an epiphytotic:

1. Host factors 2. Pathogen factors 3. Environmental factors

Factors affecting Disease Development

1. **Host factors:** The sensitivity or resistance of a host plant to a particular pathogen is determined mostly by the species and cultivar of the plant (genotype). Host plants may be resistant to pathogens at one stage of development but not at another. Other host factors as follows
   i. **Distance of susceptible plants from the source of primary inoculum**: Longer the distance from the source of survival of the pathogen, longer will be the time required for the buildup of an epiphytotic in a susceptible crop.
   ii. **Abundance and distribution of susceptible hosts**: Continuous cultivation of a susceptible variety over a large contiguous area helps in the buildup of the inoculum and improves the chances of epiphytotics.
   iii. **Disease proneness in the host due to environment**: Susceptibility is genetically controlled but the disease proneness in the plant to get infected can be induced by environment and other factors (Host nutrition, excessive application of nitrogenous fertilizers, etc).
   iv. **Presence of suitable alternate or collateral hosts**: These host plants help in the survival of inoculum of different pathogens in off season. Presence of Barbery which is an alternate host to *Puccinia graminis tritici* helps in the heterogenous infection chain. Presence of grass hosts helps in the survival of *Pyricularia oryzae* in the off-season.

5. **Pathogen factors:** The capacity of a single pathogen species to infect and cause disease symptoms can vary under, some pathogens must be at a critical life stage in order to cause infection. A population of a pathogen species is comprised of numerous genotypes that respond differently on various host genotypes or species; these genotypes may be referred to as races, pathovars, biovars, etc.

Their ability to infect a specific plant and produce disease can vary. The ability to produce disease is known as pathogenicity.

   i. **Presence of virulent/aggressive isolate of a pathogen**: For any epiphytotic, rapid cycle of infection is essential, and successful infection can be caused only by virulent isolates of the pathogen.

ii. **High birth rate**: The fungi that assume epiphytotic form invariably have the capacity to produce enormous quantity of spores that are adapted to long distance dissemination in a short time.

iii. **Low death rate of the pathogen**: Epiphytotics is attributed to low death rate of the pathogens in those in which the causal agent is systemic and protected by the plant tissues.

iv. **Easy and rapid dispersal of the pathogen**: The ability of a pathogen to cause epiphytotics is much more dependent on its dispersal rate. The units of propagation need to be dispersed by external agencies, if epiphytotics are to develop. E.g., Fungal spores disseminated by wind, water, etc. Bacteria dispersed through rain splashes and water. Viruses disseminated by insect vectors

v. **Adaptability of the pathogen**: Most of the pathogens causing epiphytotics adapt themselves to various adverse conditions.

6. **Environmental factors**: There are numerous variables in the environment that influence disease incidence and severity including temperature, sunlight, moisture, relative humidity, and time of year. Pathogens are typically restricted to an area based on the conditions of the macroclimate. A microclimate is the prevailing climatic conditions in a certain geographical area. Assuming that a particular pathogen meets all the above requirements for causing an epidemic, the infection, invasion and development of epidemic may not occur if weather is unfavourable for the pathogenesis.

   » Congenial environmental conditions, *viz.*, optimum weather conditions for sporulation, dispersal, infection and survival of pathogen, are very important for disease occurance.

   » Weather conditions such as, optimum temperature, moisture, light, etc., are very essential for the development of an epidemics.

   » Unfavorable conditions for the host can increase plant susceptibility (to hemibiotrophic and necrotrophs) and may thus be favourable for some diseases.

   » Science which deals with the relationship between weather and epiphytotics is called metereopathology.

a. Temperature

» Powdery mildew favored in summer

» Root rot (Dry) is severe in high temperature

» Rice blast is low temperature disease (Night temp<20°C)

» *Verticillium* wilt of cotton is severe in low temperature

- Loose smut of wheat is severe in temp around 19-20°C
- Bacterial blight will occur temperature around 25-30°C (not below this temp) in rice
- Apple scab will develop as epidemic temperature is near 20°C and blossom remain wet for 18 hrs.

b. Relative humidity

- Sorghum downy mildew pathogen maximize sporulation at 100% RH (if 80% no sporulation)
- Late blight of potato is severe in temperature of 10°C & RH not less 75%

c. Rainfall

- Frequent drizzling favours cumbu ergot
- Bacterial blight will occur of rainy weather, strong wind & temp of 22-26°C
- Brown spot of rice will occur under heavy rainfall accompanied by temp of 25-30°C, cloud weather and low solar radiation

d. Dew

- Sorghum downy mildew pathogen maximize sporulation at leaves at wet period of 4-5 hr.
- Blast is severe in dew for 6-8 hrs., favour disease

e. Light

- Heavy shade increases brown spot of rice, blister blight of tea
- More sporulation of cumbu downy mildew at increased light duration

f. Soil Moisture

- Low moisture favours potato scab, root rot (dry)
- High moisture favours damping off, root rot (wet)

g. Soil Temperature

- Verticillium wilt of cotton is severe in low soil moisture (<20°C)
- Fusarial wilt is severe in high soil moisture (>32°C)

h. Soil pH

- Acid soil favours club root of cabbage

» Alkaline soil favours fusarial wilt of cotton, common scab of potato

Course of epidemic / Stages of epidemic cycle

1. **Lag phase**: The rate of spread of disease is slow
2. **Logarithmic / Exponential / Progressively destructive phase**: Rate of spread is rapid, availability of susceptible host, favourable environment leads to vigorous spread of disease.
3. **Post-logarithmic phase / Decline phase**: Non availability of susceptible stage of the host, unfavourable environment leads to declining phase. In North India, wheat crop attacked by rust during Jan – March. Epidemics develop during these months. After March month disease decline due to unfavourable temperature (rise in temperature), non availability of susceptible host and aggressiveness of pathogen is reduced.

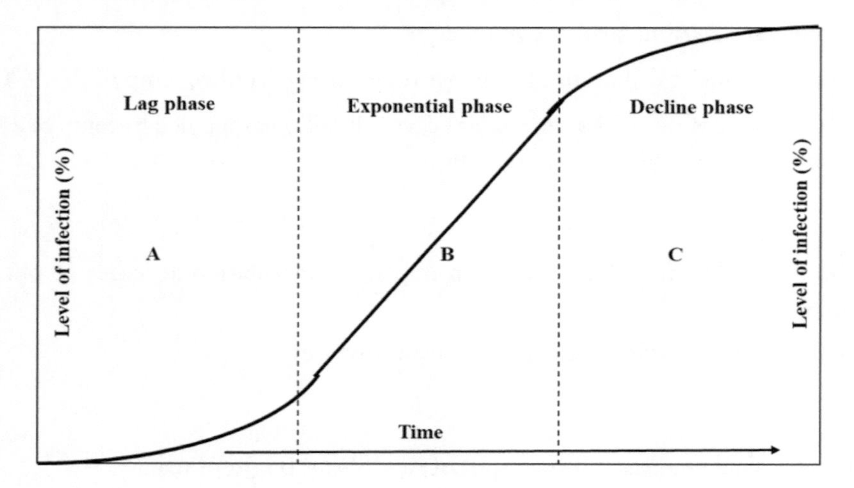

Disease surveillance is an epidemiological practice by which the spread of disease is monitored in order to establish patterns of progression. The main role of disease surveillance is to predict, observe, and minimize the harm caused by outbreak, epidemic, and pandemic situations, as well as increase knowledge about which factors contribute to such circumstances.

Survey on the plant diseases is one of the important aspects in the disease management. It is useful

i. To know the prevalence of disease in a season in a crop of a particular area.

ii. To correlate the disease incidence with the weather factors, biological and soil factors.

iii. To forecast and inform about the outbreak of the diseases to the crop growers.

iv. To assess the damages caused by diseases.

v. To find out the occurrence of new diseases.

vi. To work out the management practices in the control of the diseases.

Pest and disease surveillance programmes are in existence and the Indian Government has set of staff for disease surveillance. Many states, organized pest and disease surveillance programmes which is functioning effectively throughout the state. In this programme the officials in the Departments of agriculture and Department of Horticulture and Plantation and the plant pathologists and entomologists in Agricultural Universities are involved. The departmental staff collects the data from different villages in respective districts, as the scientists (plant pathologists and entomologist) collect the data in the respective research station/ Agricultural College/ Horticutural College. Weekly reports are prepared for each district and sent to the Universities along with daily weather data for analysing and forewarning the pests and disease incidence in the state through mass media.

Methods of survey

1. **Fixed plot survey:** In each agricultural division for each crop, two villages are selected and each village, two fields of one acre are fixed. In each field of one acre five sampling plots are selected. They should be three metres away from the bunds. The sampling plot should be of 1m × 1m size. In each plot twenty representative plants / leaves are selected and observed for the disease incidence utilizing the standard score chart having 0-9 grades for leaf spot, rust, leaf blight, downy mildew, powdery mildew etc. The per cent disease index (PDI) is worked out using standard formula. For systemic diseases like viral and pytoplasmal diseases, root rot, wilt, damping-off, sugarcane smut, green ear etc, percentage of disease incidence is worked out using the specific formula. In the case of vector-borne diseases, the number of vectors per tiller / plant has to be observed and included in the report. Survey on the crop diseases is done every week, the report is prepared.
2. **Roving survey:** In roving survey, for each Panchayat Union / Block four representative villages for individual crops are selected. In each village two fields are fixed. In each field, 100 plants / leaves are observed and scored for the disease intensity by walking across starting from South-West corner to North-East corner. The percentage of disease / per cent disease index is calculated by using the standard formula. As in the fixed plot roving survey is also made at every week and report is prepared.

Disease surveillance report

The weekly data on disease incidence (including vectors and their population) along with the weather data collected from fixed plot survey and roving survey by the Assistant Director of Agriculture are sent to the Joint Director of Agriculture in respective districts with any one of the following.

a. White report: It contains disease / pest surveillance details of a particular block / division / district.

b. Yellow card: It should be sent when the disease / pest occurrence reached half of the level of economic threshold level (ETL)

c. Red card: It should be sent when the disease / pest occurrence exceeded the economic threshold level (ETL).

After analysing the weekly survey reports and the weather data, the message on the severity level of the disease and management measures to be adopted are disseminated to the farmers through mass media like newspaper and All India Radio. Leaflets are also prepared on the disease(s) and their management and distributed to the farmers for adoption. If the disease is in epidemic form, materials are mobilized by staff for immediate and effective control of disease in a short span of time.

Assessment of plant diseases

Assessment of a disease on a plant is essentially required to study quantitative epidemiology. Assessment is essential for the fungicide manufacturing firms, plant breeders and academicians for studying treatment efficacies and evaluation of resistant varieties. Assessment of a specific disease in a crop over several years provides factors governing its incidence and severity. It can be used to device forecasting system for different crops in any location.

Methods of disease assessment

1. Assessment using percent disease index - PDI: In this method assessment is done for systemic diseases like virus and phytoplasmal diseases, sugarcane smut, loose smut of wheat, green ear and root diseases like root-rot, wilt and damping-off. In this, plant diseases which kill the plant outright or which cause about the same amount of damage to the infected plants is assessed. They include all viral, phytoplasmal diseasesvascular wilt and root rot of different crops.

The total number of plants in an area is counted along with total number of diseased plants. The percentage of disease incidence is calculated by using the following formula,

$$\text{Per cent disease incidence} = \frac{\text{No. of diseased plants}}{\text{Total number of plants observed}} \times 100$$

2. Assessment using disease grades: This method is used for downy mildews, powdery mildews, rusts, grain smuts, cankers, anthracnose, leaf spots, leaf blights, blast, ergot, etc.

For assessing most of the foliar diseases the standard disease score charts are available with pictorial diagrams. The diagrams for each disease will be based on the percentage of leaf area affected. In earhead diseases the number of grains affected are indicated. For the leaf or earhead diseases, the grade ranging from 0 to 9 are internationally used. The grades and description for each grade for important diseases are given below.

a. Leaf spot, leaf blight, anthracnose, blast, downy mildew, powdery mildew, rust, ergot and sooty mould

Grade	Description
0	No visible symptoms
1	<1% of leaf area affected
3	1-10% of leaf area affected
5	11-25% of leaf area affected
7	26-50% of leaf area affected
9	> 50% of leaf area affected

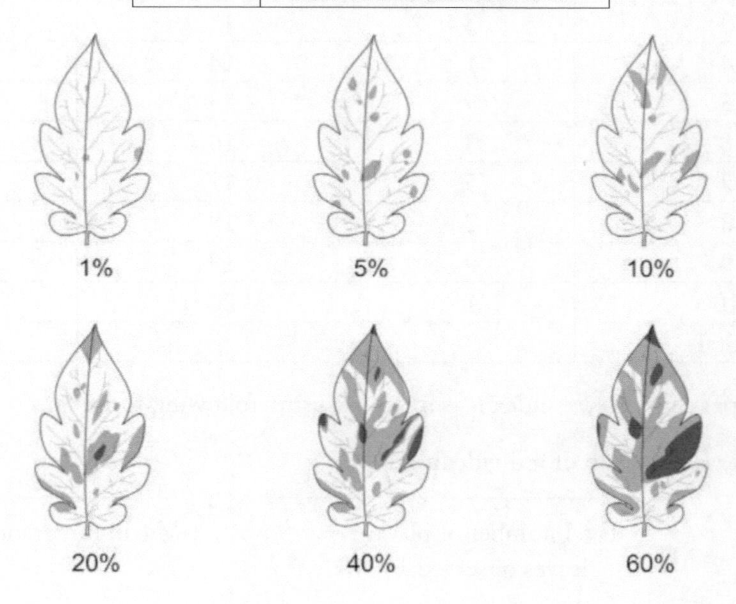

b. Ergot

Grade	Description
0	Earheads free from infection
1	<1% grains in earhead replaced by sclerotia
3	1-10% grains in earhead replaced by sclerotia

5	11-25% grains in earhead replaced by sclerotia
7	26-50% grains in earhead replaced by sclerotia
9	>75% grains in earhead replaced by sclerotia

For assessment of the above diseases, plants or leaves are observed individually for the area or number (for ergot) affected and severity grades are assigned to each diseased plant or leaf. Sampling unit and sample size vary with the host and disease involved. In assessing foliar diseases in a field, it is not possible to assess all the plants in a plot or field. Hence usually 10 per cent of the representative population is selected at random for assessing the disease. In the case of tree crops five trees are selected at random and 20 representative leaf or fruit samples are assessed. For example, if 20 representative samples are to be taken assign the grade for individual leaf or plant or fruit and enter the grade as detailed below.

Plant / Leaf number	Severity grade	Plant / Leaf number	Severity grade
1	9	11	0
2	9	12	3
3	3	13	1
4	1	14	5
5	9	15	9
6	3	16	7
7	5	17	0
8	7	18	3
9	9	19	1
10	1	20	3
Total			88

The per cent disease index is worked out using following formula:

$$\text{Per cent Disease Index (PDI)} = \frac{\text{Sum of individual ratings}}{\text{Total number of plants / leaves observed}} \times \frac{100}{\text{Maximum grade}}$$

Disease forecasting

Crop diseases have the potential to cause devastating epidemics that threaten the world's food supply and vary widely in their dispersal pattern, prevalence, and severity. Disease dynamics itself involves a complex interaction between a host, a pathogen, and their environment, representing one of the largest integrated risks facing the long-term sustainability of agriculture. Genetic factors (e.g., emergence of new diseases and of new races), environmental-driven influence (e.g., global climate change impacts on disease spread), and management-intervention driven agroecosystem interactions (e.g., crop breeding and monitoring technologies) are all important considerations in disease risk mitigation. It is an advance warning on the occurrence or outbreak of disease in a particular area, so that suitable control measures can be undertaken in advance to avoid losses.

a. Long range forecasting.

b. Short range forecasting.

Long range forecasting is helpful to enable the farmer to decide the crop cultivar for the season. In the case of simple interest diseases, farmer can choose disease free seed or planting material. It is more useful for the farmers than short range forecasting. In the case of short-range forecasting, it is helpful to take up protective spraying well in advance to avoid yield loss.

Since forecasting is a part of applied epidemiology, the disease triangle informations (elated to epidemic development are required for forecasting.

Forecasting systems developed by different countries for the following diseases

1. Grapevine downy mildew (*Plasmopara viticola*) - Australia, France, Germany, Italy, USA

2. Potato late blight (*Phytophthora infestans*) – Australia, Brazil, Germany, USA

3. Apple scab (*Venturia inaequalis*) – Australia, USA

4. Cucumber downy mildew (*Pseudoperonospora cubensis*) - USA

5. Rice blast (*Pyricularia oryzae*) – India and Japan

6. Rice tungro virus - India

7. Wheat brown rust (*Puccinia recondita*) – India and USA

Important forecasting models

1. Blitecast - computerized version for late blight of potato (USA)
2. Blast L - first leaf blast model developed in Japan
3. Blastcast, Pyricularia - model developed in Japan for rice blast
4. MARYBLYT - Model developed in USA for apple blossom blight symptoms
5. CougerBlight - Model developed in USA for apple fire blight, predicts blossom infections using temperature data to estimate the growth rate of fire blight bacteria on the stigma

during over the approximately 96 hours prior to a wetting period, which is required for flower infections

6. Billing's Integrated System (BIS95) - Model developed in USA for apple fire blight
7. FAST - Model developed for tomato early blight
8. *Xcv* infection model - Model developed for bacterial spot-on hot pepper
9. *Septoria* - Timer - Wheat septoria
10. *Psa* risk model - Model developed for bacterial canker of kiwifruit
11. Early leaf spot Adison model developed in USA for groundnut early leafspot
12. PYRNEW - model developed for rice blast in Indonesia
13. LEAF BLAST, EPIBLAST – rice blast model developed in Korea
14. BSP Cast - Pear brown spot

Disease prediction models

Disease prediction models using advanced statistical methods integrating weather and aerobiological monitoring data have been successfully developed and validated for white blister on Brassica crops (Brassica spot™). This has the potential to improve the timeliness, effectiveness and foresight for controlling crop diseases, while minimizing crop loss.

A prototype integrated, model-based framework for forecasting disease risk. Risk components for the major disease progression stages (i.e., immigration, deposition, germination, infection, incubation/residency, multiplication, re-dispersal, perennation) are included, alongside major risk variables, e.g., weather trigger events and conditions, host exposure and susceptibility, pathogen survival, residency, latency, endemic infection, cycling, epidemic transmission and dispersal, and evolved virulence associated with the risk of resistance breakdown in the field. Full boxes are components of the framework that are considered in the current disease risk modeling, while dashed boxes will be considered in future extended modeling.

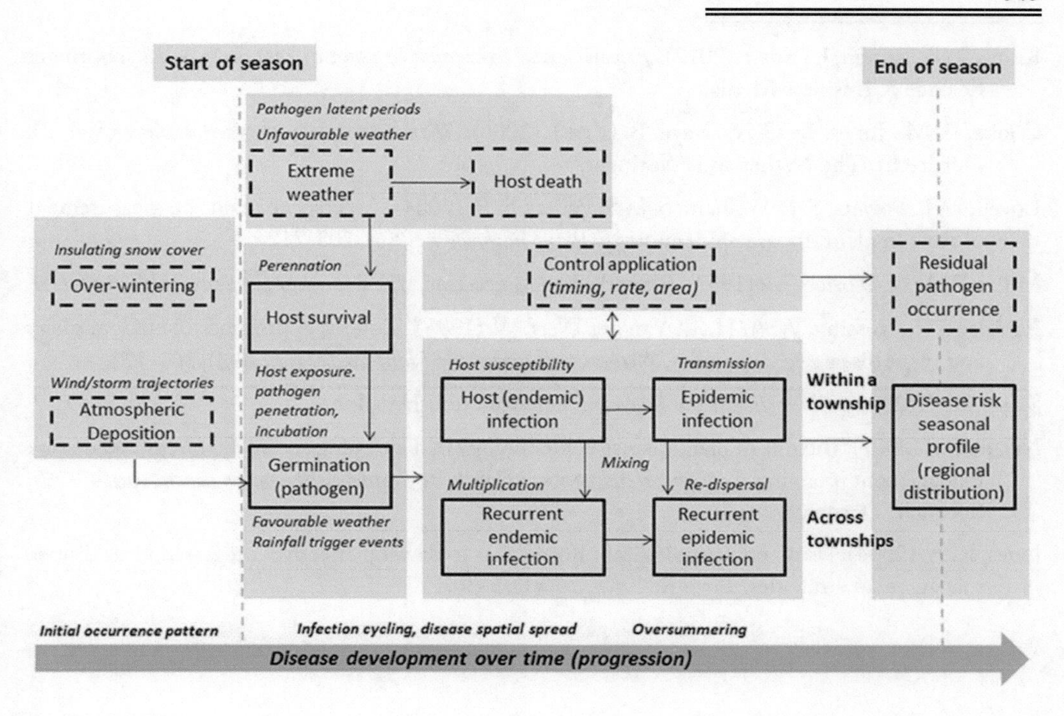

Practice Questions

1. Write the factors involved in disease development
2. Give definition and examples of the following types of disease:
    i. Epidemic or Epiphytotic disease
    ii. Sporadic disease
    iii. Endemic disease
    iv. Pandemic disease
3. How does plant disease forecasting help to control disease?
4. What are BLIGHTCAST, CERCOS, MYCOS, EPICORN, EPIVEN, and EPIDEMIC?
5. Write the differences between percent disease index and incidence.

Suggested Readings

Campbell, C. L., & Madden, L. V. (1990). Introduction to plant disease epidemiology. John Wiley & Sons.

Arneson, P. A. (2001). Plant disease epidemiology. *The Plant Health Instructor*.

Zadoks, J. C., & Schein, R. D. (1979). Epidemiology and plant disease management. *Elsivier.*

Kranz, J., & Rotem, J. (Eds.). (2012). *Experimental techniques in plant disease epidemiology*. Springer Science & Business Media.

Cooke, B. M., Jones, D. G., & Kaye, B. (Eds.). (2006). *The epidemiology of plant diseases* (Vol. 2). Dordrecht, The Netherlands: Springer.

Lovell, D. J., Powers, S. J., Welham, S. J., & Parker, S. R. (2004). A perspective on the measurement of time in plant disease epidemiology. *Plant Pathology*, *53*(6), 705-712.

Miller, P. R., & O'brien, M. (1952). Plant disease forecasting. *The Botanical Review*, *18*(8), 547-601.

Rabbinge, R., Rossing, W. A. H., & Van der Werf, W. (1993). Systems approaches in epidemiology and plant disease management. *Netherlands Journal of Plant Pathology*, *99*(3), 161-171.

Singh, R. S. (2001). *Plant disease management*. Science Publishers, Inc..

Nutter, F. F. (2007). The role of plant disease epidemiology in developing successful integrated disease management programs. In *General concepts in integrated pest and disease management* (pp. 45-79). Springer, Dordrecht.

Jones, R. A. (2004). Using epidemiological information to develop effective integrated virus disease management strategies. *Virus Research*, *100*(1), 5-30.

Chapter - 32

Defense Mechanism in Plants

In general plants defend themselves against pathogens by two ways: structural or morphological characteristics that act as physical barriers and biochemical reactions that take place in cells and tissues that are either toxic to the pathogen or create conditions that inhibit the growth of the pathogen in the plant.

1. Structural defense mechanisms: These may be pre-existing, which exist in the plant even before the pathogen comes in contact with the plant or induced, *i.e*, even after the pathogen has penetrated the preformed defense structures, one or more type of structures is formed to protect the plant from further pathogen invasion.

Pre-existing structural defense structures

These include the amount and quality of wax and cuticle that cover the epidermal cells and the size, location and shapes of natural openings (stomata and lenticels) and presence of thick walled cells in the tissues of the plant that hinder the advance of the pathogen.

i) Waxes: Waxes on leaf and fruit surfaces form a hydrophobic or water repellent surface preventing the germination of fungi and multiplication of bacteria.

ii) Cuticle and epidermal cells: A thick cuticle and tough outer wall of epidermal cells may increase resistance to infection in diseases in which the pathogen enters its host only through direct penetration. E.g., Disease resistance in Barbery species infected with *Puccinia graminis tritici* has been attributed to the tough outer epidermal cells with a thick cuticle. In linseed, cuticle acts as a barrier against *Melampsora lini*.

The silicification and lignifications of epidermal cells offers protection against *Pyricularia oryzae* and *Streptomyces scabies* in paddy and potato, respectively.

iii) Sclerenchyma cells: The sclerenchyma cells in stems and leaf veins effectively blocks the spread of some fungal and bacterial pathogens that cause angular leaf spots.

iv) Structure of natural openings:

a) Stomata: Most of the pathogens enter plants through natural openings. Some pathogens like stem rust of wheat can enter its host only when the stomata are open. The wheat varieties (Hope) in which stomata open late in the day are resistant as the germ tubes of the spores germinating in the night dew desiccate owing to evaporation of the dew before stomata begin to open. This can also be called as functional resistance. The structure of stomata provides resistance to penetration by certain plant pathogenic bacteria.

E.g., The citrus variety, szinkum, is resistant to citrus canker because it posses a broad cuticular ridge projecting over the stomata and a narrow slit leading to the stomatal cavity thus preventing the entry of bacterial and fungal spores into the interior of the leaf.

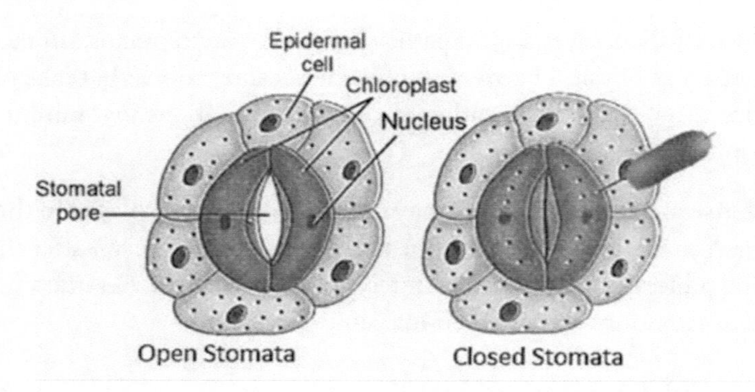

Stomatal closing during pathogen (bacteria) entry

b) Lenticels: The shape and internal structure of lenticels can increase or decrease the incidence of fruit diseases. Small and suberised lenticels will offer resistance to potato scab pathogen, *Streptomyces scabies.*

Post-infectional structural defense mechanisms/Induced structural barriers: These may be regarded as histological defense barriers (cork layer, abscission layers and tyloses) and cellular defense structures (hyphal sheathing).

i) Histological defense structures

a) Cork layer: Infection by fungi, bacteria, some viruses and nematodes induce plants to form several layers of cork cells beyond the point of infection and inhibits the further invasion by the pathogen beyond the initial lesion and also blocks the spread of toxin substances secreted by the pathogen. Furthermore, cork layers stop the flow of nutrients and water from the healthy to the infected area and deprive the pathogen of nourishment. E.g., Potato tubers infected by *Rhizoctonia*; *Prunus domestica* leaves attacked by *Coccomyces pruniphorae*.

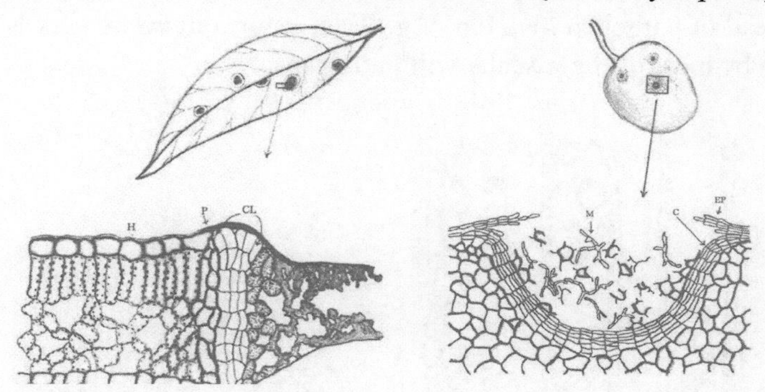

(Key: H = healthy area, P = pathogen, CL = cork layer, I = infected area, M = mycelium of the fungal pathogen, C = cork layer, Ep = epidermis.)

b) Abscission layers: An abscission layer consists of a gap formed between infected and healthy cells of a leaf surrounding the locus of infection due to the disintegration of the middle lamella of parenchymatous tissue. Gradually, infected area shrivels, dies, and sloughs off, carrying with it the pathogen. Abscission layers are formed on young active leaves of stone fruits infected by fungi, bacteria or viruses. E.g., *Xanthomonas pruni,* and *Closterosporium carpophylum* on peach leaves.

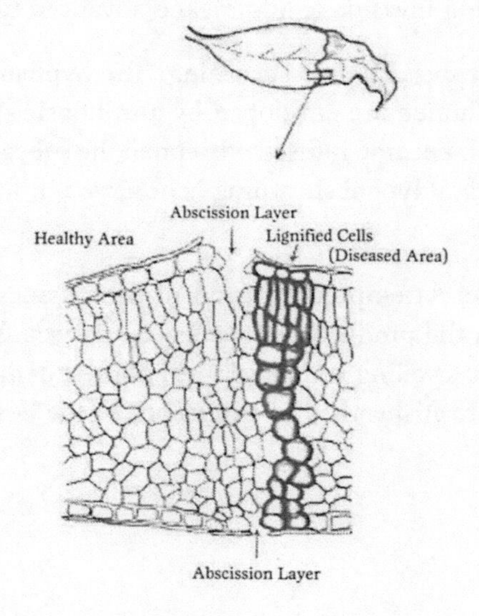

c) Tyloses: Tyloses are the cellulosic overgrowths of the protoplast of adjacent living parenchymatous cells, which protrude into xylem vessels through pits. Tyloses have cellulosic walls and are formed quickly ahead of the pathogen and may clog the xylem vessels completely interrupting the active and passive transport of the pathogen from the site of local invasion to outlying areas of host tissue in resistant varieties. In susceptible varieties, few or no tyloses are formed ahead of pathogen invasion. E.g., Tyloses form in xylem vessels of most plants under invasion by most of the vascular wilt pathogens.

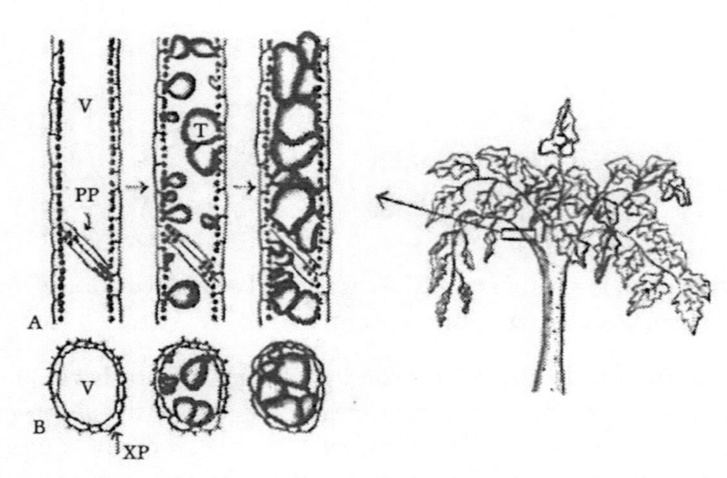

(Key: V = xylem vessel, PP = perforated vessel plate, T = tylose, XP = xylem parenchyma Cell.)

(A) depicts longitudinal sections of a healthy, noninvaded xylem vessel (far left), the beginning of tylose formation following invasion (middle), and advanced tylose formation

ii) Cellular defense structures: Hyphal sheathing: The hyphae penetrating the cell wall and growing into the cell lumen are enveloped by a cellulosic sheath (callose) formed by extension of cell wall, which become infused with phenolic substances and prevents further spread of the pathogen. E.g., Hyphal sheathing is observed in flax infected with *Fusarium oxysporum* f.sp. *lini.*

d) Sheathening of hyphae: Attempted invasion of host tissues by a potential pathogen is often responded to with the production of 'adjusted' tissues. At the localized regions of invasion, cell walls thicken, swelling to many times normal density. Furthermore, invaded portions of the pathogen are ensheathed by extensions of the host cell wall.

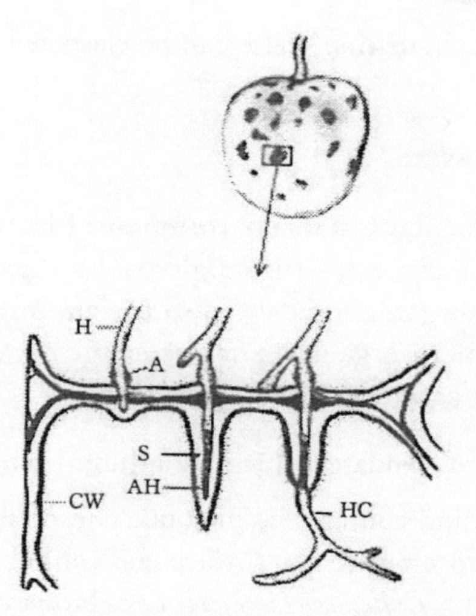

(Key: H = pathogen hyphae, A = pathogen appresorium or "pad," HC = hypha in the host cytoplasm, AH = advancing hypha, S = sheath formed by the thickened cell wall of the host, CW = cell wall of the host.)

e) Defensive gummosis: When woody tissues are invaded, isolation or wailing off the intruding pathogen oftentimes takes the form of gum production and lignification. Gums are complex polysaccharides with colloidal properties which allow them to expand and contract as well as to form impenetrable barriers against the advancing pathogen. Intercellular spaces surrounding the zone of infection are quickly filled with gums, thereby forming an impenetrable barrier. Adjacent tissues eventually become lignified, thereby completing the mature walled-off barrier.

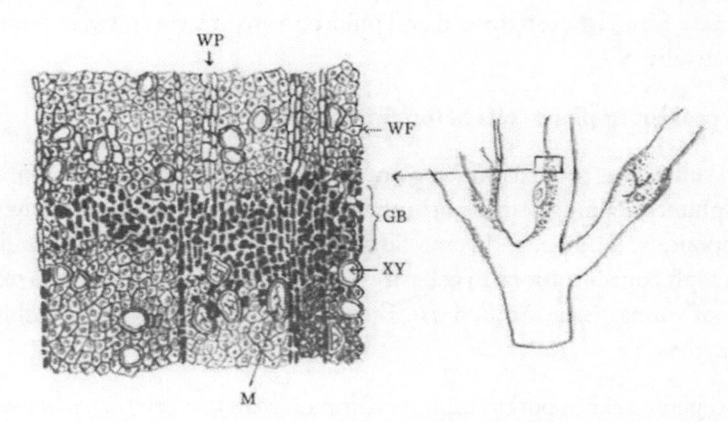

(Key: WP = wood parenchyma tissue, WF = wood fiber, GB = gum barrier, XV = xylem vessels, M = mycelia of the fungal pathogen.)

2) Biochemical defense mechanisms: These can be classified as pre-existing and induced biochemical defenses.

Pre-existing chemical defenses:

a) Inhibitors released by the plant in its environment: Plants exude a variety of leaf and root exudates which contain aminoacids, sugars, glycosides, organic acids, enzymes, alkaloids, flavones, toxic materials, inorganic ions and also certain growth factors. The inhibitory substances directly affect micro-organisms or encourage certain groups to dominate the environment which may act as antagonists to pathogen.

» Tomato leaves secrete exudates which are inhibitory to *Botrytis cinerea*

» Red scales of red onion contain the phenolic compounds, protocatechuic acid and catechol, which diffuse out to the surface and inhibit the conidial germination of onion smudge fungus, *Colletotrichum circinans*. However, these fungitoxic phenolic compounds are missing in white scaled onions.

» Resistant varieties of apple secrete waxes on the leaf surface which prevents the germination of *Podosphaera leucotricha* (powdery mildew of apples).

» In *Cicer arietinum* (chickpea), the *Ascochyta* blight resistant varieties have more glandular hairs which have maleic acid which inhibit spore germination.

» Resistant varieties of linseed secrete HCN in roots which are inhibitory to linseed wilt pathogen, *Fusarium oxysporum* f.sp. *lini*.

» Root exudates of marigold contain a-terthinyl which is inhibitory to nematodes.

» Chlorogenic acid present in sweet potato, potato and carrot inhibits *Ceratocystis fimbriata*. Similarly, caffeic acid and phloretin are present in sweet potato and apple, respectively.

b) Inhibitors present in plant cells before infection:

Antimicrobial substances pre-existing in plant cells include unsaturated lactones, cyanogenic glycosides, sulphur containing compounds, phenols, phenolic glycosides and saponins. Several phenolic compounds, tannins, and some fatty acid like compounds such as dienes, which are present in high concentrations in cells of young fruits, leaves or seeds are responsible for the resistance of young tissues to *Botrytis*. These compounds are potent inhibitors of many hydrolytic enzymes.

» Chlorogenic acid in potato inhibits common scab bacteria, *Streptomyces scabies,* and to wilt pathogen, *Verticillium alboatrum*

» Saponins have antifungal membranolytic activity which excludes fungal pathogens that lack saponinases. E.g., Tomatine in tomato and Avenacin in oats.

- Lectins, which are proteins that bind specifically to certain sugars and occur in large concentrations in many types of seeds, cause lysis and growth inhibition of many fungi.
- Plant surface cells also contain variable amounts of hydrolytic enzymes such as glucanases and chitinases which may cause breakdown of pathogen cell wall.

Post inflectional or induced defense mechanisms: A plant's resistance response to pathogen intrusion depends upon how quickly **the plant can mount a defense. The speed of the response determines whether a plant remains healthy or becomes afflicted with disease.**

The rise in defense metabolism following pathogen challenge, it will become clear that rapidity of the resistance response is commensurate with the health status of the plant and its ability to quickly prioritize production of various 'secondary metabolites' and compounds necessary for resisting disease.

All such reactions require a plant to be metabolizing nutrients efficiently, an ample supply of the molecules necessary for building tissue structures, and a supply of chemically stored energy to drive these reactions.

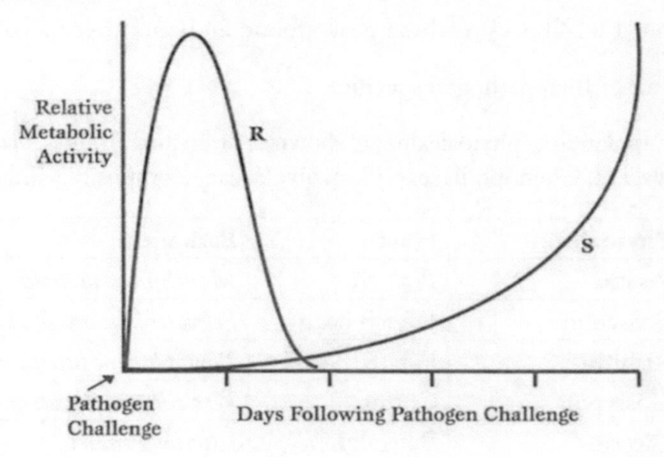

S = a susceptible plant, R = a resistant plant.

a) Phytoalexins (*Phyton* = plant; *alexin* = to ward off) Muller and Borger (1940) first used the term phytoalexins for fungistatic compounds produced by plants in response to injury (mechanical or chemical) or infection. Phytoalexins are toxic antimicrobial substances produced in appreciable amounts in plants only after stimulation by phytopathogenic micro-organisms or by chemical or mechanical injury.

Phytoalexins are not produced by uninfected healthy plants, but produced by healthy cells adjacent to localized damaged or necrotic cells in response to materials diffusing from the infected cells. These are not produced during compatible biotrophic infections.

Phytoalexins accumulate around both resistant and susceptible necrotic tissues. However, resistance occurs when one or more phytoalexins reach a concentration sufficient to restrict pathogen development.

Characteristics of phytoalexins

1. Fungitoxic and bacteriostatic at low concentrations.

2. Produced in host plants in response to stimulus (elicitors) and metabolic products.

3. Absent in healthy plants

4. Remain close to the site of infection.

5. Produced in quantities proportionate to the size of inoculum.

6. Produced in response to the weak or non-pathogens than pathogens

7. Produced within 12-14 hours reaching peak around 24 hours after inoculation.

8. Host specific rather than pathogen specific.

Synthesis and accumulation of phytoalexins are shown in diversified families, *viz.*, Leguminosae, Solanaceae, Malvaceae, Chenopodiaceae, Convolvulaceae, Compositae and Graminaceae.

S. No.	Phytoalexin	Host	Pathogen
1	Pisatin	Pea	*Monilinia fructicola*
2	Phaseolin	French bean	*Sclerotinia fructigena*
3	Rishitin	Potato	*Phytophthora infestans*
4	Gossypol	Cotton	*Verticillium alboatrum*
5	Cicerin	Bengalgram	*Ascochyta rabiei*
6	Ipomeamarone	Sweet potato	*Ceratocystis fimbriata*
7	Capsidol	Pepper	Colletotrichum capsici

b) Hypersensitive response (HR)

The term hypersensitivity was first used by Stakman (1915) in wheat infected by rust fungus, *Puccinia graminis* var. *tritici*. The hypersensitive response is a localized induced cell death in the host plant at the site of infection by a pathogen, thus limiting the growth of pathogen. In the infected plant part, HR is seen as water-soaked large sectors which subsequently become necrotic and collapsed. HR occurs only in incompatible host-pathogen combinations. HR may occur whenever virulent strains or races of pathogens are injected into non-host plants or into resistant varieties, and when avirulent strains or races of pathogens are injected into susceptible cultivars. HR is initiated by the recognition of specific pathogen-produced signal molecules, known as elicitors. Recognition of the elicitors by the host results in altered cell functions leading to the production of defense related compounds.

The most common new cell functions and compounds include:

» A rapid burst of oxidative reactions.

» Increased ion movement, especially of K+ and H+ through cell membrane.

» Disruption of membranes and loss of cell compartmentalization.

» Cross-linking of phenolics with cell wall components and strengthening of plant cell wall

» Production of antimicrobial substances such as phytoalexins and pathogenesis-related proteins (such as chitinases).

Cellular responses during HR

In many host-pathogen combinations, as soon as the pathogen establishes contact with the cell, the nucleus moves toward the invading pathogen and soon disintegrates. Brown resin like granules form in the cytoplasm, first around the point of penetration of pathogen and then throughout the cytoplasm. As the browning discolouration of the cytoplasm continues and death sets in, the invading hypha begins to degenerate and further invasion is stopped.

c) Plantibodies: Transgenic plants have been produced which are genetically engineered to incorporate into their genome, and to express foreign genes, such as mouse genes that produce antibodies against certain plant pathogens. Such antibodies, encoded by animal genes, but produced in and by the plant, are called plantibodies. E.g., Transgenic plants producing plantibodies against coat protein of viruses, such as, artichoke mottle crinkle virus have been produced.

Practice Questions

1. What are the different morphological defense mechanisms in plants? Give suitable examples.
2. What is biochemical defense mechanism?
3. What are phytoalexins? Are phytoalexins responsible for host specificity?
4. Define plantibodies
5. Write the differences between phytoalexins and phytoanticipins

Suggested Readings

Bari, R., & Jones, J. D. (2009). Role of plant hormones in plant defence responses. *Plant molecular biology*, *69*(4), 473-488.

Bennett, R. N., & Wallsgrove, R. M. (1994). Secondary metabolites in plant defence mechanisms. *New phytologist*, *127*(4), 617-633.

Agrawal, A. A. (2011). Current trends in the evolutionary ecology of plant defence. *Functional Ecology*, *25*(2), 420-432.

Ebel, J., & Mithöfer, A. (1998). Early events in the elicitation of plant defence. *Planta*, *206*(3), 335-348.

Goyal, S., Lambert, C., Cluzet, S., Mérillon, J. M., & Ramawat, K. G. (2012). Secondary metabolites and plant defence. In *Plant defence: biological control* (pp. 109-138). Springer, Dordrecht.

Chapter - 33

Principles of Plant Disease Management

The word 'control' is a complete term where permanent 'control' of a disease is rarely achieved whereas, 'management' of a disease is a continuous process and is more practical in influencing adverse affect caused by a disease. Understanding the disease cycle is important when considering control options. Learning the chain of events that contribute to a disease helps point out the weakest links. Control measures can then be used to break the cycle. Disease management requires a detail understanding of all aspects of crop production, economics, environmental, cultural, genetics and epidemiological information upon which the management decisions are made.

Management: Management of plant disease is continuous cycle based on the principle of eradication of the pathogen and on the principle of minimizing the damage or loss due to disease below economic injury level.

Importance: Plant diseases are important because of the losses (qualitative and quantitative) they cause. Loss may occur at any time between sowing of the crop and consumption of the produce. Measures taken to prevent the incidence of the disease, reduce the amount of inoculum that initiates and spreads the disease and finally minimize the loss caused by the disease are called as management practices.

Essential considerations in plant disease management

1. Benefit-cost ratio
2. Procedures for disease control should fit into general schedule of operations of crop production
3. Control measures should be adopted on a co-operative basis over large adjoining areas. This reduces frequency of applications, cost of control and increases chances of success of control measures

4. Knowledge aspects of disease development are essential for effective economical control. Information is needed on the following aspects:
   a. Cause of a disease
   b. Mode of survival and dissemination of the pathogen
   c. Host parasite relationship
   d. Effect of environment on pathogenesis in the plant or spread in plant population
5. Prevention of disease depends on management of primary inoculum
6. Integration of different approaches of disease management is always recommended

General principles of plant disease management

Management of pathogen: Avoidance, Exclusion and Eradication

Management of host: Protection, Immunization and Therapy

The major priciples in plant disease management are

1. **Avoidance**: Avoiding disease by planting at times when, or in areas where, inoculum is ineffective due to environmental conditions, or is rare or absent

2. **Exclusion of inoculum**: Preventing the inoculum from entering or establishing in the field or area where it does not exist.

3. **Eradication**: Reducing, inactivating, eliminating or destroying inoculum at the source, either from a region or from an individual plant in which it is already established.

4. **Protection**: Preventing infection by creating a chemical toxic barrier between the plant surface and the pathogen.

5. **Disease resistance (Immunization)**: Preventing infection or reducing effect of infection by managing the host through improvement of resistance in it by genetic manipulation or by chemical therapy.

1. Avoidance of the pathogen: These methods aim at avoiding the contact between the pathogen and susceptible stage of the crop. This is achieved by

a. Proper selection of geographical area

b. Proper selection of the field

c. Adjusting time of sowing

d. Disease escaping varieties

e. Proper selection of seed and planting material

a) Proper selection of geographical area: Many fungal and bacterial diseases are more severe in wet areas than in dry areas. Cultivation of pearl millet in wet areas is not profitable due to the diseases, smut (*Tolyposporium penicillariae*) and ergot (*Claviceps microcephala*). Growing of pearl millet in dry areas with irrigation facilities can be avoided from smut and ergot diseases.

b) Proper selection of the field: Proper selection of field will help in the management of many diseases, especially the soil borne diseases. Raising of a particular crop year after year in the same field makes the soil sick, where disease incidence and severity may be more. E.g., Wilt of redgram, late blight of potato (*Phytophthora infestans*), green ear of bajra (*Sclerospora graminicola*), etc.

c) Time of sowing: Generally, pathogens are able to infect the susceptible plants under certain environmental conditions. Alteration of date of sowing can help in avoidance of favourable conditions for pathogen. E.g., *Rhizoctonia* root rot of redgram is more severe in the crop sown immediately after the rains. Delayed sowing will help in reducing the incidence of disease. E.g., Infection of black stem rust of wheat (*Puccinia graminis* var. *tritici*) is more in late sowing, hence, early sowing helps in reduction of stem rust incidence.

d) Disease escaping varieties: Certain varieties of crops escape the disease damage because of their growth characteristics.

E.g., Early maturing varieties of wheat or pea escape the damage due to *Puccinia graminis* var. *tritici* and *Erysiphe polygoni*, respectively.

Early maturing varieties of groundnut escape early leaf spot infection (*Cercospora arachidicola*) and early varieties of wheat escape rust and loose smut infection.

e) Proper selection of seed and planting material: Selection of seed and seedling material from healthy sources will effectively manage the diseases such as loose smut of wheat (*Ustilago nuda tritici*), bunchy top of banana (*BBTV*), Panama wilt of banana (*Fusarium oxysporum* f.sp. *cubense*) and whip smut of sugarcane (*Ustilago scitaminae*). Potato seed certification or tuber indexing is followed for obtaining virus free seed tubers. Citrus bud wood certification programme will help in obtaining virus free planting material.

2. Exclusion of the pathogen: These measures aim at preventing the inoculum from entering or establishing in the field or area where it does not exist. Different methods of exclusion are seed treatment, seed inspection and certification, and plant quarantine regulation.

a) Seed inspection and certification: Crops grown for seed purpose are inspected periodically for the presence of diseases that are disseminated by seed. Necessary precautions are to be taken to remove the diseased plants in early stages, and then the crop is certified as disease free. This practice will help in the prevention of inter and intra regional spread of seed borne diseases.

b) Plant quarantine regulation: Plant quarantine is defined as "a legal restriction on the movement of agricultural commodities for the purpose of exclusion, prevention or delaying the spread of the plant pests and diseases in uninfected areas".

Plant quarantine laws were first enacted in France (1660), followed by Denmark (1903) and USA (1912). These rules were aimed at the rapid destruction or eradication of barbery bush which is an alternate host of *Puccinia graminis* var. *tritici*.

In India, plant quarantine rules and regulations were issued under Destructive Insects and Pests Act (DIPA) in 1914. In India, 16 plant quarantine stations are in operation by the "Directorate of Plant Protection and Quarantine" under the Ministry of Food and Agriculture, Government of India.

Pest: "Any species, strain or biotype of plant or animal or any pathogenic agent, injurious to plants or plant products" It includes insects, mites, nematodes, fungi, bacteria, viruses, phytoplasma, ricketssia like organisms and weeds.

Quarantine pest: "A pest of potential economic importance to the area endangered thereby and present there but not widely distributed and being officially controlled".

Diseases	Host	Country	Introduced from	Losses caused
Canker	Citrus	USA	Japan	13 million; 19.5 million trees destroyed
Dutch elm	Elm	USA	Holland	25-50,000 million
Powdery mildew	Grapevine	France	USA	80% in wine production
Downy mildew	Grapevine	France	USA	$50,000 million
Bunchy top	Banana	India	Sri Lanka	Rs. 4 crores
Wart	Potato	India	Netherlands	1000 ha infected
Blight	Chestnut	USA	Eastern Asia	$ 100 – 1000 million
Blue mould	Tobacco	Europe	UK	$ 50 million

Diseases believed to have been introduced into India from other countries

Diseases	Host	First record	Introduction from
Leaf rust (*Hemileia vastatrix*)	Coffee	1879	Sri Lanka
Late blight (*Phytophthora infestans*)	Potato, Tomato	1883	Europe

Rust (*Puccinia carthami*)	Chrysanthemum	1904	Japan (or) Europe
Flag smut (*Urocystis tritici*)	Wheat	1906	Australia
Downy mildew (*Plasmopara viticola*)	Grapevine	1910	Europe
Downy mildew (*Pseudoperonospora cubensis*)	Cucurbits	1910	Sri Lanka
Downy mildew (*Sclerospora philippinensis*)	Maize	1912	Jawa
Foot rot (*Fusarium moniliforme*)	Rice	1930	S.E.Asia
Leaf spot (*Phyllachora sorghi*)	Sorghum	1934	S. Africa
Powdery mildew (*Oidium heveae*)	Rubber	1938	Malaya
Black shank (*Phytophthora parasitica* var. *nicotianae*)	Tobacco	1938	Dutch east
Fire blight (*Erwinia amylovora*)	Pear, Pomes	1940	England
Crown gall (*Agrobacterium tumefaciens*) and Hairy root (*Agrobacterium rhizogenes*)	Apple, Pear	1940	England
Banana bunchy top virus (BBTV)	Banana	1940	Sri Lanka
Canker (*Sphaeropsis malorum*)	Apple	1943	Australia
Wart (*Synchytrium endobioticum*)	Potato	1943	Netherlands
Foot rot (*Fusarium moniliforme*)	Rice	1930	S.E.Asia

Plant quarantine measures are three types.

a. Domestic quarantine: Rules and regulations issued prohibiting the movement of insects and diseases and their hosts from one state to another state in India is called domestic quarantine. Domestic quarantine in India exists for two pests (Rooted scale and Sanjose scale) and three diseases (Bunchy top of banana, banana mosaic and wart of potato).

- **Bunchy top of banana**: It is present in Kerala, Assam, Bihar, West Bengal and Orissa. Transport of any part of *Musa* species excluding the fruit is prohibited from these states to other states in India.
- **Banana mosaic**: It is present in Maharashtra and Gujarat. Transport of any part of *Musa* species excluding the fruit is prohibited from these states to other states in India.
- **Wart of potato**: It is endemic in Darjeeling area of West Bengal, therefore seed tubers are not to be imported from West Bengal to other states.

b. Foreign quarantine: Rules and regulations issued prohibiting the import of plants, plant materials, insects and fungi into India from foreign countries by air, sea and land. Foreign quarantine rules may be general or specific. General rules aim at prevention of introduction of pests and diseases into a country, where as the specific rules aim at specific diseases and insect pests. The plant materials are to be imported only through the prescribed airport and seaports of entry.

c. Total embargoes: Total restriction on import and export of agricultural commodities.

Phytosanitary certificate: It is an official certificate from the country of origin, which should accompany the consignment without which the material may be refused from entry.

Diseases not entered into India: Swollen shoot of cocoa, leaf blight of rubber and many viral diseases.

3. Eradication: These methods aim at breaking the infection chain by removing the foci of infection and starvation of the pathogen (i.e., elimination of the pathogen from the area by destruction of sources of primary and secondary inoculum). It is achieved by

a) Rouging: Removal of diseased plants or their affected organs from field, which prevent the dissemination of plant pathogens.

E.g., Loose smut of wheat and barley, whip smut of sugarcane, red rot of sugarcane, ergot of bajra, yellow vein mosaic of bhendi, khatte disease of cardamom, etc. During 1927-1935, to eradicate citus canker bacterium in USA, 3 million trees were cut down and burnt.

b) Eradication of alternate and collateral hosts: Eradication of alternate hosts will help in management of many plant diseases.

E.g., Barbery eradication programme in France and USA reduced the severity of black stem rust of wheat.

E.g., Eradication of *Thalictrum* species in USA to manage leaf rust of wheat caused by *Puccinia recondita*.

Eradication of collateral hosts, such as *Panicum repens*, *Digitaria marginata* will help in the management of rice blast disease (*Pyricularia oryzae*).

c) Crop rotation: Continuous cultivation of the same crop in the same field helps in the perpetuation of the pathogen in the soil. Soils which are saturated by the pathogen are often referred as sick soils. To reduce the incidence and severity of many soil borne diseases, crop rotation is adopted. Crop rotation is applicable to only root inhabitants and facultative saprophytes, and may not work with soil inhabitants.

E.g., Panama wilt of banana (long crop rotation), wheat soil borne mosaic (6 yrs) and club root of cabbage (6-10 yrs), etc.

Growing paddy followed by tomato can suppress *Fusarium* wilt in tomato.

d) Field sanitation: Collection and destruction of plant debris from soil will help in the management of soil borne facultative saprophytes as most of these survive in plant debris. Use of clean seed or disease free seed is effective in checking different diseases (e.g. potato viruses and bean anthracnose). Eradication of host and pathogen or rouging and elimination of over summering or over wintering host or control of weeds also will in elimination viral and soil borne pathogens.

e) Tillage: Tillage help in loosening of soil hence provide aeration which is effective in preventing/ against soil borne pathogen like *Fusarium* root rot of beans and *Sclerotium rolfsii*.

f) Manures and fertilizers: The deficiency or excess of a nutrient may predispose a plant to some diseases. Excessive nitrogen application aggravates diseases like stem rot, bacterial leaf blight and blast of rice. Nitrate form of nitrogen increases many diseases, whereas, phosphorous and potash application increases the resistance of the host. Addition of farm yard manure or organic manures such as green manure, 60-100 t/ha, helps to manage the diseases like cotton wilt, *Ganoderma* root rot of citrus, coconut, etc.

Bacterial disease of pomegranate caused by *Xanthomonas axonopodis* pv. *punicae* is appeared with higher severity when plants grown on high pH soil was mainly associated with lower concentration of Mg, Ca, Mn, and Cu and higher concentration of N in leaves.

g) Altering the soil pH: In certain soil borne diseases adjustment of soil reaction helps in the reduction of inoculum level of the pathogens. The altered pH of the environment forms a barrier against the pathogen.

- A very low pH less than 5.2 is unfavourable to common scab bacterium on potato (*Streptomyces scabies*). Thus, use of acid forming fertilizers (like sulphur) and avoiding lime and calcium ammonium nitrate application are effective in controlling the common scab disease.
- The club root pathogen of cabbage (*Plasmodiophora brassicae*) cannot live and infect when the soil pH is 7.0 or more. Hence liming which increases the soil pH gives satisfactory control of club root disease.
- In Punjab, root rot of tobacco (*Macrophomina phaseolina*) has been overcome by application of 2.5 to 5.0 tons of lime /ha to the soil.

h) Mixed cropping: Root rot of cotton (*Phymatotrichum omnivorum*) is reduced when cotton is grown along with sorghum. Fusarium wilt of pigeon pea is reduced when pigeon pea is grown along with sorghum due to secretion of Hydrogen cyanide (HCN) by sorghum root and this compound harm to wilt pathogen.

i) Intercropping: Intercropping is a practice of growing two or more crops within a field in order to derive the benefits of unutilized inter-row spaces, complementation, and competition among the crops. Intercropping sorghum in cluster bean reduces the incidence of root rot and wilt (*Rhizoctonia solani*). Intercropping watermelon with upland rice suppressed Fusarium wilt in rice.

j) Barrier cropping: Taller crops can be grown to protect a crop of lesser height from virus vectors. The insects may land at the taller crops (barrier crops) and the dwarf crop may escape from virus diseases by those insects. Barrier cropping with 3 rows of maize or sorghum or pearlmillet around the main crop namely blackgram or greengram is effective in reducing the vector population and incidence of yellow mosaic.

k) Decoy crop and trap crop: Decoy crops (hostile crops) are non-host crops sown with the purpose of making soilborne pathogens waste their infection potential. This is effected by activating dormant propagules of fungi, seeds of parasitic plants, etc. in absence of the host. A list of pathogens that can be decoyed is given in table

Host	Pathogen	Decoy crops
Sorghum	*Striga asiatica*	Sudan grass
Cabbage	*Plasmodiophora brassicae*	Rye grass, *Papaver rhoeas, Reseda odorata*
Potato	*Spongospora subterranean*	*Datura stramonium*
Tomato, Tobacco	*Orobanche* spp.	Sunflower, safflower, lucerne, chickpea etc.

l) Summer ploughing: Ploughing the soil during summer months expose soil to hot weather which will eradicate heat sensitive soil borne pathogens.

m) Soil amendments:

» Application of organic amendments like saw dust, straw, oil cake, etc., will effectively manage the diseases caused by *Pythium, Phytophthora, Verticillium, Macrophomina, Phymatotrichum* and *Aphanomyces*. Beneficial micro-organisms increases in soil and helps in suppression of pathogenic microbes.

» Application of lime (2500 kg/ha) reduces the club root of cabbage by increasing soil pH to 8.5.

» Application of Sulphur (900 kg/ha) to soil brings the soil pH to 5.2 and reduces the incidence of common scab of potato (*Streptomyces scabies*).

» Addition of sphagnum peat to soil was able to inhibit the disease caused by *Pythium* spp.

» Compost treatment of soil up to 3-4 years improved cotton stand and also significantly reduced the inoculum population of *Rhizoctonia solani* in soil.

» Application of vermin-compost to a conductive pot culture resulted in suppression of Fusarium wilt (*F. oxysporum*) in tomato.

n) Changing time of sowing: Pathogens are able to infect susceptible plants under certain environmental conditions. Alternation in date of sowing can help avoidance of favourable conditions for the pathogens. E.g., Rice blast can be managed by changing planting season.

o) Method of sowing: Planting method is effective in increasing and reducing disease severity, deep sowing of wheat seed increased bunt and flag smut diseases. When ginger planted in raised bed, then incidence of *Pythium myriotylum* infection is less.

p) Seed rate and plant density: Close spacing raises atmospheric humidity and favours sporulation by many pathogenic fungi. A spacing of 8'×8' instead of 7'×7' reduces sigatoka disease of banana due to better ventilation and reduced humidity. High density planting in chillies leads to high incidence of damping off in nurseries. Disease caused by *Cercospora* spp. on pulses spread very fast under crowdy conditions.

q) Irrigation and drainage: The amount, frequency and method of irrigation may affect the dissemination of certain plant pathogens. Many pathogens, including, *Pseudomonas solanacearum, X. campestris* pv. *oryzae* and *Colletotrichum falcatum* are readily disseminated through irrigation water. High soil moisture favours root knot and other nematodes and the root rots caused by species of *Sclerotium, Rhizoctonia, Pythium, Phytophthora, Phymatotrichum*, etc.

r) Flooding of field: If flooded the field up to 30cm deep for 8-10 weeks will help in eradicating several soil borne pathogens due to creations of anaerobic and low oxygen conditions in the soil.

Physical Methods: Physical methods include soil solarization and hot water treatments.

a. **Soil solarization**: Soil solarization or slow soil pasteurization is the hydro/thermal soil heating accomplished by covering moist soil with polyethylene sheets as soil mulch during summer months for 4-6 weeks. Soil solarization was developed for the first time in Israel (Exley and Katan) for the management of plant pathogenic pests, diseases and weeds. Polythene mulching enhances the soil surface temperature to 54.5°C up to 5 cm depth; this temperature was found to be lethal to the Bunt spores.

b. **Soil sterilization:** Soil can be sterilized in green houses and sometimes in seed beds by aerated steam or hot water. At about 50°C, nematodes, some oomycetous fungi and other water molds are killed. At about 60 and 72°C, most of the plant pathogenic fungi and bacteria are killed. At about 82°C, most weeds, plant pathogenic bacteria and insects are killed. Heat tolerant weed seeds and some plant viruses, such as TMV are killed at or near the boiling point (95-100°C).

c. **Hot water or Hot air treatment**: Hot water treatment or hot air treatment will prevent the seed borne and sett borne infectious diseases. Hot water treatment of certain seeds, bulbs and nursery stock is done to kill many pathogens present in or on the seed and other propagating materials. Hot water treatment is used for controlling sett borne diseases of sugarcane [whip smut, grassy shoot and red rot of sugarcane (52°C for 30min)] and loose smut of wheat (52°C for 10 min).

Host Plant Resistance (Immunization)

Disease resistance: It is the ability of a plant to overcome completely or in some degree the effect of a pathogen or damaging factor.

Susceptibility: The inability of a plant to resist the effect of a pathogen or other damaging factor.

Advantages of resistant varieties:

» Resistant varieties can be the simplest, practical, effective and economical method of plant disease management.

» They not only ensure protection against plant diseases but also save the time, energy and money spent on other measures of control

» Resistant varieties, if evolved can be the only practical method of control of diseases such as wilts, viral diseases, rusts, etc.

» They are non-toxic to human beings, animals and wild life and do not pollute the environment

» They are effective only against the target organisms, whereas, chemical methods are not only effective against target organisms but also effective against non-target organisms.

» The resistance gene, once introduced, is inherited and therefore permanent at no extra cost.

Disadvantages:

1. Breeding of resistant varieties is a slow and expensive process

2. Resistance of the cultivar may be broken down with the evolution of the pathogen

Types of resistance:

1. **Vertical resistance**: When a variety is more resistant to some races of the pathogen than others, the resistance is called vertical resistance (race-specific resistance, qualitative resistance, discriminatory resistance). Vertical resistance is usually governed by single gene and is unstable.
2. **Horizontal resistance**: When the resistance is uniformly spread against all the races of a pathogen, then it is called horizontal/generalized/non-specific/field/qualitative resistance. Horizontal resistance is usually governed by several genes and is more stable.
3. **Monogenic resistance**: When the defense mechanism is controlled by a single gene pair, it is called monogenic resistance.
4. **Oligogenic resistance**: when the defense mechanism is governed by a few gene pairs, it is called oligogenic resistance.
5. **Polygenic resistance**: When the defense mechanism is controlled by many genes or more groups of supplementary genes, it is called polygenic resistance.

Cross protection: The phenomenon in which plant tissues infected with mild strain of a virus are protected from infection by other severe strains of the same virus. This strategy is used in the management of severe strains of *Citrus tristeza virus.*

Development of resistance in host: Development and utilization of host plant resistance remains the most viable, environmentally safe, ecologically sound and also less expensive technique for the management of diseases.

- » Selection and hybridization for disease resistance
- » Resistance through chemotherapy
- » Resistance through host nutrition

Selection and hybridization for disease resistance: Selection of cultivar having disease resistant character and hybridizing them with susceptible cultivar having high yielding character is the aim of developing resistance through hybridization. Biotechnological tools are also being used for developing disease resistant varieties.

Vertifolia effect: Van der Plank introduced the term Vertifolia effect derived from the German potato variety 'Vertifolia' having the late blight resistance genes R3 and R4. The variety became susceptible when the appropriate Pathotypes P (3,4) evolved in the pathogen leading to a nearly complete failure of the crop; such a total failure of vertical resistance leading to a disease epidemic is known as Vertifolia effect.

Vertifolia effect refers to an epidemic development in a variety carrying vertical resistance genes and a low level of HR, leading to heavy economic losses.

Boom and burst cycle: In varietial improvement programmes, it is easy to incorporate the monogenic vertical resistance genes. But the success of exploiting the monogenic host resistance invariably does not last long. Whenever a single gene-based resistant variety is widely adopted, the impact would be the arrival of new matching pathotypes. These pathotypes soon build up in population to create epidemics and eventually the variety is withdrawn. This phenomenon is generally called - boom and burst.

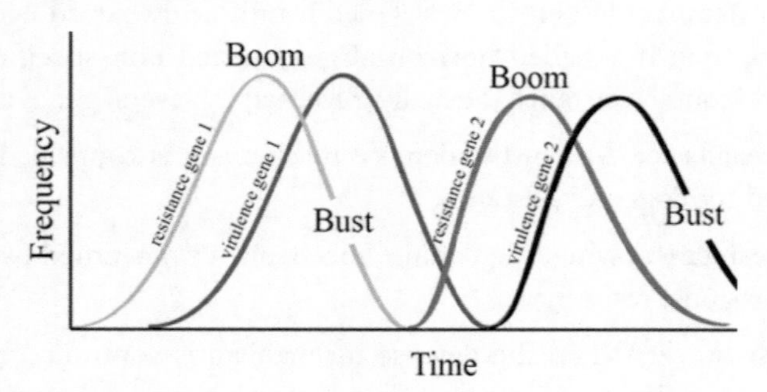

The avoid the implications of boom and burst phenomenon, use of durable host resistance is advocated in several crops. Durable resistance remains effective even though it may be widely grown ovoer a long period of time, in an environment that favours the disease.

Resistant varieties

Diseases	**Important Resistant Varieties in India**
Rice	
Blast	CR1009, ADT36, Co37, IR64, Ponmani, ASD16, Tulsi, Vikas
Brown spot	Co20, Bala, Bhavani, Jagannath, Padma, Rasi
Sheath rot	Tetop, IR50
Bacterial blight	IR20, Zenith, Pavizham, TKM6, Vaigai, Lalat, Indra
Yellow dwarf	IR64, IR69
RTV	Pusa 2-21, Vikramarya, White ponni, IR20, IR28, IR29, Co45
Grassy shoot	IR28, IR30
Sorghum	
Downy Mildew	WCC75, X6, X7, Co7, QLS, Co25, Co26, Co21
Maize	
Downy Mildew	KDMH16, EHQ16, DHM117, 900M Gold
Wheat	
Stem rust	Lerma Roja, Safed Lerma, Sonalika, Choti Lerma, Girija, Sonara 64
Yellow rust	Hira, VP262, Sonalika, Lerma Roja, DWL 5023

Brown Rust	Pratap, Arjun, HB208, Sonalika, Lerma Roja, Janak
Loose Smut	Kalyani, Kalyan Sona, Sonalika, Choti Lerma, PV18, L302
Chick pea	
Wilt	ICCC32, H82-2, BDN 1,2
Redgram	
Wilt	NP41, NP80
PPSMV	VR3, DA11, Bahar
Yellow mosaic	Co6, Vamban 2
Groundnut	
Leaf spot	ALR-1, ALR-3
Bud necrosis	Co3
Blackgram	
Yellow mosaic	VBN-1, VBN3
Tobacco	
TMV	TMVR2, RR29, RR3
Cotton	
Black arm	MCV10, Paiyur 1, Sujatha, CRH71
Alternaria leaf spot	Paiyur 1
Sugarcane	
Red rot	CoC 99061, CoC 22, Karan 2, Karan 8
Smut	Co7704, Co1111, Bo11, Bo22
Potato	
Late blight	Kufri Jyoti, Kufri Malar, Kufri Thangam
Bhendi	
BYVMV	Prabhani Kranti
Coffee	
Rust	Cauvery

Biological methods

Biological control of plant disease is a condition or practice whereby survival or activity of a pathogen is reduced through the agency of any other living organism (Except human beings), with the result that there is reduction in incidence of the disease caused by the pathogen (Garett, 1965).

Biological control is the reduction of inoculum density or disease producing activity of a pathogen or a parasite in its active or dormant state by one or more organisms accomplished naturally or through manipulation of the environment of host or antagonist by mass introduction of one or more antagonists (Baker and Cook, 1974)

Biological control has been developed and extensively used against soil-borne diseases in the nursery and in the main fields. It has also been used in the control of other diseases. Management of plant disease by chemical methods is uneconomical, less effective and harmful to the beneficial microbes. In addition, they cause residual problem in the soil and farm produce. Seed treatment chemicals enable protection only in the early stages of crop growth (about 15 days) but biological control is cheaper, highly effective and has no residual problem. They are not harmful to the beneficial microorganisms. When they are applied to the seed, they reach the soil, multiply in the organic matter in the soil and offer protection throughout the crop growth. In addition, they can be mixed safely with the biofertilizers and sown immediately after the seed treatment.

Advantages of biological control

1. Biological control is less costly and cheaper than any other methods.
2. Biocontrol agents give protection to the crop throughout the crop period.
3. Biocontrol agents not only control the disease but also enhance the growth by way of encouraging the beneficial soil microflora. It increases the crop yield. It helps in the availability of certain nutrients. For example, *Bacillus subtilis* solubilizes phosphorus.
4. Application of biocontrol agents is safer to the environment and the person applying them.
5. They are highly effective against specific disease.
6. Biological agents can eliminate pathogens from the site of infection.
7. Biocontrol agents are very easy to handle and apply to the targets.
8. Biocontrol agents can be combined with biofertilizers.
9. They are easy to manufacture.
10. They do not cause toxicity to the plants.
11. They multiply easily in the soil and leave no residual problem.

Disadvantages

1. Biocontrol agents can be used only against specific diseases.
2. They are less effective than the fungicides
3. Only few biocontrol is available for use and are available in few places.
4. They are unavailable in larger quantities at present.

5. They have slow effect in the control of plant disease
6. This method is only a preventive measure and not curative.
7. Biocontrol agents should be multiplied and supplied without contamination and requires skilled person.
8. Shelf life of biocontrol agents is short. *Trichoderma viride* is viable for 4 months and *P. fluorescens* is viable only for three months.
9. The required amount of population of biocontrol agents should be checked at periodical intervals and should be maintained at required level.
10. Their efficiency is decided by the climatic conditions.
11. Under certain circumstances a biocontrol agent may become a pathogen.

Characters of biocontrol agents

Biocontrol agents (fungi and bacteria) are used in the control of plant diseases and are highly effective in the nursery diseases and in fields where commercial crops are grown.

A good biocontrol agent should have the following features / characteristics.

- » It should not be pathogenic to plants
- » It should have broad spectrum of its activity in controlling many types of diseases.
- » It should have fast growth and sporulation.
- » It should be amenable for mass multiplication.
- » It should be compatible with bio-fertilizers
- » It should have least susceptibility to the action by the seed treating chemicals.
- » It should not be toxic to human beings and animals
- » It should not be toxic to beneficial organisms in or on the target area.
- » It should be easily formulated.
- » It should have more shelf life

Mechanisms of biological control

1. Competition: Most of the biocontrol agents are fast growing and they compete with plant pathogens for space, organic nutrients and minerals. E.g. (i) the competition for nutrients between *Pythium aphanidermatum*, *P. ultimum* and bacteria suppress the damping off disease in cucumbers.

Most aerobic and facultative anaerobic micro-organisms respond to low iron stress by producing extracellular, low molecular weight (500-1000 daltons) iron transport agents, designated as siderophores, which selectively make compete with iron (Fe 3+) with very high affinity. Siderophore producing strains are able to utilize Fe 3+ - siderophore compete and restrict the growth of deleterious micro-organisms mostly at the plant roots. Iron starvation prevents the germination of spores of fungal pathogens in rhizosphere as well as rhizoplane. Fluorescent siderophores (iron chelaters) such as pseudobactinis and pyoverdins produced by *P. fluorescens* chelates iron available in the soil, thereby depriving the pathogen of its Fe requirements, helps in the control of soft rot bacterium, *Erwinia caratovora.*

2. Antibiosis: Antagonism mediated by specific or non-specific metabolites of microbial origin, by lytic agents, enzymes, volatile compounds or other toxic substances is known as antibiosis.

3. Hyperparasitism: Direct parasitism or lysis and death of the pathogen by another micro-organism when the pathogen is in parasitic phase is known as hyperparasitism.

E.g., *T. harzianum* parasitize and lyse the mycelia of *Rhizoctonia* and *Sclerotium*.

4. Lysis: The biocontrol against parasitizes the pathogen by coiling around the hyphae, E.g., *Trichoderma viride*; various bacteria and fungi secrete hydrolytic about the degradation of cell wall of pathogens.

E.g., (i) *Bacillus* sp. cause hyphal lysis of *Gaeumanomyces graminis*

(ii) The chitinolytic enzymes of *Serratia marcescens* caused cell wall lysis of *Sclerotium rolfsii.*

(iii) *Trichoderma* sp. produces chitinases and β-1,3 glucanases which lyses the cell wall of *Rhizoctonia solani*

5. Induced Systemic Resistance (ISR)

The phenomenon of induced systemic resistance is at present widely accepted and even exploited for the biocontrol of plant diseases. Induced Systemic Resistance is the ability of an agent (a fungus, bacterium, virus, chemical, *etc.*) to induce plant defense mechanisms that lead to systemic resistance to a number of pathogens. Inoculation of plants with weak pathogens or non-pathogens leads to induced systemic plant resistance against subsequent challenge by pathogens. The mechanisms remain largely unknown but typically the induced resistance operates against a wide range of pathogens and can persist for 3 - 6 weeks. Then a booster treatment is required. Rhizobacteria mediated induced systemic resistance (ISR) has been demonstrated against fungi, bacteria and viruses in *Arabidiopsis*, bean, carnation, cucumber, radish, tobacco and tomato under conditions in which the inducing bacteria and the challenging pathogen remain spatially separated. Rhizobacteria known to induce resistance can be found among *Pseudomons aeruginosa, P. aureofaciens, P. corrugata, P. fluorescens,*

P. putida, and *Serratia marcescens*. Also, some fungi, including *Trichoderma harzianum* T 39 are able to induce resistance.

E.g., Phenazine-1-carboxylic acid by *P. fluorescens* plays an important role in suppressing the take all disease of wheat.

Biocontrol agents for the management of plant pathogens

Biocontrol agent	Pathogen/ Disease
Fungal biocontrol agent	
Ampelomyces quisqualis	Powdery mildew fungi
Darluca filum	Rust
Pichia gulliermondii	*Botrytis, Penicillium*
Most of the species of *Trichoderma*, viz., *T. harzianum, T. viride, T. virens* (*Gliocladium virens*)	Soil borne diseases, such as, root rots, seedling rots, collar rots, damping off and wilts caused by the species of *Pythium, Fusarium, Rhizoctonia, Macrophomina, Sclerotium, Verticillium*, etc.
Verticillium leccani	*Uromyces dianthi, Hemellia vastatrix*
Gliocladium virens	*Rhizoctonia solani*
Laetisaria arvalis	*Sclerotinia sclerotiarum*, *Rhizoctonia solani* and *Pythium* spp.
Paecilomyces lilacinus	Egg parasite of *Meloidogyne incognita*
Bacterial biocontrol agent	
Pseudomonas fluorescens	Seed borne, soil borne and foliar diseases (*Pythium ultimum, Pythium aphanidermatum, Rhizoctonia solani, Fusarium oxysporum* f.sp. *udum, Fusarium oxysporum* f.sp. *cubense* and *Fusarium oxysporum* f.sp. *ciceri*)
Bacillus subtilis	Seed borne, soil borne and foliar diseases (*Fusarium oxysporum* f.sp. *dianthi, Macrophomina phaseolina, Rhizoctonia solani* and *Sclerotium rolfsii*)
Pasteuria penetrans	Juvenile parasite of root knot nematode

Plant growth promoting rhizobacteria (PGPR):

Rhizosphere bacteria that favourably affect plant growth and yield of commercially important crops are designated as plant growth promoting rhizobacteria. The growth promoting ability of PGPR is due to their ability to produce phytohormones, Siderophores, Hydrogen cyanide (HCN), chitinases, volatile compounds or antibiotics which will reduce infection of host through phyto-pathogenic micro-organisms. Many bacterial species, *viz.*, *Bacillus subtilis, Pseudomonas fluorescens*, etc., are usually used for the management of plant pathogenic microbes. Bacillus has ecological advantages as it produces endospores that are tolerant to extreme environmental conditions. *Pseudomonas fluorescens* have been extensively used to manage soil borne plant pathogenic fungi due to their ability to use many carbon sources that exude from the roots and to compete with microflora by the production of antibiotics, HCN and Siderophores that suppress plant root pathogens.

Plant products and antiviral principles in plant disease management

» The neem tree (*Azadirachta indica*), popularly called as china berry, crackjack, Nim, Indian lilac, margosa and paradise tree, contains several active principles in various parts.

Neem products:

» **Neem Seed Kernel Extract** (NSKE-5%): It is prepared by soaking 5 kg of powdered neem seed kernel (in a gunny bag) in 100 litres of water for 8 hours for sheath blight of rice.

» **Neem oil solution**: 100 ml of teepol is mixed first with 100 litres of water. Then, 3 litres of neem oil is slowly added to this solution with constant shaking. The spray volume is 500 litres/ha.

» **Neem cake extract**: Ten kg of powdered neem cake in a gunny bag is soaked in 100 litres of water for 8 hours. The gunny bags removed after thorough shaking.

» **Neem cake**: Powdered neem cake is directly applied to the field at the time of last ploughing. The quantity applied is 150 kg/ha.

Diseases controlled by neem products

» **Paddy**: Tungro (virus) (Vector: *Nephotettix virescens*): Neem cake is applied at 150 kg/ ha as basal dose. In addition, 3% neem oil or 5% NSKE at) 500 l/ ha can be sprayed.

» **Paddy**: Sheath rot (*Sarocladium oryzae*), five per cent NSKE or 3% neem oil can be sprayed at 500 lit/ ha at the time of grain emergence.

» **Paddy**: Blast (*Pyricularia oryzae*), spraying 5% neem oil is effective.

» **Paddy**: Sheath blight (*Rhizoctonia solani*) Application of 150 kg of neem cake/ha or NSKE 5%.

» **Groundnut**: Rust (*Puccinia arachidis*)

» **Groundnut**: Foot rot (*Sclerotium rolfsii*)

» **Coconut**: Wilt (*Ganoderma lucidum*), application of 5 kg of neem cake/ tree/ year during the rainy season.

» **Black gram**: Powdery mildew (*Erysiphe polygoni*)

» **Black gram**: Root rot (*Macrophomina phaseolina*)

» **Black gram**: Yellow mosaic (Virus), application of 3% neem oil is effective.

» **Soybean**: Root rot (*M. phaseolina*), application of neem cake at 150 kg/ha.

Other Plant Products

» The leaf extract of tulsi (*Ocimum sanctum*) is found effective against *Helminthosporium* oryzae (paddy brown spot).

» The leaf and pollen extracts of vilvam (*Aegle marmolos*) early blight of tomato (*Altenaria solani*) and blight of onion (*A. porri*).

» solani is checked by flower extract of periwinkle (*Catheranthus roseus*) and bulb extract of garlic (*Allium sativum*).

Anti-Viral Principle (AVP)

» Plants are also known to contain some compounds which are inhibitory to virus. They are called Anti-Viral Principles (AVP) or Anti-Viral Factors (AVF).

» The leaf extracts of sorghum, coconut, bougainvillea, *Prosopis juliflora*, *Mirabilis jalapa* and *Cyanodon dactylon* are known to contain virus inhibiting principles.

Practice Questions

1. Write the components of integrated disease management?
2. What are the cultural methods of plant disease control?
3. What are plant quarantines?
4. When and where was the first quarantine act enacted?
5. What is the D.I.P. Act? When it was first enacted?
6. Name, some plant diseases, which have been introduced before and after the enforcement of plant quarantines in India?
7. Is quarantine necessary for all types of diseases?
8. Name some plant diseases, which can be controlled by crop rotation.
9. What is seed certification?
10. What is biological control of plant pathogens? Give some suitable examples where it has been successfully used to control disease

11. Breeding for disease resistance is the best method of controlling plant diseases. Why?
12. What is integrated Pest/Disease Management (IPM or IDM).
13. What is solarization?
14. What are the innovative methods of plant disease control? Give few examples.
15. Name some plant pathogens of plant quarantine importance to India.

Suggested Readings

Gururani, M. A., Venkatesh, J., Upadhyaya, C. P., Nookaraju, A., Pandey, S. K., & Park, S. W. (2012). Plant disease resistance genes: current status and future directions. *Physiological and molecular plant pathology*, *78*, 51-65.

Chaube, H. (2018). *Plant disease management: principles and practices*. CRC Press.

Thurston, H. D. (2019). *Sustainable practices for plant disease management in traditional farming systems*. CRC Press.

van Bruggen, A. H., Gamliel, A., & Finckh, M. R. (2016). Plant disease management in organic farming systems. *Pest Management Science*, *72*(1), 30-44.

Thurston, H. D. (1990). Plant disease management practices of traditional farmers. *Plant disease*, *74*(2), 96-102.

Muthaiyan, M. C. (2009). *Principles and practices of plant quarantine* **(Vol. 1). Allied Publishers.**

Fry, W. E. (2012). *Principles of plant disease management*. Academic Press.

Gutterson, N. (1990). Microbial fungicides: recent approaches to elucidating mechanisms. *Critical reviews in biotechnology*, *10*(1), 69-91.

Chapter - 34

Chemical Fungicides and Antibiotics

Protection: Use of chemicals for the control of plant diseases is generally referred to as protection or therapy. The prevention of the pathogen from entering the host or checking the further development in already infected plants by the application of chemicals is called protection and the chemicals used are called protectants.

Therapy means cure of a disease, in which fungicide is applied after the pathogen is in contact with the host. Chemicals used are called therapeutants.

Fungicide: Any agent (chemical) that kills the fungus.

Fungistat: Some chemicals which do not kill fungi, but simply inhibit the fungus growth temporarily.

Antisporulant: The chemical which inhibits spore production without affecting vegetative growth of the fungus.

Classification of fungicides based on

a. **Mode of action**: Protectant, Eradicant, Therapeutant.
b. **General use**: Seed protectant, soil fungicides (Pre-plant, Post plant), Foliage and blossoms protectant, Fruit protectant and Tree wound dresser.
c. **Chemical composition**: Mercury fungicides, sulphur fungicides, copper fungicides, quinons, heterocyclic nitrogenous compounds, Aromatic compounds, miscellaneous compounds and Antibiotics.

Fungicides are classified into three categories based on mode of action

a. **Protectants**: These are the chemicals which are effective only when used before infection (prophylactic in behavior). Contact fungicides which kill the pathogen present on the host surface when it comes in contact with the host are called protectants. These are applied to seeds, plant surfaces or soil. These are non-systemic in action (i.e, they cannot penetrate plant tissues). E.g., Zineb, sulphur, captan, Thiram, etc.

b. **Eradicants**: Those chemicals which eradicate the dormant or active pathogen from the host. They can remain on/in the host for some time. E.g., Lime sulphur, Dodine.

c. **Therapeutants**: These are the agents that inhibit the development of a disease syndrome in a plant when applied after infection by a pathogen. Therapy can be by physical means (solar and hot water treatment) and chemical means (by use of systemic fungicides, i.e., chemotherapy).

History of fungicides

1000 BC – Homer, Greek poet mentioned sulphur with pest control properties.

1761 – Schulthess first suggested $CuSO_4$ to control stinking smut on wheat seed.

1807 – Prevost (France) found effectivity of $CuSO_4$ against stinking smut of wheat.

1821 – Robertson found sulphur – effective against peach mildew.

1882 – Prof. P. M. A. Millardet developed Boredeaux mixture.

1887 – Mason (France) introduced Burgundy mixture

1934 – Tisdale and Williams reported fungitoxic activity of dithiocarbamates.

1952 – Kittleson introduced captan as fungicide (kittleson's killer)

1966 – von Schmeling and Marshal Kulka discovered systemic fungicides (Oxathiin compounds)

1973 – Trizoles were introduced

1978 – Fosetyl-Al was introduced

India

1885 – Ozanne – first person to use $CuSO_4$ as a fungicide to control sorghum smut

1904 – Lawrence (Pune) used Bordeaux mixture first time to control of Groundnut *Cercospora* spp.

1906 – Sulphur was first tested in grapes

1914 – Burns used $HgCl_2$ first time to control *R. solani* in potato.

1925 – Hilton tested organomercurials first time to control of sorghum smut.

1964 – Discovery of Aureofungin by Thirumalachar *et al.*

Classification of fungicides based on chemical nature

Many fungicides have been developed for purpose of managing crop diseases which may be used as sprays, dusts, paints, pastes, fumigants, etc. The discovery of Bordeaux mixture in 1882 by Professor Millardet, University of Bordeaux, France led to the development of fungicides. Major group of fungicides used include salts of toxic metals and organic acids, organic compounds of sulphur and mercury, quinones and heterocyclic nitrogenous compounds. Copper, mercury, zinc, tin and nickel are some of the metals used as base for inorganic and organic fungicides. The non metal substances include, sulphur, chlorine, phosphorous etc. The fungicides have been classified based on their chemical nature as follows

Copper fungicides: copper fungicides can be classified as preparatory and proprietory copper compounds.

Common name	Chemical composition	Diseases managed
1. Bordeaux mixture	It is prepared by suspending 5 kg of copper sulphate and 5 kg of lime in 500 liters of water (1%)	Downy mildew of grapes, Coffee rust, Tikka leaf spot of groundnut, citrus canker, citrus scab, etc.
2.Bordeaux paste	It is prepared by mixing 1 kg of copper sulphate and 1 kg of lime in 10 liters of water	It is a wound dressing fungicide and can be applied to the pruned parts of the host plants such as fruit crops and ornamentals. E.g., Citrus gummosis, Stem bleeding of coconut, Bud rot of coconut, etc.
3. Burgundy mixture	Sodium carbonate is used in place of lime. It is prepared by mixing 1 kg of copper sulphate and 1 kg of sodium carbonate in 100 liters of water	Downy mildew of grapes, Coffee rust, Tikka leaf spot of groundnut, citrus canker, citrus scab
4.Cheshunt compound	It is a compound prepared by mixing 2 parts of copper sulphate and 11 parts of ammonium carbonate	It is used for soil drenching only. Sclerotial wilt diseases of chilli, tomato and groundnut. Fusarial wilt diseases. Damping-off diseases of solanaceous crops.
5.Chaubattia paste	It is a compound prepared by mixing 800g of copper sulphate and 800g of red lead in 1 liter of lanolin or linseed oil	Pink disease of citrus, stem canker and collar rot of apple and pears

Proprietary copper fungicides or Fixed or insoluble copper fungicides: In the fixed or insoluble copper compounds, the copper ion is less soluble than in Bordeaux mixture. So, these are less phytotoxic than Bordeaux mixture but are effective as fungicides.

Common name	Trade name	Dosage	Disease managed
1. Copper oxy chloride	Blitox-50, Blue copper-50, Cupramar-50	0.3 to 0.5% for foliar application, 25 to 35 Kg/ha for dusting	Anthracnose of grapevine, Tikka leaf spot of groundnut, Sigatoka leaf spot of banana, citrus canker, black arm of cotton
2.Cuprous oxide	Fungimar and Perenox	0.3% for foliar spray	Anthracnose of grapevine, Tikka leaf spot of groundnut, Sigatoka leaf spot of banana, citrus canker, black arm of cotton
3.Copper hydroxide	Kocide	0.3% for foliar spray	Blister blight of tea, False smut of rice, Tikka leaf spot of groundnut

Sulphur Fungicides: Sulphur is probably the oldest chemical used in plant disease management for the control of powdery mildews and can be classified as inorganic sulphur and organic sulphur. Inorganic sulphur fungicides include lime sulphur and elemental sulphur fungicides. Organic sulphur fungicides, also called as carbamate fungicides, are the derivatives of dithiocarbamic acid.

Inorganic Sulphur Fungicides

Common name	Trade name	Dosage	Disease managed
Preparatory sulphur compounds			
1. Lime sulphur	It is prepared by mixing 20 Kg of rock lime and 15 Kg of sulphur in 500 liters of water	10-15 liters in 500 liters of water	Powdery mildew of apple, Apple scab, bean rust
2. Sulphur dust	Kolo dust, Mico999	4-5 g/kg seed for ST, 10-30 kg/ha for dusting on crops, 100 kg /ha for soil application in tobacco, 500 Kg/ha for furrow application in potato	Common scab of potato, Grain smut of jowar, Powdery mildew of tobacco, chilli, rose, mango, grapes, etc.
3.Wettable sulphur	Sulfex, Thiovit, Cosan	0.2-0.4 % for foliar spray	Powdery mildews of various crops

Organic sulphur compounds: Organic sulphur compounds are derived from dithiocarbamic acid and are widely used as spray fungicides. In 1931, Tisdale and Williams were the first to describe the fungicidal nature of Dithiocarbamates. Dithiocarbamates can be categorized into two groups, viz., dialkyl dithiocarbamates (ziram, ferbam and thiram) and monoalkyl dithiocarbamates (nabam, zineb, vapam and maneb).

Common name	Trade name	Dosage	Disease managed
Dialkyl Dithiocarbamates			
1. Ziram	Ziride, Hexazir, Milbam, Zerlate	0.15 to 0.25% for foliar spray	Anthracnose of pulses, tomato, beans, tobacco, etc., bean rust
2. Ferbam	Coromet, Ferbam, Fermate, Fermocide, Hexaferb, Karbam Black	0.15 to 0.25% for foliar spray	Fungal pathogens of fruits and vegetables, leaf curl of peaches, apple scab, downy mildew of tobacco
3. Thiram	Arasan, Hexathir, Tersan, Thiram, Thiride	0.15 to 0.2% as foliar spray, 0.2 to 0.3% as dry seed treatment, 15 - 25 kg/ha as soil application	Soil borne diseases caused by *Pythium, Rhizoctonia solani, Fusarium*, etc. Rust of ornamental crops, Scab on pears and *Botrytis* spp. on lettuce
Monoalkyl dithyiocarbamates			
1. Nabam	Chembam, Dithane D-14, Dithane A-40 and Parzate liquid	0.2% as foliar spray	Used as foliar spray against leaf spot diseases of fruits and vegetables. Also used against soil borne pathogens, *Fusarium, Pythium* and *Phytophthora*
2. Zineb	Dithane Z-78, Hexathane, Lanocol and Parzate	0.1 to 0.3% for foliar application	Chilli die-back and fruit rot, Apple scab, Maize leaf blight, early blight of potato
3.Vapam or Metham sodium	Chem-vape, vapam, vitafume, VPM	1.5 to 2.5 liters per 10 m^2 area	Fungicide with fungicidal, nematicidal and insecticidal properties. Soil fungal pathogens like *Fusarium, Pythium, Sclerotium* and *Rhizoctonia.*

4. Maneb	Dithane M22, Manzate and MEB. Mancozeb (78% Maneb + 2% zinc ion): Dithane M 45, Indofil M 45	0.2% to 0.3% as foliar application	Early and late blight of potato and tomato, rust diseases of field and fruit crops

Hetrocyclic Nitrogenous Compounds: The group of heterogeneous fungicides includes some of the best fungicides like captan, folpet, captafol, vinclozoline and Iprodione. Captan, folpet and captafol belong to dicarboximides and are known as pthalamide fungicides. The new members of dicarboximide group are Iprodione, vinclozolin, etc.

Common name	Trade name	Dosage	Disease managed
1.Captan (Kittleson's killer)	Captan 50W, Captan 75 W, Esso fungicide, Orthocide 406, Hexacap, Vancide 89	0.2 to 0.3% for dry seed treatment, 0.2 to 0.3% for foliar spray, 25 to 30 Kg/ha for furrow application	Onion smut, Chilli die-back and fruit rot, Damping off of beans, chilli and tomato, seed rots and seedling blights of maize
2. Folpet	Phaltan	0.1 to 0.2% for spraying	Apple scab, tobacco brown spot, rose black spot
3.Captafol	Difosan, Difolaton, Sanspor, Foltaf	0.15 to 0.2% for spraying, 0.25% for seed treatment, 0.15% for soil drenching	Sorghum anthracnose, cotton seedling diseases, seed rot and seedling diseases of rice, downy mildew of crucifers, apple scab
4. Iprodione	Rovral, Glycophene	0.1 to 0.2% for foliar application	Diseases caused by *Botrytis*, *Monilinia*, *Alternaria*, *Sclerotinia*, *Helminthosporium* and *Rhizoctonia*
5. Vinclozolin	Ornalin, Ronilan, Vorlan	0.1 to 0.2% for foliar application	Effective against sclerotia forming fungi like *Botrytis*, *Monilinia* and *Sclerotinia*

Miscellaneous Fungicides

Common name	Trade name	Dosage	Disease managed
1. Chlorothalonil	Bravo, Daconil, Kavach, Thermil, Egotherm, Safeguard	0.2 to 0.3% for Foliar application	A broad spectrum contact fungicide often used in greenhouses for control of *Botrytis* on ornamentals and for several molds and blights of tomato. Also used for the control of sigatoka leaf spot of banana, onion purple blotch, tikka leaf spot and rust of groundnut
2. Dinocap	Karathane, Arathane, Capryl, Mildex, Mildont and crotothane	0.1 to 0.2% for spraying	It is a good acaricide and contact fungicide and it controls powdery mildews of fruits and ornamentals effectively. This can be safely used on sulphur sensitive crops like cucurbits and apple varieties for control of powdery mildews
3. Dodine	Cyprex, Melprex, and Syllit	0.075% for spraying	Apple scab, black spot of roses and cherry leaf spot

Disadvantages of non-systemic fungicides

- Lack of appreciable surface redistribution
- It gives protection only at the point, where it falls.
- The new flushes are not protected.
- Efficacy is subjected to weather parameters.
- Hence it needs repeated application.
- Inaccessible parts are difficult to protect.

Systemic Fungicides: A chemical that can able to translocate freely in the plant system after absorption and can control plant pathogens remote from the point of application and can be detected or identified is known as systemic fungicides. The systemic fungicides were first introduced by Von Schelming and Marshall Kulka in 1966. The discovery of Oxathiin fungicides was soon followed by confirmation of systemic activity of pyrimidines and benzimidazoles. A systemic fungicide is capable of managing a pathogen remote from the point of application.

Characteristics of an ideal systemic fungicide

- The compound may either be directly toxic to the pathogen or be converted in the plant to become a fungitoxicant.
- The compound / derivative formed in the plant may alter the host metabolism or induce host resistance.
- It should not adversely affect the host plant and reduce the quality and quantity of the produce.
- The compound must be absorbed by the plant and stable within the plant for considerable period.
- It should be having low mammalian toxicity to avoid residual problem at consumer level.
- It should be cheap and stable under normal conditions.
- It should have wide spectrum activity and safe to beneficial microorganisms.
- It should be safer to the user and consumer.

Disadvantages of systemic fungicides

- The uptake and translocation of compound is passive most of the compounds have got only apoplastic (Upward) movement via the xylem vessels.
- In herbaceous plants, the translocation is quick. Whereas in woody plants it is slow and little.
- Most of them have broad spectrum activity but are not effective against Mastigomycotina fungi.
- They are generally costly.
- Development of resistant strains occurs.

General mechanism of action

- Inhibition of nucleic acid synthesis.
- Inhibition of respiration
- Interfere with biosynthesis of new cell material required for growth.
- Disruption of cell structure, permeability of cell membrane results in leakage of cell contents.

Common name	Trade name	Dosage	Disease managed
Acylalanines			
1. Metalaxyl	Ridomil 25% WP, Apron 35 SD, Subdue, Ridomil MZ-72WP	3-6 g/Kg seed for seed treatment, 1 to 1.5 Kg a.i/ ha for soil application, 0.1 to 0.2% for foliar spray	It is highly effective against *Pythium, Phytophthora* and many downy mildew fungi
2. Benalaxyl	Galben 25% WP and 5% G	0.1 to 0.2% for foliar spray, 1 to 1.5 Kg a.i/ha for soil application	Blue mold of tobacco, late blight of potato and tomato, downy mildew of grapevine
Aromatic Hydrocarbons			
1. Chloroneb	Demosan	0.2% for seed treatment	Seedling diseases of cotton, peanut, peas and cucurbits caused by species of *Pythium, Phytophthora, Rhizoctonia* and *Sclerotium*
Benzimidazoles			
1. Carbendazim	Bavistin 50WP, MBC, Derosol 60WP, Agrozim, Zoom	0.1% for foliar spray, 0.1% for soil drench, 0.25% for ST, 500-1000ppm for post-harvest dip of fruits.	Effectively controls anthracnose, powdery mildews and rusts caused by various fungi. It is also used as a soil drench against wilt diseases and for post harvest treatment of fruits
2. Benomyl	Benlate 50WP	0.1 to 0.2% for ST, 50-60g/100 L for foliar spray, 50-200ppm for soil drenching, 12-45 kg a.i/ ha for soil broadcast, 100-500 ppm for post harvest fruit dip	Effective against powdery mildews of cucurbits, cereals and legumes. It is highly effective against diseases caused by the species of *Rhizoctonia, Theilaviopsis* and *Cephalosporium*. Benomyl has no effect against Oomycetes and some dark coloured fungi such as *Alternaria* and *Helminthosporium*
3. Thiabendazole	Mertect 60WP, Mycozol, Arbotect, Tecto and Storite	0.2 to 0.3% for spraying, 1000 ppm for fruit dip	Blue and green molds of citrus, loose smut of wheat, Tikka leaf spot of groundnut
Aliphatics			
1. Prothiocarb	Previcur, Dynone	5.6 kg a.i/ha for soil application	Highly active against soil borne Oomycetes like *Pythium* and *Phytophthora*

2. Propamocarb	Previcur-N, Dynone-N, Prevex, Benol	3.4 and 4.8 kg a.i/ha for soil application	Effective against against soil borne Oomycetes like *Pythium* and *Phytophthora*
Oxathins or Carboximides			
1. Carboxin	Vitavax 75WP, Vitaflow	0.15 to 0.2% for seed treatment, 0.5% for spraying	Highly effective against smut diseases. Commonly used for the control of loose smut of wheat, onion smut, grain smut of sorghum. As a soil drench it is used for the control of diseases caused by *Rhizoctonia solani* and *Macrophomina phaseolina*.
2. Oxycarboxin	Plantavax 75 WP, Plantavax 20EC, Plantavax 5% liquid	0.1 to 0.2% for foliar spray, 0.2 to 0.5% for ST	Highly effective against rust diseases. Commonly used for the control of rusts of wheat, sorghum, safflower, legumes, etc.
Imidazoles			
1. Imazalil	Fungaflor, Bromazil and Nuzone	0.1 % as post harvest dip	Blue and green molds of citrus
2. Fanapanil	Sistane 25 EC	0.05% foliar spray	Spot blotch of barley, loose and covered smut of barley
Morpholines			
1. Tridemorph	Calixin 75EC, Bardew, Beacon	0.1% for foliar spray	Powdery mildew of cereals, vegetables and ornamentals. Rusts of pulses, groundnut and coffee, Sigatoka leaf spot of banana, pink disease of rubber, *Ganoderma* root rot and wilt
Organophosphates			
1. Iprobenphos	Kitazin 48EC, Kitazin 17G, Kitazin 2% D	30-45 kg of granules/ha, 1 to 1.5 liters of 48% EC in 1000 ml of water for foliar spray.	Fungicide with insecticidal properties. Highly specific against rice blast, stem rot and sheath blight of rice
2. Ediphenphos	Hinosan 30 and 50% EC, Hinosan 2%D	400 to 500 ppm for spraying, 30 to 40 kg/ha	Highly specific against rice blast, stem rot and sheath blight of rice
Alkyl Phosphonates			
1. Fosetyl-Al or Aluminium Tris	Aliette 80WP	0.15% for foliar spray, 0.2% for soil drench	Ambimobile fungicide. Specific against Oomycetes fungi

Pyrimidines			
1. Fenarimol	Rubigan 50% WP, 12%EC	2g/kg seed as ST, 20 to 40 ml/100 liters of water for spraying	Powdery mildew of cucurbits, apple, mango, roses, grapes and ornamental crops
Thiophanates			
1. Thiophanate	Topsin 50WP, Cercobin 50WP	0.1 to 0.2% for spraying	Powdery mildew of cuurbits and apple, club root of crucifers, rice blast
2.Thiophanate methyl	Topsin M 70WP, Cercobin M 70WP	0.1% for spraying	Blast and sheath blight of rice, sigatoka leaf spot of banana, powdery mildew of beans, chilli, peas and cucurbits
Triazoles			
1. Triadimefon	Bayleton, Amiral	0.1 to 0.2% for spraying, 0.1% for seed treatment	Highly effective against powdery mildews and rusts of several crops. Effective against diseases caused by species of *Erysiphe, Sphaerotheca, Puccinia, Uromyces, Phakopsora, Hemileia* and *Gymnosporangium*
2. Tricyclazole	Beam 75WP, Baan 75WP, Trooper 75WP	2g/kg seed for ST, 0.06% for spraying	Highly effective against blast of rice
3. Bitertanol	Baycor and Sibutol	0.05 to 0.1% for foliar spray	Powdery mildews and rusts of various crops, apple scab, *Monilinia* on fruit crops, late leaf spot of groundnut and sigatoka leaf spot of banana
4. Hexaconazole	Contaf 5%EC, Anvil	0.2% for spraying	Sheath blight of rice, powdery mildew and rust of apple, rust and tikka leaf spot of groundnut
5. Propiconazole	Tilt, 25% EC, Desmel	0.1% for foliar application	Sheath blight of rice, Sigatoka leaf spot of banana, brown rust of wheat
6. Myclobutanil	Systhane 10WP	0.1 to 0.2% for spraying	Apple scab, cedar apple rust and powdery mildew of apple
Strobilurins			
1. Azoxystrobin	Amistar, Quadris	0.1% for spraying	Broad spectrum fungicide
2.Kresoxim methyl	Ergon, Discus, Stroby	0.1% for spraying	Commonly used for control of ornamental diseases

Host defense inducing /SAR inducing chemicals

Acibenzolar-*S*-methyl - Trade name: Bion, Actigard, Blockade

Phosporic acid - Trade name: Agri-Fos, Alude, Phospho Jet

Harpin - Trade name: Messenger

Probenazole - Trade name: Oryzemate

Fosetyl-Al - Trade name: Allite, Signature, Legion, Flanker

Antibiotics: Metabolites of microorganisms, which in very low or dilute concentration have the capacity to inhibit the growth or destroy other microorganisms.

Most of the antibiotics are the products of actinomycetes. Few of them are produced by fungi and bacteria.

a. Antibacterial (Streptomycin, Tetracycline, Penicillin, etc.)
b. Antifungal (Aureofungin, Griseofulvin, Cyclohexamide, Kasugamycin, Blasticidin, etc.,)

» Penicillin is the first antibiotic discovered by Alexander Fleming (1928)
» Streptomycin is the first antibiotic used in plant disease control
» Chloramphenicol is the first antibiotic synthesized on a commercial scale.

Advantages

» Systemic in action (eradicant and therapeutants)
» Effectively control at very low concentration
» Easy to manufacture
» Easily decompose and protect the crop.
» Specific in their action against pathogen
» No operation hazards
» It moves both Upward and downward directions and kills the pathogen directly or indirectly by altered physiology of host.

Disadvantages

» Most of them are unstable.
» Phytotoxic if they are used at higher doses

» Narrow spectrum of antipathogenic action (antifungal are not effective against bacteria and vice-versa. Even in antibacterial antibiotics, some are effective against Gram +ve bacteria. Some are effective against Gram –ve bacteria)

» Development of resistance in fungi / bacteria (*Erwinia amylovora* developed resistance to streptomycin).

Antibacterial antibiotics

a. **Streptomycin sulphate**: Streptomycin is an antibacterial, antibiotic produced by *Streptomyces griseus*. This group of antibiotics act against a broad range of bacterial pathogens causing blights, wilt, rots etc. This antibiotic is used at concentrations of 100-500 ppm. Some important diseases controlled are blight of apple and pear (*Erwinia amylovora*), Citrus canker (*Xanthomonas campestris* pv. *citri*), Cotton black arm (*X.c.* pv. *malvacearum*), bacterial leaf spot of tomato (*Pseudomonas solanacearum*), wild fire of tobacco (*Pseudomonas tabaci*) and soft rot of vegetables (*Erwinia carotovora*). In addition, it is used as a dip for potato seed pieces against various bacterial rots and as a disinfectant in bacterial pathogens of beans, cotton, crucifers, cereals and vegetables. Although it is an antibacterial antibiotic, it is also effective against some diseases caused by Oomycetous fungi, especially foot-rot and leaf rot of betelvine caused by *Phytophthora parasitica* var. *piperina*.

b. **Tetracyclines**: Antibiotics belonging to this group are produced by many species of *Streptomyces*. This group includes Terramycin or Oxymicin (Oxytetracycline). All these antibiotics are bacteriostatic, bactericidal and mycoplasmastatic. These are very effective against seed-borne bacteria. This group of antibiotic is very effective in managing MLO diseases of a wide range of crops. These are mostly used as combination products with Streptomycin sulphate in controlling a wide range of bacterial diseases. Oxytetracyclines are effectively used as soil drench or as root dip controlling crown gall diseases in rosaceous plants caused by *Agrobacterium tumefaciens*.

Antifungal antibiotics

a. **Aureofungin**: It is a hepataene antibiotic produced in sub-merged culture of *Streptoverticillium cinnamomeum* var. *terricola*. It is absorbed and translocated to other parts of the plants when applied as spray or given to roots as drench. The diseases controlled are citrus gummosis caused by several species of *Phytophthora*, powdery mildew of apple caused by *Podosphaera leucotricha* and apple scab (*Venturia inaequalis*), groundnut tikka leaf spot, downy mildew, powdery mildew and anthracnose of grapes, potato early and late blight. As seed treatment it effectively checked are *Diplodia* rot of mango, *Alternaria* rot of tomato, *Pythium* rot of cucurbits and *Penicillium* rot of apples and citrus. As a truck application/root feeding, 2 g of aureofunginsol + 1g of copper sulphate in 100 ml of water effectively reduce. Thanjavur wilt of coconut.

b. **Griseofulvin:** This antifungal antibiotic was first discovered to be produced by *Penicillium griseofulvum* and now by several species of *Penicillium*. It is highly toxic to powdery mildew of beans and roses, downy mildew of cucumber. It is also used to control *Alternaria solani* in tomato *Sclerotinia fructigena* in apple and *Botrytis cinerea* in lettuce.

c. **Cycloheximide**: It is obtained as a by-product in streptomycin manufacture. It is produced by different species of Streptomyces, including *S. griseus* and *S. nouresi*. It is active against a wide range of fungi and yeast. Its use is limited because it is extremely phytotoxic. It is effective against powdery mildew of beans (*Erysiphe polygoni*), Bunt of wheat (*Tilletia* spp.) brownnot of peach (*Sclerotinia fructicola*) and post harvest rots of fruits caused by *Rhizopus* and *Botrytis* spp.

d. **Blasticidin**: It is a product of *Streptomyces griseochromogenes* and specifically used against blast disease of rice caused by *Pyricularia oryzae*.

e. **Antimycin:** It is produced by several species of *Streptomyces*, especially *S. griseus* and *S. kitasawensis*. It is effectively used against early blight of tomato, rice blast and seedling blight of oats.

f. **Kasugamycin**: It is obtained from *Streptomyces kasugaensis*. It is also very specific antibiotic against rice blast disease. It is commercially available as Kasumin.

g. **Thiolution**: It is produced by *Streptomyces albus* and effectively used to control late blight of potato and downy mildew of cruciferous vegetables.

h. **Endomycin**: It is a product of *Streptomyces endus* and effectively used against leaf rust of wheat and fruit rot of strawberry (*Botrytis cinerea*).

i. **Bulbiformin**: It is produced by a bacterium, *Bacillus subtills* and is very effectively used against wilt diseases, particularaly redgram wilt.

j. **Nystatin**: It is also produced by *Streptomyces noursei*. It is successfully used against anthracnose disease of banana and beans. It also checks downy mildew of cucuribits. As a post harvest dip, it effectively reduces brown rot of peach and anthracnose of banana in stroage rooms.

k. **Eurocidin**: It is a pentaene antibiotic produced by *Streptomyces anandii*. It is effectively used against diseases caused by several species of *Colletotrichum* and *Helminthosporium*.

Some of the fungicides and antibiotics included in the previous lists are prohibited for manufacture, import and use, for agriculture purpose in India, in this book it was included for knowledge purpose only.

Fungicide banned for manufacture, import and use

» Aldicarb (vide S.O. 682 (E) dated 17th July 2001)

» Benomyl (vide S.O 3951(E) dated 8th August, 2018)

» Fenarimol (vide S.O 3951(E) dated 8th August, 2018)

» Methoxy Ethyl Mercury Chloride (vide S.O 3951(E) dated 8th August, 2018)

» Pentachloro Nitrobenzene (PCNB) (vide S.O. 569 (E) dated 25th July 1989)

» Tridemorph (vide S.O 3951(E) dated 8th August, 2018)

» Pentachlorophenol

» Phenyl Mercury Acetate

Fungicide formulations banned for import, manufacture and use

» Carbofuron 50% SP (vide S.O. 678 (E) dated 17th July 2001)

Fungicide formulations banned for use but continued to manufacture for export

» Captafol 80% Powder (vide S.O. 679 (E) dated 17th July 2001)

Fungicide formulations withdrawn

» Ferbam (S.O. 915(E) dated 15th Jun, 2006)

Fungicide formulations refused registration in India

» Fentin Acetate

» Fentin Hydroxide

» Chinomethionate (Morestan)

» Leptophos (Phosvel)

Prohibition of Streptomycin + Tetracycline in agriculture order, 2021 (S.O. 5295(E) dated 17th Dec, 2021)

» Streptomycin and Tetracycline, Streptocycline (90:10 combinations of streptomycin and tetracycline) for use in agriculture beginning February 1, 2022. The order ensures a complete ban on the use of the two antibiotics in agriculture from January 1, 2024, onwards.

Practice Questions

1. What are the different chemical systemic fungicides for plant disease control?
2. How do the fungicides act in the pathogen?
3. What is chemotherapy?
4. What is the mechanism of the action of various groups of fungicides?
5. Name some important antibiotics used for plant disease control.

Suggested Readings

Bartlett, D. W., Clough, J. M., Godwin, J. R., Hall, A. A., Hamer, M., & Parr-Dobrzanski, B. (2002). The strobilurin fungicides. *Pest Management Science: formerly Pesticide Science*, *58*(7), 649-662.

Morton, V., & Staub, T. (2008). A short history of fungicides. *APSnet Features*, *308*.

McCallan, S. E. A. (2012). History of fungicides. *Fungicides, an advanced treatise*, *1*, 1-37.

Thomson, W. T. (1973). Agricultural chemicals book IV-fungicides. *Agricultural chemicals book IV-fungicides.*

Oliver, R. P., & Hewitt, H. G. (2014). *Fungicides in crop protection*. Cabi.

Yamaguchi, I., & Fujimura, M. (2005). Recent topics on action mechanisms of fungicides. *Journal of pesticide science*, *30*(2), 67-74.

Knight, S. C., Anthony, V. M., Brady, A. M., Greenland, A. J., Heaney, S. P., Murray, D. C., & Youle, A. D. (1997). Rationale and perspectives on the development of fungicides. *Annual review of phytopathology*, *35*(1), 349-372.

Cremlyn, R. J. W. (1961). Systemic fungicides. *Journal of the Science of Food and Agriculture*, *12*(12), 805-812.

Sharvelle, E. G. (1961). The nature and uses of modern fungicides. *The nature and uses of modern fungicides.*

McManus, P. S., & Stockwell, V. O. (2001). Antibiotic use for plant disease management in the United States. *Plant Health Progress*, *2*(1), 14.

Brian, P. W. (1952). Antibiotics as systemic fungicides and bactericides. *Annals of Applied Biology*, *39*(3), 434-438.

Oliver, R. P., & Beckerman, J. L. (2022). *Fungicides in Practice*. CABI.

Sinha, A. P., Kishan, S., & Mukhopadhyay, A. N. (1988). *Soil fungicides. Volume I.* CRC Press, Inc.

Chapter - 35

Application of Fungicides

Characters of Fungicides

- » High toxicity to pathogen at low concentration.
- » It should not be toxic to plant, man, animal, beneficial microbes, earthworms, etc.,
- » Slow or no loss of toxicity in storage (shelf life).
- » It should retain toxicity on dilution
- » It should have good spreading quality on host surface
- » High toxicity or retention on the host surface
- » It should have broad spectrum activity
- » It should be compatible with commonly used insecticides and acaricides without any deleterious effect.
- » It should be cheap and available easily in the market
- » It should not cause environmental pollution.
- » It should be easily transportable

Classification of Fungicides Based on Method of Application

The fungicides can also be classified based on the nature of their use in managing the diseases.

1. Seed protectants: E.g., Captan, Thiram, Carbendazim, Carboxin etc.
2. Soil fungicides (preplant): E.g., Bordeaux mixture, Copper oxy chloride, Chloropicrin, Formaldehyde, Vapam, etc.

3. Soil fungicides: E.g., Bordeaux mixture, Copper oxy chloride, Captan, PCNB, Thiram etc.
4. Foliage and blossom: E.g., Capton, Ferbam, Zineb, Mancozeb, Chlorothalonil etc.
5. Fruit protectants: E.g., Captan, Maneb, Carbendazim, Mancozeb etc.
6. Eradicants: E.g., Lime sulphur
7. Tree wound dressers: E.g., Boreaux paste, Chaubattia paste, etc.

Mode of action of fungicides:

The killing of fungal cells or spores after solubilization of fungicides on the host surface may be brought about by

1. General disruption of cell functions and inactivation of enzymes. E.g., Copper fungicides, Phthalamides, Triazoles, Quinones, Chlorothalonil, Chinomethionate, Aromatic and heterocyclic compounds.
2. Disruption of membrane functions and inhibition of sterol biosynthesis. E.g., Triazoles, Imidazoles.
3. Inhibition of glycerophopholipid biosynthesis. E.g., Iprofenphos, Edifenphos, Validamycin-A.
4. Disruption of nuclear process and nucleic acid metabolism. E.g., Benzimidazole, acylalanines, pyrimidines.
5. Inhibition of tubilin biosynthesis. E.g., Benzimidazoles.
6. Effects on cell wall function

   Inhibition of chitin biosynthesis. E.g., Organophosphate compounds (Iprofenphos, Edifenphos).

   Inhibition of melanin biosynthesis. E.g., Tricyclazole

   Inhibition of protein synthesis. E.g., Blasticidin, Kasugamycin, Sulphur, Copper.

   Inhibition of respiration. E.g., Dinocap, Binapacryl, Carboxin, Oxycarboxin, Sulphur

   Non-specific disruption of cell membrane integrity. E.g., Dodine.

7. The protectant fungicides (Copper, Mercury, Sulphur, etc.) have unspecific modes of action. These fungicides can not penetrate the host and exhibit chemicals reactions with sulphydryl, amino, hydroxylic or carboxylic groups in proteins and nucleic acids or their precursors.

Application of the fungicides

Seed treatment: (a) Mechanical, (b) Chemical and (c) Physical.

Mechanical method by sieving or flotation, the infected grains separated. Eg. tundu disease of wheat.

Removal of ergot in cumbu seeds: Dissolve 2 kg of common salt in 10 litres of water (20% solution).

Chemical methods: Using fungicides on seed is one of the most efficient and economical methods of chemical disease control. The seed dressing chemicals may be applied by (i) Dry treatment (ii) Wet treatment and (iii) Slurry.

Dry seed treatment

» Fungicide adheres in a fine from on the surface of the seeds.

» Fungicides treated with the seeds of small lots using simple rotary seed dresser (Seed treating drum) or of large seed lots at seed processing plants using grain treating machines. E.g., Dry seed treatment in paddy.

» Any one of the following chemical may be used for treatment at the rate of 2g/ kg: E.g., Thiram or Captan or Carboxin or Tricyclazole.

Wet seed treatment

» Dipping the seeds or seedlings or propagative materials for a specified time fungicide suspension. Vegetatively propagative materials like cuttings, tubers, corms, setts rhizomes, bulbs etc., which are not amenable to dry or slurry treatment.

» **Seed dip / seed soaking:** E.g., Seed dip treatment in paddy.

» **Seedling dip / root dip**: The seedlings of vegetables and fruits are normally dipped in 0.25% copper oxychloride or 0.1% carbendazin solution for 5 minutes to protect against seedling blight and rots.

» **Rhizome dip:** The rhizomes of cardamom, ginger and turmeric are treated with 0.1% carbendazim + mancozeb.

» **Sett dip / Sucker dip**: The sets of sugarcane and tapioca are dipped in 0.1% carbendazim solution for 30 minutes.

» **Slurry treatment (Seed pelleting)**: Chemical is applied in the form of a thin paste (active material is dissolved in small quantity of water). The slurry treatment is more efficient than the rotary seed dressers. E.g., Seed pelleting in ragi.

Special method of seed treatment

Acid-delinting in cotton: To kill the seed-borne fungi and bacteria. The seeds are treated with concentrated sulphuric acid at 100 ml/kg of seed for 2-3 minutes.

Soil treatment

» Soil harbours a large number of plant pathogens and the primary sources of many plant pathogens happens to be in soil where dead organic matter supports active or dormant stages of pathogens.

» Soil treatment is largely curative in nature as it mainly aims at killing the pathogens in soil and making the soil safe for the growth of the plant. The soil treatment methods: Soil drenching, broadcasting, furrow application, fumigation and chemigation.

Soil drenching

» Followed for controlling damping off and root rot infections at the ground level.

» Requisite quantity of fungicide suspension is applied per unit area so that the fungicide reaches to a depth of atleast 10-15 cm. Eg. Emisan, PCNB, Carbendazim, Copper fungicides, etc.

Broadcasting: Followed in granular fungicides wherein the pellets broadcasted near the plant.

Furrow application: practiced in the control of powdery mildew of tobacco where the sulphur dust is applied in the furrows.

Fumigation:

» Volatile toxicants (fumigants) such as methyl bromide, chloropicrin, formaldehyde and vapam are the best chemical sterilants for soil to kill fungi and nematodes as they penetrate the soil efficiently.

» Fumigations are normally done in nursery areas and in glass houses.

» The fumigant is applied to the soil and covered by thin polythene sheets for 5-7 days and removed. For example, Formaldehyde is applied at 400 ml/100 Sq.m. The treated soil was irrigated and used 1 or 2 weeks later.

» Vapam is normally sprinkled on the soil surface and covered.

» Volatile liquid fumigants are also injected to a depth of 15-20 cm, using sub-soil injectors.

Chemigation: Fungicides are directly mixed in the irrigation water. It is normally adopted using sprinkler or drip irrigation system.

Spraying: This is the most commonly followed method. Spraying of fungicides is done on leaves, stems and fruits. Wettable powders are most commonly used for preparing spray solutions. The most common diluent or carrier is water. The dispersion of the spray is usually achieved by its passage under pressure through nozzle of the sprayer.

Depending on the volume of fluid used for coverage, the sprays are categorised into high volume, medium volume, low volume, very high volume and ultra low volume. The different equipments used for spray application are: Foot-operated sprayer, rocking sprayer, knapsack sprayer, motorised knapsack sprayer (Power sprayer), tractor mounted sprayer, mist blower and aircraft or helicopter (aerial spray).

Dusting: Dusts are applied to all aerial parts of a plant as an alternative to spraying. Dry powders are used for covering host surface. The equipments employed for the dusting operation are: Bellow duster, rotary duster, motorised knapsack duster and aircraft (aerial application).

Post-harvest application

» Fruits and vegetables are largely damaged after harvest by fungi and bacteria. Many chemicals have been used as spray or dip or fumigation.

» Systemic fungicides, particularly thiabendazole, benomyl, carbendazim, metalaxyl, fosety-AI have been found to be very effective against storage diseases. In addition, dithiocarbamates and antibiotics are also applied to control the post-harvest diseases.

» Wrapping the harvested products with fungicide impregnated wax paper is the latest method available.

Special method of applications

a. Trunk Application / Trunk Injection: Adopted in coconut gardens to control Thanjavur wilt (*Ganoderma lucidum*). In the infected plant, a downward hole is made to a depth of 3-4" at an angle of 45°C at the height of 3" from the ground level with the help of an auger. The solution containing 2 g of Aureofungin sol and 1 g of copper sulphate in 100 ml of water is taken in a saline bottle and the bottle is tied with the tree.

b. Root feeding: This is for the control of Thanjavur wilt of coconut instead of trunk application. The root region is exposed; actively growing young root is selected and given a slanting cut at the tip. The root is inserted into a polythene bag containing 100 ml of the fungicidal solution. The mouth of the bag is tied tightly with the root.

c. Pseudostem injection: Very effective in controlling the aphid vector (*Pentalonia nigronervosa*) of bunchy top of banana. The banana injector is used for injecting the insecticide. Banana injector is nothing but an Aspee baby sprayer of 500 ml capacity. In which, the nozzle is replaced by leurlock system and aspirator needle No. 16. The tip of the needle is closed and two small holes are made in opposite direction. It is for free flow of fluid and the lock system prevents the needle from dropping from the sprayer. One ml of monocrotophos mixed with water at 1:4 ratios is injected into the pseudostem of 3 months old crop and repeated twice at monthly intervals.

The same injector can also be used to kill the bunchy top infected plants by injecting 2 ml of 2, 4-D (Femoxone) mixed in water at 1:8 ratios.

d. Corm injection: Effective method used to control Panama wilt of banana caused by *Fusarium oxysporum* f. sp. *cubense*. Capsule applicator is used for this purpose. It is nothing but an iron rod of 7 mm thickness to which a handle is attached at one end. The length of the rod is 45 cm and an iron plate is fixed at a distance of 7 cm from the tip. The corm is exposed by removing the soil and a hole is made at 45) angle to a depth of 5 cm. One or two gelatin capsules containing 50-60 mg of carbendazim is pushed in slowly and covered with soil. Instead of capsule, 3 ml of 2 % carbendazim solution can also be injected into the hole.

e. Paring and Pralinage: Used to control Fusarium wilt and burrowing nematode (*Radopholus similis*) of banana. The roots as well as a small portion of corm is removed or chopped off with a sharp knife and the sucker is dipped in 0.1% carbendazim solution for 5 minutes. Then, the sucker is dipped in clay slurry and furadan granules are sprinkled over the corm at 40 g/corm.

Practice Questions

1. Attempt a short note on Integrated Disease Management (IDM).
2. Describe IDM for Nursery Disease.
3. Attempt short note on 'Soil solarization'.
4. List different cultural practices used for the management of diseases of crops.
5. Define Biological management of crop diseases and name common fungal and bacterial bio- agents used for Seed priming and soil application.

Suggested Readings

Oliver, R. P., & Beckerman, J. L. (2022). *Fungicides in Practice*. CABI.

Nita, M. (Ed.). (2013). *Fungicides: Showcases of Integrated Plant Disease Management from Around the World*. BoD–Books on Demand.

Panda, H. (2003). *The Complete Technology Book on Pesticides, Insecticides, Fungicides and Herbicides with Formulae & Processes*. NIIR Project Consultancy Services.

Nene, Y. L., & Thapliyal, P. N. (1993). *Fungicides in plant disease control* (No. Ed. 3). International Science Publisher.

Knowles, D. A. (1998). Formulation of agrochemicals. In *Chemistry and technology of agrochemical formulations* (pp. 41-79). Springer, Dordrecht.

Erwin, D. C. (1973). Systemic fungicides: disease control, translocation, and mode of action. *Annual Review of Phytopathology, 11*(1), 389-422.

Damicone, J., & Smith, D. (2009). *Fungicide resistance management.* Oklahoma Cooperative Extension Service.

Bartett, D. W., Clough, J. M., Godfrey, C. R., Godwin, J. R., Hall, A. A., Heaney, S. P., & Maund, S. J. (2001). Understanding the strobilurin fungicides. *Pesticide Outlook, 12*(4), 143-148.

Domsch, K. H. (1964). Soil fungicides. *Annual Review of Phytopathology, 2*(1), 293-320.

Hewitt, G. (2000). New modes of action of fungicides. *Pesticide Outlook, 11*(1), 28-32.

Chapter - 36

Principles of Integrated Nematode Management

Plant parasitic nematodes can be controlled by several methods. The nematode control aims to improve growth, quality and yield by keeping the nematode population below the economical threshold level. The control measures to be adopted should be profitable and cost effective. It is essential to calculate the cost benefit ratio before adopting control measures.

The nematode control methods are

1. Regulatory (Legal) control

2. Cultural control

3. Physical control

4. Biological control

5. Chemical control.

1. Regulatory control: Regulatory control of pests and diseases is the legal enforcement of measures to prevent them from spreading or having spread, from multiplying sufficiently to become intolerably troublesome. The principle involved in enacting quarantine is exclusion of nematodes from entering into an area which is not infested, in order to avoid spread of the nematode. Quarantine principles are traditionally employed to restrict the movement of infected plant materials and contaminated soil into a state or country. Many countries maintain elaborate organizations to intercept plant shipments containing nematodes and other pests. Diseased and contaminated plant material may be treated to kill the nematodes or their entry may be avoided.

Quarantine also prevents the movement of infected plant and soil to move out to other nematodes free areas.

Plant Quarantine in India: The Destructive Insects and Pests Act, 1914 (DIP) was passed by the Government of India which restricts introduction of Egotic pests and disease into the country from abroad. The agricultural pests and disease acts of the various states prevent interstate spread of pests within the country. The rules permit the plant protection advisor to the government of India or any authorizes officer to undertake inspection and treatments. Strict regulations have been made against *G. rostochiensis*, the potato cyst nematode and *Rhadinaphelenchus cocophilus*, the red ring nematode of coconut. Domestic quarantine regulations have also been imposed to restrict the movement of potato both for seed and table purposes in order to present the spread of potato cyst nematode from Tamil Nadu to other states in India.

2. Cultural control: Cultural nematode control methods are agronomical practices employed in order to minimize nematode problem in the crops.

» Use of cover crop like *Calapagonium.*
» Use of resistant variety like.
» Sow sunnhemp or marigold in and around the basin of plants at early crop stage and incorporate their biomass one month later. It reduces nematode build up.

a. **Selection of healthy seed material:** In plants, propagated by vegetative means we can eliminate nematodes by selecting the vegetative part from healthy plants. The golden nematode of potato, the burrowing, spiral and lesion nematodes of banana can be eliminated by selecting nematode free plant materials. The wheat seed gall nematode and rice white tip nematode can be controlled by using nematode free seeds.

b. **Adjusting the time of planting:** Nematode life cycle depends on the climatic factors. Adjusting the time of planting helps to avoid nematode damage. In some cases crops may be planed in winter when soil temperature is low ad at that time the nematodes cannot be active at low temperature. Early potatoes and sugar beets grow in soil during cold season and escapes cyst nematode damage since the nematodes are not that much active, to cause damage to the crop during cold season.

c. **Fallowing:** Leaving the field without cultivation, preferably after ploughing helps to expose the nematodes to sunlight and the nematodes die due to starvation without host plant. This method is not economical.

d. **Deep summer ploughing:** During the onset of summer, the infested field is ploughed with disc plough and exposed to hot sun, which in turn enhances the soil temperature and kills the nematodes. For raising small nursery beds for vegetable crops like tomato and brinjal seed beds can be prepared during summer, covered with polythene sheets which enhance soil temperature by 5 to 10°C which kills the nematodes in the seed bed. This method is very effective and nematode free seedling can be raised by soil solarization using polythene sheets.

e. **Manuring:** Raising green manure crops and addition of more amount of farm yard manure, oil cakes of neem and castor, pressmud and poultry manure etc enriches the soil and further encourages the development of predacious nematodes like *mononchus* spp. and also other nematode antagonistic microbs in the soil which checks the parasitic nematodes in the filed.

f. **Flooding:** Flooding can be adopted where there is an enormous availability of water. Under submerged conditions, anaerobic condition develops in the soil which kills the nematodes by asphyxiation. Chemicals lethal to nematodes such as hydrogen sulphide and ammonia are released in flooded condition which kills the nematodes.

g. **Trap cropping:** Two crops are grown in the field, out of which one crops is highly susceptible to the nematode. The nematode attacks the susceptible crop. By careful planning, the susceptible crop can be grown first and then removed and burnt.

Thus, the main crop escapes from the nematode damage. Cowpea is highly susceptible crop can be grown first and then removed and burnt. Cowpea is highly susceptible to root – knot nematode and the crop can be destroyed before the nematodes mature.

h. **Antagonistic crops:** Certain crops like mustard, marigold and neem etc have chemicals or alkaloids as root Egudates which repell or suppress the plant parasitic nematodes. In marigold (*Tagetes* spp.) plants the – terthinyl and bithinyl compounds are present throughout the plant from root to shoot tips. This chemical kills the nematodes. In mustard allyl isothiocyanate and in pangola grase pryrocaterchol are present which kills the nematodes. Such enemy plants can be grown along with main crop or included in crop rotation.

i. **Removal and destruction of infected plants:** Early detection of infested plants and removal helps to educe nematode spread. After harvest the stubbles of infested plants are to be removed. In tobacco, the root system is left in the field after harvest. This will serve as a inoculum or the next season crops. Similarly, in *D. angstus* the nematode remains in the left-out stubbles in the field after harvest of rice grains. Such stubbles are to be removed and destroyed and land needs to be ploughed to expose the soil.

j. **Use of resistant varieties:** Nematode resistant varieties have been reported from time to time in different crops. Use of resistant varieties is a very effective method to avoid nematode damage. Nemared, Nematex, Hisar Lalit and Atkinson are tomato varieties resistant to *M. incognita*. The potato variety Kufriswarna is resistant to *G. rostochiensis*.

k. Crop rotation: Rotation to non-host crops such as corn, cucurbits, potatoes, and tomatoes is an effective control for the cyst nematode (*Heterodera* spp.), **but is less likely to** control the root knot nematode because of its wider host range. All species of *Meloidogyne* are called ‹root knot› nematode, but each species has a different host range, causing confusion over which crops or cultivars are resistant or tolerant to which species of root knot nematode. Rotations to non-host crops for more than a year reduce populations below damaging levels but will not eliminate them. It is generally agreed that asparagus, corn, onions, garlic, small grains, cahaba white vetch, and ‹nova› vetch are sufficiently resistant that they can be grown as a rotation crop in soils infested with root knot nematodes. Crotalaria, velvet bean, soybean and grasses such as rye are also resistant to root knot nematode and are good rotation crops.

l. Cover crops: Cover crops of winter rapeseed significantly reduced subsequent damage to potatoes by the Columbia root knot nematode.

» Sudangrass reduced soil nematode populations, but suppression may not be of sufficient duration for commercial potato production.

» Using sesame as a cover crop has been reported to decrease nematodes.

m. Increasing soil organic matter: Higher soil organic matter content protects plants against nematodes by increasing soil water-holding capacity and enhancing the activity of naturally-occurring biological organisms that compete with nematodes in the soil. Low soil moisture puts even more stress on plants with nematode-damaged root systems.

3. Physical control: It is very easy to kill the nematodes in laboratory by exposing the nematodes to heat, irradiation and osmotic pressure etc., but it is extremely difficult to adopt these methods in fieldconditions. These physical treatments maybe hazardous to plant or the men working with the treatments and the radiation treatments may have residual effects. Sun drying of banana sucker for 2 weeks or Hot water treatment at 55°C for 10 min.

a. Heat treatment of soil: Sterilization of soil by allowing steam is a practice in soil used in green house, seed beds and also for small area cultivation. Insects, weed seeds, nematodes, bacteria and fungi are killed by steam sterilization. In such cases steam is introduced into the lower level of soil by means of perforated iron pipes buried in the soil. The soil surface needs to be covered during steaming operation. Plastic sheets are used for covering. In the laboratory and for pot culture experiments autoclaves are used to sterilize the soil.

b. Hot water treatment of planting material: Hot water treatment is commonly used for controlling nematodes. Prior to planting the seed materials such as banana corms, onion bulbs, tubers seeds and roots of seedlings can be dipped in hot water at 50–55°C for 10 minutes and then planted.

c. **Irradiation:** Irradiation also kills the nematode. Cysts of *G. rostochiensis* exposed to 20,000 Ỵ contained only dead eggs and at 40,000 Ỵ exposure, the eggs lost their contents. *Ditylenchus myceliphagus* in mushroom compost exposed to Ỵ rays between 48,000 to 96,000 Ỵ inactivated the nematodes. UV light also kills the nematodes. But this irradiation is not practically feasible under field conditions.

d. **Osmotic pressure:** Complete nematode mortality when sucrose or dextrose was added to nematode infested soil at 1 to 5% by weight. But these methods are not practical and economical.

e. **Washing process:** Plant parasitic nematodes are often spread by soil adhering to potato tubers, bulbs and other planting materials. Careful washing of such planting material helps to avoid the nematodes in spreading in new planting field. Washing apparatus for cleaning potato and sugarbeet tubers are commercially developed and are being used in many countries.

f. **Seed cleaning:** Modern mechanical seed cleaning methods have been developed remove the seed galls from normal healthy wheat seeds.

g. **Ultrasonics:** Ultrasonic have little effect on *Heterodera* spp. The use of this ultrasonics is not practically feasible.

4. Biological control aims to manipulate the parasites, predators and pathogens of nematodes in the rhizosphere in order to control the plantparasitic nematodes. Addition of organic amendments such as farm yard manure, oil cakes, green manure and pressmud etc encourages the multiplication of nematode antagonistic microbes which inturn checks the plant parasitic nematodes.

The addition of organic amendments acts in several ways against the plant parasitic nematodes. Organic acid such as formic, acetic propionic and butric acids are released in soil during microbial decomposition of organic amendments. Ammonia and hydrogen sulphide gases are also released in soil during decomposition. These organic acids and gases are toxic to nematodes.

a. **Crab meal:** Crab meal compost is potentially nematode-suppressive. Blue crab compost applied at a 10 to 20 percent ratio (weight compost: weight soil being treated) suppressed root galling and egg mass production by *Meloidogyne javanica* on tomato. Raw crab scrap at 0.05 percent was even more effective in suppressing root galling than the 20-percent compost. Crab meal, like other nematode control practices, must be applied before planting because these materials need to penetrate as much of the rooting zone as possible to be effective. This is only possible if they can be incorporated before planting.

Nematode antagonistic microbes multiply rapidly due to addition of organic matter. Organic amendements improve soil conditions and helps the plants to grow. The organic matter also provides nutrition for the crop plants.

b. **Predacious fungi:** Most of the predacious fungi come under the order Moniliales and Phycomycetes. There are two types of predacious activities among these fungi. They are nematode trapping fungi and endozoic fungi.

c. **Non-constricting rings:** The trap is formed similar to the constricting ring. It is a non – adhesive trap. The reing becomes an infective structure and kills the nematode E.g., *Daclylaria candida*. In addition to formation of traps and adhesive secretions, the predacious fungi may also produce toxin which kills the nematodes.

d. **Endozoic fungi:** The endozoic fungi usually enter the nematode by a germ tuber that penetrates the cuticle from a sticky spore. The fungal hyphae ramify the nematode body, absorb the contents and multiply. The hyphae then emerge from dead nematode. *Catenaria vermicola* often attacks sugarcane nematodes. *Pasteuria penetrans* was found to be very effective against the root-knot nematodes in many crops. The *P. penetrans* infested J2 of root knot nematodes ca be seen attached with spores throughout the cuticle.

5. Chemical management: The development of chemical controls for plant-parasitic nematodes is a formidable challenge. Because most phytoparasitic nematodes spend their lives confined to the soil or within plant roots, delivery of a chemical to the immediate surroundings of a nematode is difficult. The outer surface of nematodes is a poor biochemical target and is impermeable to many organic molecules. Delivery of a toxic compound by an oral route is nearly impossible because most phytoparasitic species ingest material only when feeding on plant roots. Therefore, nematicides have tended to be broad-spectrum toxicants possessing high volatility or other properties promoting migration through the soil. The resulting record of less-than-perfect environmental or human health safety has resulted in the widespread deregistration of several agronomically important nematicides (e.g., ethylene dibromide and dibromochloropropane). The most important remaining fumigant nematicide, methyl bromide, faces immediate severe restrictions and future prohibition because of concerns about atmospheric ozone depletion.

History of chemical control:

- Kuhn (1881) first tested CS_2 to control sugarbeet nematode in Germany and he could not get encouraging results.
- In South Carolina State, U.S.A. Bessey (1911) treated CS_2 or the control of root – knot nematodes but the method proved impractical.
- Latter on the chemicals like formaldehyde, cyanide and quick line were observed to have nematicidal properties, but all these chemicals were found to be highly expensive.
- Mathews (1919) observed the effect of chloropicrin (tear gas) against plant parasitic nematodes in England.

- Carter (1943) an entomologist of Hawai, Pineapple Research Institute, reported the efficacy of 1,3 dichoropropene 1,2 dichloropropane (DD) mixtureat 250 1b/acre, against the plant parasitic nematodes.
- In 1944, scientists from California and Florida states of USA reported the efficacy of ethylene dibromide (EDB). In the same year the Dow Chemical company, USA introduced the chemical as a soil fumigant for the management of nematodes.
- The introduction of these two nematicides *viz.*, DD and EDB paved way for the chemical control of nematodes.

Although the discovery of nematicidal activity in a synthetic chemical dates from the use of carbon disulfide as a soil fumigant in the second half of the nineteenth century, research on the use of nematicides languished until surplus nerve gas (chloropicrin) became readily available following World War I. In the 1940s, the discovery that D-D (a mixture of 1,3-dichloropropene and 1,2-dichloropropane) controlled soil populations of phytoparasitic nematodes and led to substantial increases in crop yield provided a great impetus to the development of other nematicides, as well as the growth of the science of nematology.

A nematicide is a type of chemical pesticide used to kill plant-parasitic nematodes. Nematicides have tended to be broad-spectrum toxicants possessing high volatility or other properties promoting migration through the soil. Most soil nematicides are also registered as insecticides or fungicides.

FUMIGANTS	CARBAMATES	ORGANOPHOSPHATES
D-D mixture	Aldoxycarb	Fosthiazate
1,3-Dichloropropene	Carbofuran	Cadusafos
Ethylene Dibromide	Oxamyl	Fenamiphos
1,2-Dibromo-3-Chloropropane		Ethoprop
Methyl Bromide	-	-
Chloropicrin	-	-
Metam Sodium, Dazomet, and Methyl Isothiocyanate (MITC)	-	-
Sodium Tetrathiocarbonate	-	-

Description of some important nematicides

a. **Ethylene dibromide (EDB):** 1.2 Dibromonehane. It is a colourless liquid and the gas in non-inflammable. It is available ad 83% liquid formulation containing 1.2 kg active ingredient per litre and as 35% granules. It is injected or dibbled into the soil for the control of nematodes at 60 to 120 1 or 200 kg ai/ha but it is not very effective against cyst nematodes. *Heterodera* spp. and soil fungi. Crops like onion, garlic and other bulbs should not be planted after soil treatment with EDB. It is available as Bromofume and Dowfume.

b. **Dibromochloropropane (DBCP):** 1,2 – dipromo – 3 – chloropropane. It is a straw coloured liquid, a litre of it weighing 1.7 kg. It can be used s soil treatment before planting, at the time of planting or as post when the soil temperature is above 20°C. It is applied as a sprinkle depending upon the crop and stage. Certain crops like tobacco and potato are sensitive due to high bromine content in the chemical. It functions more efficiently than other fumigant at high soil temperature due to its high boiling point (195.6°C). trade names are Nemagon and fumazone.

c. **D-D mixture:** It is the trade name of the mixture of compounds, chief of them contain thecis and the trans isomers in equal quantities of 1,3-dichloropropene 30.35%, and a few other chlorinated compounds up to about 5%. Of these, dichloropropene is the most toxic component and among its two isomers, the transisomer is twice as toxic as the cis-isomer. It is a black liquid of 100% formulation and a litre of it contains approximately 1 kg of technical compounds. It is used in the control of soil insects and nematodes and injected into the soil at a depth of 15-20 cm at 25 × 30 cm spacing. Since it is highly phytotoxic, it is used for preplant soil application at least 2 -3 weeks before planting. It is used for preplant soil application at least 2 -3 weeks before planting. It is used as such at 225 – 280 1/ha, but in clay and peaty soils a higher dosage is required. It taints potato tubers and carrots grown in treated soil. Dichloropropence is available under the trade name Telone andin mixture with dibromoetane under the name Dorlane.

d. **Methylbromide or Bromomethane** It boils at 4.5°C. At ordinary temperatures it is a gas and therefore, confined in containers under pressure as a liquid. The gas is 1.5 times as heavy as air. Its insecticidal properties were described by Le Goupil in 19-32. Its power of penetration into packed foodstuffs such as flour is remarkable. As it kills insects slowly a longer period of exposure to gas may be required. For control of stored grain pests, it is used at 24 – 32 g.m, exposure period being 48 h. In tent fumigation for the control of termites and powder post beetles, the dosage recommended is 32 0-64 g/ m^3. For fumigating live plants, the dosage is 16 32. Some plants are likely to be injured. In soil application for the control of nematodes.

e. **Phorate**: 0,0 – diethyl S – (ethylthiomethyl) phosphorodithioate. Trade name is Thimet. It is systemic insecticide cum nematicide, available as 10% granule. It has got both contact and fumigant action.

f. **Aldicarb:** 2 – methyl – 2 (methilthio) propinaldehyde 0 (methylcarbomy) oxime. Trade name is Temik. The sulphur atom in the molecule is oxidized to sulfoxide and then to sulfone. It is a systemic 10% granule. The residues remain in plants for 30 -35 days as a lethal dose. It also acts as repellent, contact nematicide and interferes with reproduction of the nematodes by way of sex reversal.

g. **Carbofuran:** It is 2,3 – dihydro – 2, 2, dimethyl 7 benzofuranyl methyl carbamate. Trade name Furadon. It is a systemic insecticide cum nematicide. It is formulated as 3% granule and also as 40F. The residual effect last for 30 – 60 days. It has also got phytotonic effect. This systemic chemical has got acropetal action and applied at 1 -2 kg ai. /ha.

h. **Sodium Tetrathiocarbonate:** Sodium tetrathiocarbonate is more recently registered preplant soil fumigant active against fungi, insects, and nematodes. It is supplied as a liquid formulation and may be applied via drip or surface irrigation. Sodium tetrathiocarbonate rapidly degrades in soil into carbon disulfide, sodium hydroxide, hydrogen sulfide, and sulfur. Carbon disulfide is the active principle. Although carbon disulfide has a long history as a fumigant, its flammability is legendary. Carbonates and sulfates are the terminal degradation products. Unlike other commonly used fumigants, sodium tetrathiocarbonate does not readily move through soil air and requires a high level of soil moisture when applied in order to be distributed throughout the soil.

Note: Many of these nematicides are banned for use, manufacturing, import in India. It is given here only for educational purpose.

Newer Molecules in the Market

a. Fluensulfone 2% GR: It belongs to a new chemical class fluoroalkenyle (thioether), non-fumigant type exhibits systemic activity and causes irreversible paralysis within 24 hours and subsequent death in 48 hours. Mainly for managing root-knot nematode in tomato (1 g/ plant), cucumber (1 g/ plant), okra (1 g/ plant), pomegranate (10 g/ dipper), capsicum (1.5 g/ plant).

b. Fluopyrum 34.48% SC: The active ingredient fluopyrum selectively inhibits Complex II of the mitochondrial respiratory chain of nematodes. Following application, nematodes cannot generate energy to sustain hence initially take on the shape of a needle, becomes immobile and eventually dies. It is used at 0.1% to manage tomato root knot nematode (*Meloidogyne incognita*), yellow potato cyst nematode (*Globodera rostochiensis*) and white potato cyst nematode (*Globodera pallida*), and cyst nematode (*Heterodera carotae*), root lesion nematode (*Pratylenchus penetrans*), stunt nematode (*Tylenchorhynchus* spp.), stubby root nematode (*Trichodorus* spp.) in carrot.

Methods of Applications

The methods for treating agricultural soils with Nematicides are similar to those used for other pesticides examined in this volume. Nematicide application research is being driven by the need to maximize efficacy while minimizing groundwater and atmospheric contamination.

a. Fumigation

Soil fumigation requires prior preparation to be effective. Prior to fumigant or nonfumigant application, soil is often turned or tilled to increase porosity and uniformity and promote decomposition of residual plant roots, which can serve as hiding places for nematodes or interfere with fumigant movement. Adequate but not excessive soil moisture is critically important to the success of some fumigants.

- » Fumigants are typically injected with chisels or shanks into the upper 15–40 cm of soil, with the actual depth a function of compound, soil structure, and crop. Although deep injection is often required to minimize the escape of fumigant into the surrounding air, inadequate levels of nematicide in the upper soil layers may result in some situations.
- » Experimental chisels angled to the side 45° in order to eliminate chisel trace formation have provided control of root-knot nematodes on tomato equivalent to conventional chisels.
- » Fumigation usually involves the use of plastic tarpaulins to minimize atmospheric losses and deliver nematicide to the target organism. Sometimes, tarpaulins must be in place for 10 days.
- » Impervious sheeting, warm temperatures, and deep injection often enhance nematicidal activity and permit the use of much smaller quantities of fumigant.

b. Irrigation

Liquid and emulsifiable formulations of nematicides can often be applied through surface or drip irrigation systems. Drip irrigation in particular offers a means of precisely controlling the amount of active ingredient delivered to a field, as well as regulating the amount of water, so that leaching of active ingredient beyond the root zone and into groundwater can be eliminated. Drip irrigation also is useful for postplant applications, and it avoids the use of granular materials that may pose risks to birds. Use of drip irrigation also reduces the amount of personal protective equipment required for field workers.

c. Granules and Broadcast Sprays

The most widely practiced method of applying nonfumigant nematicides is with granular formulations. In some cases, adequate control can be achieved by band application of nematicides at or before sowing.

- » In band application, plant roots may eventually grow beyond the treated area at a time when the root system will be sufficiently vigorous to not suffer serious damage.
- » In-furrow application sometimes is practiced but may result in lack of delivery to

the root zone; in other cases, in-furrow application may be preferable.

- » In some cases, sidedress applications of nematicides are useful replacements or additions to at-plant applications.
- » In other cases, broadcast application of granules or sprays followed by a thorough mixing of the soil may be effective.
- » Tillage is necessary to distribute nematicide to a broad enough area to provide control, and a thorough mixing is particularly important for nematicides with poor soil mobility characteristics.
- » Use of broadcast sprays instead of granules often promotes greater uniformity in distribution. For many annual crops, incorporating nematicides into the upper 10–15 cm of soil provides the best balance of efficacy, expense, ease, and safety to wildlife.
- » Paring and pralinge: Removal of lesioned portion + dipping in mud slurry (clay soil: water in the ratio of 1:5) + sprinkling of Carbofuran 3G at 40g/ rhizome).
- » Dip the corm with 0.5% Monocrotophos for 15 minutes, shade dry and then plant.

d. Seed Dressing and Bare Root Dip

Few nematicides have been registered as seed coatings include the difficulty in applying a sufficient quantity of nematicide needed to provide control beyond the seedling stage.

Practice Questions

1. Name some important nematicides used for controlling plant nematode diseases.
2. What is the rationale in growing trap crops? How could it be utilized for controlling plant nematodes?
3. Name some important trap crops used to control plant nematodes.
4. Give some suitable examples of controlling nematode diseases by means of biological agents.
5. What is soil solarization? How can it be used to control nematode population in soil?
6. Why are some of the earlier recommended nematicides abandoned now?

Suggested Readings

Wang, K. H., Sipes, B. S., & Schmitt, D. P. (2002). Crotalaria as a cover crop for nematode management: a review. *Nematropica*, 35-58.

Bird, G. W. (1981). Integrated nematode management for plant protection. *Plant parasitic nematodes*, *3*, 355-375.

Sikora, R. A., Desaeger, J., & Molendijk, L. (Eds.). (2021). *Integrated nematode management: state-of-the-art and visions for the future*. CABI.

Sikora, R. A., Bridge, J., & Starr, J. L. (2005). Management practices: an overview of integrated nematode management technologies. *Plant parasitic nematodes in subtropical and tropical agriculture*, (Ed. 2), 793-825.

Abd-Elgawad, M. M., & Askary, T. H. (2018). Fungal and bacterial nematicides in integrated nematode management strategies. *Egyptian journal of biological pest control*, *28*(1), 1-24.

Mashela, P. W., De Waele, D., Dube, Z., Khosa, M. C., Pofu, K. M., Tefu, G., ... & Fourie, H. (2017). Alternative nematode management strategies. In *Nematology in South Africa: a view from the 21st century* (pp. 151-181). Springer, Cham.

Thoden, T. C., Korthals, G. W., & Termorshuizen, A. J. (2011). Organic amendments and their influences on plant-parasitic and free-living nematodes: a promising method for nematode management?. *Nematology*, *13*(2), 133-153.

Barker, K. R., & Koenning, S. R. (1998). Developing sustainable systems for nematode management. *Annual review of phytopathology*, *36*(1), 165-205.

Duncan, L. W. (1991). Current options for nematode management. *Annual Review of Phytopathology*, *29*(1), 469-490.

Glossary

- Anamorph: the asexual reproductive or propagative manifestation of a fungus.
- Apomorphy: a character state which is unique to a single, terminal taxon. Example: among primates, complex grammar is an apomorphy of human beings. It is quite diagnostic of humans, but useless in determining phylogenetic relationships because it is not a shared, derived characteristic, or synapomorphy, of any larger group.
- Apothecium: a disk-shaped or cup-shaped ascocarp (fruiting body) of a lichen or non-lichenized ascomycete. In an apothecium, the asci are born in a single, orderly layer on an open, fairly flat surface.
- Arthrospore: A spore resulting from the fragmentation of a hypha.
- Ascocarp: the fruiting body of an ascomycete; the multicellular structure that produces asci, and acts as the platform from which the spores are launched.
- Ascohymenial: relating to an ascocarp which forms after nuclear pairing. The ascocarps produced by ascohymenial development extend out into the medium, as in the image at apothecium, and the asci are unitunicate.
- Ascoma: (pl. *ascomata*) (1) same as ascocarp (2) "an ascocarp having the spore-bearing layer of cells (the hymenium) on a broad disklike receptacle." This latter definition is frequently cited but appears to be either incorrect or useless. For our purposes, *ascoma* and *ascocarp* will be treated as synonymous.
- Ascospore: A meiospore borne in an ascus
- Ascus: a sac-like cell containing the ascospores cleaved from within by free cell formation after karyogamy (nuclear fusion) and meiosis. Eight ascospores typically are formed within the ascus, but this number may vary considerably.
- Autotroph: an organism which obtains energy from inorganic sources, sunlight or the oxidation of inorganic chemicals.
- Ballistospore: the ability to launch spores into the air. The mechanism of ballistospore is not fully understood. This ability is common in Basiodiomycota and is also known in some Ascomycota.
- Basidiospore: a meiospore borne on the outside of a basidium.
- Basidium: structure produced by basidiomycetes on which sexual spore formation occurs.

- Binding hyphae: thick walled, typically aseptate, highly branched vegetative hyphae.
- Bitunicate: an ascus with differentiated inner and outer walls is said to be *bitunicate*.
- Budding: reproduction by binary fission, a characteristic form of propagation in yeasts.
- Chitin: a polymer of repeating sugar molecules (a slightly modified glucose, poly-N-acetyl-D-glucosamine). See image. Chitin is the material which makes up the exoskeleton of insects and, in more or less modified form, in almost all arthropods. In arthropods, chitin occurs in a crosslinked form, α-chitin. Significantly, it is also found in the radular «teeth» of molluscs, the setae (bristles) and jaws of annelid worms, and the cell walls of Fungi. So, this is exceedingly ancient stuff, possibly predating the split between bacteria and metazoans.
- Chitosan: Chitosan (poly-D-glucosamine) is one of the most common polymers found in nature. Structurally, it is related to cellulose, which consists of long chains of glucose molecules linked to each other. Chitosan is chitin with acetic acid amide linkage hydrolysed from the 2-amino group. In fact, this is exactly how chitosan is commercially prepared.
- Clade: a "natural group" or phylogenetically defined group of organisms. A clade is usually defined as a particular organism «and all of its descendants.»
- Clamp connection: a bridge-like hyphal connection involved in maintaining the dikaryotic condition.
- Cleistothecial: of an ascocarp (especially Aspergillaceae and Erysiphaceae), have a closed spore-bearing structure without a pore (ostiole) for spore release, and from which spores are released only by decay or disintegration.
- Cleistothecium: a cleistothecial ascocarp.
- Coenocytic: only divisions (septa) are formed between the nuclei of the hypha when reproductive cells develop.
- Columella: the swollen tip of the sporangiophore (the stalk) on which mitospores develop. The image shows an intact columella, with spores attaches. When the spores have dispersed, the columella collapses into a structure that looks remarkably like a microscopic mushroom.
- Colony: coherent mycelium or mass of cells, like yeast cells, of one origin.
- Conidium: Conidia are sporangia grown on elaborate structures called conidiophores. These are usually stalked, lifting the conidia off the substrate for better dispersal and to avoid microscopic grazing animals. They are often produced hundreds or thousands of at a time.
- Deuteromycetes: Fungi that can only reproduce asexually.

- Dikaryon: a pair of closely associated, sexually compatible nuclei, may or may not be derived from a different parent hypha or cell. Occasionally, polykaryons are produced with nuclei from more than two individuals.
- Diploid: a nucleus is diploid if it contains two copies of each non-redundant gene. In Fungi, it is necessary to distinguish between diploid nuclei and diploid cells. A hypha may contain several haploid nuclei (either identical or from different individuals). Technically, the cell is diploid or polyploid. However, there may be no diploid nuclei.
- Diploidization: The process which occurs when compatible haploid hyphae from different individuals exchange nuclei. This does not necessarily result in the formation of diploid nuclei, meiosis, or genetic exchange. In fact, some fungi can maintain populations of nuclei from up to six independent individuals in a single hypha.
- Endomycorrhiza: mycorrhiza in which the fungal hyphae penetrate cell walls of host plant.
- Endophyte: a fungus living within plants, often without causing visible symptoms.
- Ergosterol: a sterol lipid which occurs in fungal cell membranes, apparently in place of cholesterol. It is found in metazoans as a precursor to vitamin D2.
- Fission: cytoplasmic division of a cell to form two cells, a form of asexual reproduction.
- Fissitunicate: A bitunicate «ascus has a distinctly bilayered wall, with the outer layer being rigid and the inner layer being expandable. As it matures, the thin outer layer splits and the thick inner layer absorbs water and expands upward. The ascus stretches up into the narrow neck of the ascoma, and the ascospores are expelled. These asci with a "jack-in-the-box› design are also called fissitunicate".
- Flagellum: (pl. *flagella*) A eukaryotic flagellum is a bundle of nine fused pairs of microtubules called "doublets" surrounding two central single microtubules (the so-called 9+1 structure of paired microtubules; also called the "axoneme"). At the base of a eukaryotic flagellum is a microtubule organizing center about 500 nm long, called the basal body or kinetosome. The flagellum is encased within the cell's plasma membrane, so that the interior of the flagellum is accessible to the cell's cytoplasm. This is necessary because the flagellum's flexing is driven by the protein dynein bridging the microtubules all along its length and forcing them to slide relative to each other, and ATP must be transported to them for them to function. This extension of the cytoplasm is called the *axosome*. Important note: The eukaryotic flagellum is completely different from the prokaryote flagellum in structure and in evolutionary origin. The only thing that the bacterial, archaeal, and eukaryotic flagella have in common is that they stick outside of the cell and wiggle to produce propulsion.
- Fragmentation: «Many fungi can reproduce by fragmentation. Any mycelium that is fragmented or disrupted, provided that the fragment contains the equivalent of the

peripheral growth zone, can grow into a new colony. Many fungi are sub-cultured using this hyphal fragment technique.

» Fruiting body: any complex fungal structure that contains or bears spores.

» Gametangium: a single-celled structure producing gametes (sex cells) or gametic nuclei.

» Ghost lineage: a period of time during which a clade is inferred to exist, but is not known in the fossil record. As discussed in the text, Ascomycota and Basidiomycota are sister groups. Basidiomycota is known from the Early Devonian. This means that Ascomycota must have diverged from Basidiomycota at least this early.

» Glucan: a polysaccharide composed of repeating units of glucose.

» Haploid: a nucleus is haploid if it contains only one copy of each non-redundant gene. In Fungi, it is necessary to distinguish between haploid nuclei and haploid cells. A hypha may contain several haploid nuclei (either identical or from different individuals). Technically, the cell is diploid or polyploid. However, there may be no diploid nuclei.

» Heterogametes: male and female gametes that are morphologically distinguishable.

» Heterothallic: describes fungi in which two genetically distinct but compatible mycelia must meet before sexual reproduction can take place

» Heterotroph: an organism which obtains energy from the oxidation of organic matter.

» Homing Endonuclease: an enzyme which catalyzes the insertion of an intron into chromosomal DNA. Homing enzymes may be coded by the intron itself, or may come from elsewhere. Homing enzymes may work in a variety of ways. For example, they may act as, or work with, a reverse transcriptase to create a DNA copy of an intron RNA. Alternatively, they may directly ligate the intron transcript (RNA) into the DNA of the «host,» relying on the host DNA repair machinery to convert the inserted sequence into double-stranded DNA.

» Homothallic: describes fungi in which a single strain can undertake sexual reproduction; self-compatible.

» Hülle Cell: in some ascomycetes, «thick-walled globose cells (Hülle cells) develop by budding at the tips of specialized hyphae. Hülle cells envelop the developing cleistothecium and may serve as nurse cells.

» Hymenium: «The hymenium is the layer of cells containing the spore-bearing cells (usually basidia or asci) of the fungus. The hymenophore is a collective term for the fleshy structures that bear the hymenium. Thus, in a gilled mushroom, all the gills constitute the hymenophore, and the hymenium is the layer of cells on the surface of those gills. Although the hymenophore may be convoluted and enclosed within the fruiting body, the hymenium still has to be, in some sense, on the outside of the hymenophore in order for either of this structure to qualify for their names. Otherwise, you have a gleba, inside a peridium.» hymenium.

» Hypha: (plural *hyphae*) is a long, branching filament that, with other hyphae, forms the feeding thallus of a fungus called the mycelium. Hyphae are also found enveloping the gonidia in lichens, making up a large part of their structure. A typical hypha consists of a tubular wall, usually made of chitin, which surrounds, supports, and protects the cells that compose a hypha. For most fungi, a cell within a hyphal filament is separated from other cells by internal cross-walls called septa (singular septum). Some forms of parasitic fungi have a portion of their hyphae modified to form haustoria that are able to penetrate the tissues of a host organism.

» Hyphal body: portion of mycelium that becomes separated from remainder of thallus.

» Imperfect state: asexual state of a fungus, also known as an anamorph in a life cycle.

» Karyogamy: fusion of two (haploid) nuclei.

» Locule: generally, in botany, a locule is a small chamber in which a reproductive structure develops. In the ascomycetes, it refers to a small chamber in the hymenium (generalized reproductive tissue) in which a perithecium develops.

» Meiosis: A process common to almost sexual reproduction in eukaryotic cells. The homologous chromosomes of a diploid nucleus first exchange homologous genes (alleles) on a roughly random basis, so that the resulting chromosomes carry a mixture of genes from each parent. The nucleus then divides normally (by mitosis) to yield two diploid daughter nuclei. Finally, the nuclei divide again, but now without DNA replication, to yield four haploid cells.

» Meiospore: a spore formed after meiosis; a spore produced by sexual reproduction.

» Microtubule: Microtubules are protein structures found within cells. They are generally long and form a structural network (the cytoskeleton) within the cell's cytoplasm, but in addition to structural support microtubules are used in many other processes as well. They form a substrate on which other cellular chemicals can interact, they are used in intracellular transport, and are involved in cell motility. The assembly and disassembly of microtubules into their subcomponent tubulin is one way in which cells can change their shape. A notable structure involving microtubules is the mitotic spindle used by eukaryotic cells to segregate their chromosomes correctly during cell division. Microtubules are also responsible for the flagella of eukaryotic cells (prokaryote flagella are entirely different).

» Monokaryon: a hypha having only a single, haploid nucleus.

» Monophyletic: a group of organisms is said to be monophyletic if it consists of only an ancestral organism and *all* descendants of that organism. If the group does not contain the last common ancestor of its members, it is *polyphyletic*. If it contains the last common ancestor of all members, but not *all* of the descendants, it is termed *paraphyletic*. A monophyletic group is called a *clade*. The significance of all this is explained at Dendrograms.

» Mushroom: fleshy, sometimes tough, umbrella like basidiocarp of certain Basidiomycota.

» Mycelium: a network of numerous hyphae which develop within or along the substrate. The spore and initial hyphae are haploid; that is, they contain only one copy of each chromosome. When a haploid mycelium meets another haploid mycelium of the same species, and they are sexually compatible, the two mycelia join together and each cell receives a nucleus from the other mycelium. This process is called diploidization. Some authors seem to want to reserve the term mycelium for hyphae that have undergone diploidization (the earlier, haploid stuff is just hyphae), but most people seem to use it for both phases.» hypha.

» Mycorrhizal association: a symbiotic relationship between fungus and plant in which the fungus interpenetrates the roots, mobilizing soil nutrients for the plant and absorbing complex organics produced by the plant.

» Necrotrophic: growing by first killing the host organism or mycelium.

» Oidium: a small, specialized haploid mating spore which fuses with a haploid hypha to produce a dikaryon.

» Operculum: a lid. Thus, for example, the asci of some pezizomycotines have a sort of lid at the distal end which opens and releases the spores at maturity. In the image, the opercula are stained blue.

» Ostiole: a pore or hole in an enclosing ascocarp for the release of spores.

» Paraphysis: (pl. *paraphyses*) a poorly-constrained term which seems to include every sort of generally elongate cell in a perithecium (or similar structure) which is not actually an ascus.

» Perfect state: sexual state of a fungus, also known as the teleomorph in a life cycle.

» Peridium: the outer membrane of a mitospore. See image at mitosporangium. In that image, the peridia are the two structures which look a bit like compound eyes in the electron micrograph. A more familiar example of a peridium may be the outer sac visible on fungal puffballs. The term is not restricted to strictly cleistothecial (completely enclosed) ascocarps (spore organs). Any outer membrane which encloses spores, even if not completely, is a peridium. The image here shows a peridium in the process of dehiscence. I have lost the source of this image.

» Periphysis: a paraphysis (elongate sterile cell) which grows into the neck of a perithecium.

- » Perithecial: relating to a perithecium.
- » Perithecium: A form of ascocarp in which the peridium (outer membrane) is almost closed but for an ostiole (pore) at the distal end.
- » Plasmodium: a naked, multinucleate mass of protoplasm that moves and feeds in an amoeboid fashion.
- » Plasmogamy: fusion of two cell cytoplasms.
- » Plectenchymatous: having strongly parallel oriented hyphae (a palisade) which rarely branch or overlap one another, as in the image.
- » Prototunicate: Some asci have no active spore-shooting mechanism. These asci are usually more or less spherical, and are found in cleistothecial (occasionally perithecial), and hypogeous ascomata.
- » Pseudoparaphysis: Pseudoparaphyses are long, hair-like cells which grow down from the roof of the locule and often attach to its base. *See* image at pseudothecium.
- » Pseudoparenchyma: Thin-walled, usually angular, randomly-arranged cells in fungi, which are tightly packed. Often used to form walls in specialized structures such as ascocarps. They are similar in appearance to the *parenchyma* cells of plants and have some of the same functions, *i.e.* as a population of undifferentiated cells competent to form various different types of specialized tissues.
- » Pseudothecium: perithecium-like fruiting body containing asci and ascospores dispersed rather than in an organized hymenium; an ascostroma with a locule or cavity and containing bitunicate asci.
- » Reverse transcriptase: an enzyme so called because it operates in a manner which reverses normal transcription. That is, a reverse transcriptase uses RNA as a template to create a complementary single-stranded DNA molecule.
- » Rhizoid: a short, thin branch of thallus, superficially resembling a root.
- » Ribozyme: an RNA molecule with catalytic properties, typically in association with a cofactor such as a neucleotide or metal ion.
- » Saprophyte: an organism which feeds on dead organic matter.
- » Septal pore: a possible synapomorphy of the Ascomycota. The hyphae of ascomycetes are partitioned off into compartments by septal walls. The septae are bridged by pores which provide controlled cytoplasmic continuity throughout the hypha.
- » Septate: with more or less regularly occurring cross-walls.
- » Septum: a cross-wall in a hypha that develops centripetally.
- » Smut fungi: fungus belonging to the Ustilaginomycetes.

- Sporangiophore: sporangia borne on stalks, or the specialized hyphae which bear sporangia.
- Sporangiospore: a spore with a nucleus formed by mitosis, an asexual spore. There seem to be an unreasonable number of names for this concept.
- Sporangium: a structure in which spores are created and released into the environment.
- Spore: «Biologically speaking, a fungal spore is a microscopic reproductive unit one or multicelled, used by fungi and other organisms on dispersal of new individuals.
- Stroma: (pl. *stromata*) fungal tissue mass of pseudoparenchyma in or on which the reproductive structures (perithecia) are formed in some sac fungi. More generally, a mass or matrix of vegetative hyphae, with or without tissue of the host or substrate, in or on which spores are produced.
- Suspensor: in Zygomycota, the remains of the gametangia that project from the sides of the zygosporangium which has developed between them.
- Synapomorphy: a character which is shared by all basal members of a clade and is derived from their common ancestor. A synapomorphy may be secondarily lost in later descendants. Only a synapomorphy may be used to infer phylogeny.
- Teleomorph: the sexual form of a fungus; unknown in many taxa.
- Thallus: Fungi have a vegetative body normally located above the substrate called a thallus or soma, composed of hyphae. In botany, a thallus is a relatively simple plant body devoid of stems, leaves and roots. In fungi, it is the somatic phase.
- Toadstool: a member of the "Agaricales" or "Boletales" (both probably polyphyletic) with an inedible fruiting body.
- Unitunicate: an ascus without differentiated inner and outer walls is said to be *unitunicate*. "The unitunicate ascus sometimes has an operculum (a small lid), which opens to liberate the ascospores when the ascus is mature. These asci are called unitunicate-operculate.
- Uredospore: dikaryotic spore of rust fungi produced in the second host and capable of reinfecting it.
- Vegetative: assimilative phase in fungi, structure or function as distinguished from the reproductive.
- Woronin body: Woronin bodies are highly refractive, electron dense, membrane bound particles found on either side of septae that divide hyphal compartments in filamentous fungi.
- Yeast: single-celled ascomycete fungus that reproduces by budding or fission.
- Zoospores: a motile, asexually produced spore.

- Zygosporangium: the teleomorph of the Zygomycetes; a usually thick-walled, often ornamented, multinucleate resting sporangium formed following anastomosis of gametangia arising from compatible mycelia (in heterothallic species) or from the same mycelium (in homothallic species).
- Zygote: a diploid cell resulting from the union of two haploid cells.